Fortschritte der chemischen Forschung
Topics in Current Chemistry

13. Band, 2. Heft

Herausgeber:

Prof. Dr. *A. Davison*   Department of Chemistry, Massachusetts Institute of Technology, Cambridge, MA 02139, USA

Prof. Dr. *M. J. S. Dewar*   Department of Chemistry, The University of Texas Austin, TX 78712, USA

Prof. Dr. *K. Hafner*   Institut für Organische Chemie der TH 6100 Darmstadt, Schloßgartenstraße 2

Prof. Dr. *E. Heilbronner*   Physikalisch-Chemisches Institut der Universität CH-4000 Basel, Klingelbergstraße 80

Prof. Dr. *U. Hofmann*   Institut für Anorganische Chemie der Universität 6900 Heidelberg 1, Tiergartenstraße

Prof. Dr. *K. Niedenzu*   University of Kentucky, College of Arts and Sciences Department of Chemistry, Lexington, KY 40506, USA

Prof. Dr. *Kl. Schäfer*   Institut für Physikalische Chemie der Universität 6900 Heidelberg 1, Tiergartenstraße

Prof. Dr. *G. Wittig*   Institut für Organische Chemie der Universität 6900 Heidelberg 1, Tiergartenstraße

Schriftleitung:

Dipl.-Chem. *F. Boschke*   Springer-Verlag, 6900 Heidelberg 1, Postfach 1780

Springer-Verlag   6900 Heidelberg 1 · Postfach 1780
Telefon (06221) 49101 · Telex 04-61723
1000 Berlin 33 · Heidelberger Platz 3
Telefon (0311) 822001 · Telex 01-83319

Springer-Verlag   New York, NY 10010 · **175**, Fifth Avenue
New York Inc.   Telefon 673-2660

Photopolymerization and Photocrosslinking of Polymers

Dr. J. L. R. Williams

Eastman Kodak Company, Research Laboratories, Rochester, N. Y., USA

Contents

I. Introduction

The fields of photopolymerization of monomers and photocross-linking of polymers are very large and their complete documentation would require many times the size of this report. The scope of this communication is restricted, therefore, to selected examples of the types of monomers and polymers which illustrate current trends.

The remarkable tendency to intermix the terms "photopolymer" and "photocrosslinkable polymer" has created inconsistencies in describing the details of the two processes of forming these polymers. In this paper, *photopolymers* signify those polymers which result from the photopolymerization of monomeric or low molecular weight compounds:

$$\left[CH_2{=}CH{-}R\right]_n \xrightarrow{\; h\nu \;} -\left[-CH_2{-}\underset{\underset{R}{|}}{CH}-\right]_n-$$

Photocrosslinkable polymers are macromolecules which can become cross-linked by formation of interchain bonds under the influence of light:

$$\begin{array}{ccc}
-M-(M)_n-M- & & -M-(M)_n-M- \\
& \xrightarrow{\text{h}\nu} & \quad | \qquad | \\
-M-(M)_n-M- & & -M-(M)_n-M-
\end{array}$$

Photopolymerization first became of interest as a preparative means of converting monomers to polymers. As early as 1860 *Hoffmann*[1] carried out the photopolymerization of vinyl bromide. During several decades following photopolymerization continued to be studied from the point of view of polymer preparation. Typical photoinitiators such as aldehydes, ketones, halogen compounds and dyes became of interest because they produced a more efficient initiation than that by the mono-mer alone.

During the 1920's, photosensitive coumarones were employed to form resists. They had, however, very low sensitivity to light. In the 1930's, *Murray*[2] showed that alketones could be photopolymerized to form resists on exposure to light. He found that a noncrystallizing layer of shellac containing dicinnamoylacetone would form photoinsolubilized images.

Photopolymerization became useful as a means of detecting the presence of free radicals, a fact which attracted many workers to study in detail the related phenomena. This period has been thoroughly review-ed from the point of view of the photoinitiators and sensitizers by *Del-zenne* [3] and recently more broadly by *Oster* and *Yang* [4].

In 1815 *Niepce* [5] discovered that the natural sensitivity of Syrian asphalt could be used to prepare a bleach-out photograph. Some time later he reported that the exposed asphalt images could be developed in turpentine. Thus Syrian asphalt constituted the first photocross-linkable polymer. Niepce went so far as to copy existing engravings by this process. For this it was necessary to print through the document. He was then able to develop the resist and subsequently engrave the surface of the metal plate in an imagewise manner. Even more than this, Niepce was able to copy a landscape "from life" with a polymer by means of the first practical camera. This was accomplished years before Daguerrotype and silver halide photography was born. Thus, Niepce has been designated "the world's first photographer". In 1852, *Lemer-cier* [5] employed asphaltum in order to sensitize litho stones. In all these cases the level of sensitivity of the resist or asphalt was very low and stemmed from their natural or inherent sensitivity to light. Exposure times of hours were neccessary.

By photocrosslinkable polymers are meant macromolecules which bear photoreactive groups capable of forming crosslinks between their respective chains. A typical photocrosslinkable polymer is poly(vinyl cinnamate) [6]. The photocrosslinking occurs when a network of crosslinks is built up between adjacent chains of poly(vinyl cinnamate). The mechanism of crosslinking will be described in detail in a later section.

Intermediate species such as macromolecules bearing acrylate-type residues represent a cross between photopolymerizable monomers and photocrosslinkable polymers. The gap between photopolymerizable systems and photocrosslinkable polymers is closed since the two systems are basically the same. A photocrosslinkable polymer can be considered the extreme case of a macromolecular photopolymerizable monomer.

Sensitivity to the near ultraviolet and visible regions of the electromagnetic spectrum has been accomplished in the above systems by use of dyes which act as spectral "sensitizers". Interest in photopolymerization and photocrosslinking of polymers has been caused by their potential value as image-forming systems. Such imaging methods find great utility in connection with photoresists for printed circuits, chemical milling and photolithography [7].

Generally the photocrosslinking of photocrosslinkable polymers is not inhibited by oxygen as is photopolymerization. For practical purposes bisacrylates or trisacrylates are uesd to insure that highly crosslinked photopolymers result by irradiation. The fact that the photocrosslinkable polymer has significantly high molecular weight *before* photocrosslinking occurs allows the attainment of insoluble crosslinked structures possessing highly stable properties. In both photocrosslinkable polymers and photopolymers, it is the caged network which allows their utilization in practical systems.

Natural polymers such as albumin, gelatin, fish glue, shellac, and gum arabic received attention up to the early 1900's, after which time other materials such as bichromate, diazo compounds, iron salts, and silver halides were added to resins in order to hasten or accomplish their crosslinking. These systems were neither photopolymerization nor photocrosslinkable polymers. Instead, the added second component, when excited by light, produced a species which itself caused the formation of radical sites on a prepolymer or polymer chain which then led to crosslinking.

Beebe and *Murray* [8], in the 1920's, sensitized condensation products of aldehydes, ketones, and amines with iodine compounds such as iodoform. Such a photoresist, called "Neokel", was sold in 1926. It consisted of "furfurane-pyrol-thiophene" resin, which could be developed in naphtha. Thus, in the strictest sense, all polymers can be considered photocrosslinkable when irradiated with sufficiently energetic light in the absence or presence of addenda. Many times photodegradation is concomitant with photocrosslinking.

II. Photopolymerization

The following photopolymerization topics have been chosen for discussion owing to current interest.

1) Image formation
2) Photografting
3) Four-center-type photopolymerization
4) Charge-transfer photopolymerization
5) p-Aminobenzoylazide photocondensation

1. Image Formation

Research into applications of photopolymerization to practical image-forming system has resulted in a multitude of patents. A limited but typical group is given in [9]. This endeavor has culminated in photopolymerization systems which involve image transfer systems for printed circuits, lithography and office copying materials. Typical of such systems are those in which a coating of a photopolymerization layer is formed on a support such that is is sandwiched between the support and a cover sheet. The cover sheet acts as an oxygen barrier and hence reduces termination of growing radical chains by oxygen.

Photopolymerizable layers usually consist of a polymeric binder, such as poly(methyl methacrylate) or poly(vinylbutal); a photopolymerizable monomer, such as methylene glycol bisacrylate; or polyethylene glycol bismethacrylate; a thermal stabilizer, p-methoxyphenol; a photoinitiator, such as an alkyl anthraquinone; and a dye, such as a cyanine dye. On exposure to light in an imagewise manner the exposed areas are photohardened by formation of caged-type photopolymerized, two-dimensional crosslink structures.

The crosslinked image areas are then insoluble in solvents. The soluble noncrosslinked areas can be removed with solvent or chemical action to provide relief patterns or images. Alternatively, the change in thermal properties on exposure can be utilized. The crosslinked areas are less easily thermally softened than are the unexposed areas. Thus the unexposed and thermally softened areas can be transferred by pressure to a receiving sheet to give a positive image. Proper choice of the polymeric binder and the monomeric acrylate or methacrylate permits control over the ultimate physical properties of the exposed and unexposed photopolymerizable areas. By modifying both the physical properties of the acrylic monomer or prepolymer and the polymeric binder, the transfer temperature can be controlled. In contrast to transfer systems, inclusion of hydrophilic groups into the binder can lead to systems which are developable in aqueous alkali washout solvents to give relief images useful for lithographic purposes. Typical of such binders are succinoylated cellulose esters. Various patents [9] disclose the changes of processability and image formation with changes of the physical and chemical properties of the photopolymerizable layer. Image transfer rheology of photopolymers has been discussed by *Berkower, Beutel* and *Walker* [10].

2. Photografting

Photografting is a technique utilizing photopolymerization such that the monomer grafts become attached to macromolecules and the surface of other materials. Several techniques exist: The macromolecule can be made to initiate polymerization at points along its chain by generation of radical sites. The sites on the chain may be formed by photoreactions that initiate or by transfer with radicals or radical formers generated in the medium. Such techniques offer an opportunity to prepare grafted polymers from monomers and polymers of radically different physical properties. An example would be the photografting of a hydrophilic monomer onto a hydrophobic macromolecule. Such polymers are of great theoretical interest as well as practical benefit to those interested in surface characteristics, dyability, etc., of fibers and films. Photografting is also important in practical photopolymerization image-forming systems involving binders.

Photoinitiation of acrylamide by eosin proceeds through the semiquinone of eosin to yield a growing chain terminated at one end by eosin [11].

Smets, DeWinter and *Delzenne* [11] attached eosin units to poly(vinylamine) through the amino groups. Photopolymerization was initiated at the eosin sites, yielding grafts of poly(vinylamine-eosin-poly-(acrylamide) where X is $-CONH_2$.

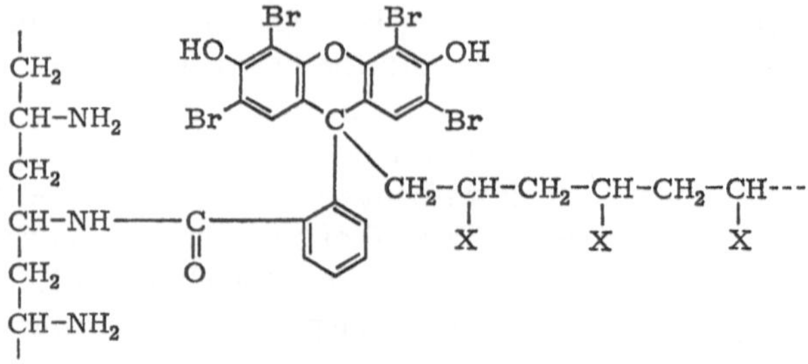

$$CH_2 - CH - CH_2 - CH - CH_2 - CH ---$$

The polymerized acrylamide was completely bound in the graft since no homopolyacrylamide was found. Block polymers have been prepared by linear grafting [11]. Eosin lactone was appended to poly(methyl methacrylate) chains by means of a terminal amino group to provide a polymeric initiator which was then used to photoinitiate polymerization of styrene and acrylamide.

By irradiation of natural rubber containing benzophenone, and in contact with aqueous acrylamide, *Oster* and *Shibata* [12] showed that grafts of poly(acrylamide) could be formed along the rubber backbone chain.

Cooper and *Fielden* [13] carried out the photografting of methyl methacrylate to natural rubber by using natural rubber latex containing l-chloranthraquinone as the photoinitiator. Ninety-eight percent of the polymerized methyl methacrylate was found as grafted poly(methyl methacrylate). *Oster* [14] reported the grafting of acrylamide onto polyethylene using benzophenone.

Grafting of acrylonitrile and acrylamide to cellophane was studied by *Tsunoda, Tanaka* and *Murata* [15] who used methyl vinyl ketone as the photoinitiator.

Cooper and co-workers [16] made a comparative study of the grafting of methyl methacrylate upon natural rubber latex using visible, ultraviolet and γ-radiations as energy sources. It was found that the grafting took place with equal efficiency whether the radiation source was light or γ-rays.

The fact that many anthraquinone dyes, when adsorbed on cellulose, undergo a photochemical reaction with the polymer was used by *Geacintov, Stannett, Abramson* and *Hermans* [17] as a means photografting upon cellulose.

Anthraquinone-2,7-disulfonic acid was used as the photoinitiator to photograft acrylonitrile, methyl methacrylate, styrene and vinyl acetate upon cellulose. The radical site formed on the cellulose chain initiated polymerization of quantities of monomers of up to three-and-one-half times the weight of the cellulose [18].

Needles [19] reported that riboflavin-5-phosphate photoinitiation of low concentrations of acrylic monomer (1—3%) proceeded with difficulty.

However, when proteins or amino acids are present in such monomer solutions, photoinitiation proceeded well. Fibrous proteins such as wool keratin and silk fibroin showed grafted polymer uptakes despite their lack of solubility [20].

Recently *Dhandraj* and *Guillet* [21] reported a new photochemical synthesis of block copolymers by irradiation of poly(tetramethylene sebacate-CO-γ-ketopimelate) in the presence of methyl methacrylate. The chain radicals formed from the Norrish Type I cleavage of the γ-ketopimelate unit initiated polymerization of methyl methacrylate leading to blocks the length of which depend on the number of scissions.

3. Four-Center-Type Photopolymerization

The photodimerization of olefinic systems such as cinnamic acid and stilbenes has been studied and reported for many years [22].

A related method for the photopolymerization of doubled-ended dimerization-type structures has been reported by *Hasegawa* and *Suzuki* for distyrylpyrazine [23] and *trans, trans*-1,4-bis(β-pyridyl-2-vinyl)benzene [24].

The fact that isomeric compounds 1,4-bis(β-pyridyl-3-vinyl)benzene and 1,4-bis(β-pyridyl-4-vinyl)benzene failed to photopolymerize under the same conditions indicates the importance of the topochemical arrangement of the double bonds. This was studied by x-ray analysis of the crystals. Topochemistry has been shown to be important in the case of the dimerization of substituted cinnamic acids [25]. p-Phenylene-diacrylic acid and its esters photopolymerize to yield polyesters containing cyclobutane units [26].

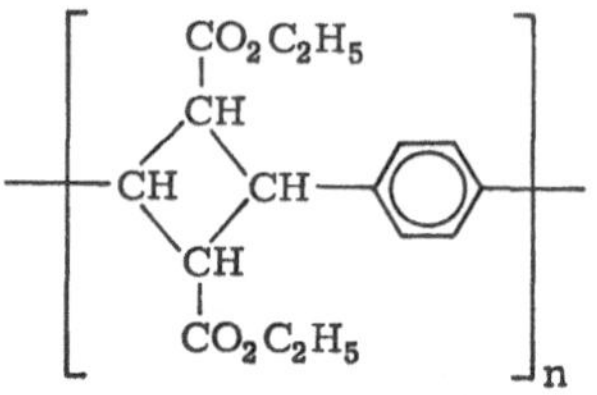

$$\left[\begin{array}{c} CO_2C_2H_5 \\ | \\ CH \\ -CH \quad CH- \bigcirc - \\ CH \\ | \\ CO_2C_2H_5 \end{array} \right]_n$$

A study with the ethyl ester showed that photopolymerization took place in the crystalline state but not in the melt [26]. A further example of this type of polymerization is reported for polymethylene dicinnamates, typical of which are tetramethylene dicinnamate and ethylene dicinnamate [27].

4. Charge-Transfer Photopolymerization

Recently a series of publications by *Tazuke* and co-workers [28,29,30] have disclosed photopolymerization systems in which ionic or charge transfer species act as the photoinitiators. It is interesting to note that the presence of oxygen in such systems causes less inhibition or retardation than in radical-type photopolymerizations. Donor-acceptor pairs such as vinylcarbazole and sodium aurochloride dihydrate typify the system:

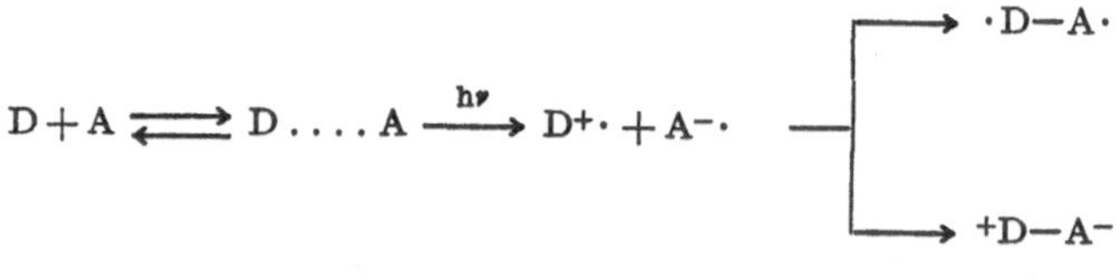

$$D + A \rightleftharpoons D \ldots A \xrightarrow{h\nu} D^+ \cdot + A^- \cdot \quad \begin{array}{l} \longrightarrow \cdot D-A \cdot \\ \\ \longrightarrow {}^+D-A^- \end{array}$$

ESR measurements indicated electron transfer from Au^{III} or Au^{II} to vinylcarbazole. Whether Au^{II} or Au^{III} is involved in the electron transfer has not been established. Slow polymerization of vinylcarbazole-sodium aurochloride dihydrate occurs at room temperature in the dark. Irradiation of the same system causes precipitous polymerization. Ammonia inhibition of the photopolymerization was indicative of the participation of a cationic propagation process.

5. *p*-Aminobenzoylazide Photopolycondensation

Reynolds, VanAllen and *Borden* [31] made use of the photochemical rearrangement of benzoyl azides to phenyl isocyanates to conduct a photopolycondensation. By incorporation of a group, reactive with isocyanates, such as amino, into the same molecule, photolysis led to formation of a polyurea structure.

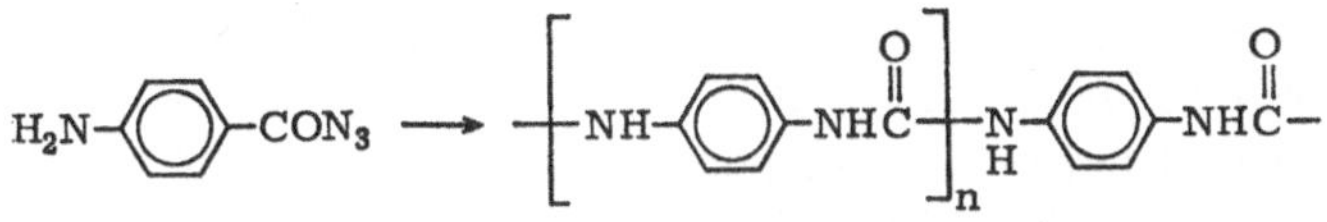

$$H_2N-\!\!\bigcirc\!\!-CON_3 \longrightarrow \left[-NH-\!\!\bigcirc\!\!-NH\overset{O}{\overset{\|}{C}}\right]_n\!\!-\underset{H}{N}-\!\!\bigcirc\!\!-NH\overset{O}{\overset{\|}{C}}-$$

Solid state irradiation of *p*-aminobenzoylazide in the absence of a binder leads to formation of the polyurea. Attempted formation of such a polymer within a matrix of a soluble polymeric binder such as poly-(vinylbutal) led to insolubilization in the exposed areas. Hydrogen abstraction can simultaneously produce crosslinking of the binder.

III. Photocrosslinkable Polymers

1. Structural Variations

Allen and *VanAllen* [32] synthesized a polymeric chalcone for use as a photolithographic material from polystyrene and cinnamoyl chloride.

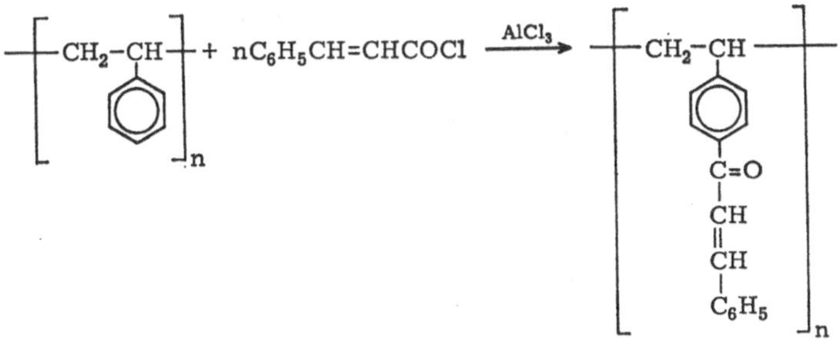

$$\left[-CH_2-CH-\right]_n + nC_6H_5CH=CHCOCl \xrightarrow{AlCl_3} \left[-CH_2-CH-\right]_n$$

with pendant groups $C=O$, CH, CH, C_6H_5

Robertson, VanDeusen and *Minsk* [33] prepared poly(vinyl cinnamate) and found that incorporation of sensitizers such as Michler's ketone into its coatings provided a means of optical sensitization. Optical sensitization decreased the required exposure time to give imagewise photocrosslinking of poly(vinyl cinnamate).

The basic unit necessary to produce crosslinking in polymers based on poly(vinyl cinnamate) is the $-C=C-CO-$ unit. This unit can be attached to many polymers in many ways and, depending on its configuration in the polymer chain or when appended to the polymer chain, the light-sensitivity and the physical properties of the polymer can be modified. If a nitro group is placed in the benzene ring of the cinnamoyl residue, the optical response of the polymer can be shifted to longer wavelengths.

Extension of the use of the cinnamoyl unit to photocrosslinkable polymers is found in the numerous chalcone-type polymers prepared by *Unruh* and *Smith* [34]. Typical is a group of chalcone polymers which were prepared by reaction of poly(4-vinylacetophenone) with aromatic aldehydes such as benzaldehyde.

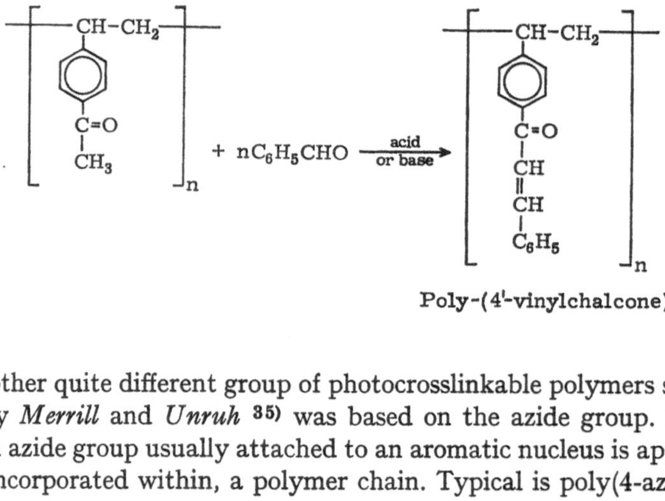

Poly-(4'-vinylchalcone)

Another quite different group of photocrosslinkable polymers synthesized by *Merrill* and *Unruh* [35] was based on the azide group. In this case, an azide group usually attached to an aromatic nucleus is appended to, or incorporated within, a polymer chain. Typical is poly(4-azidostyrene).

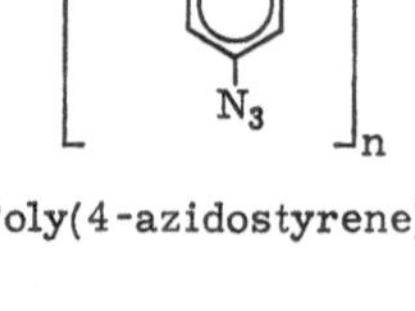

Poly(4-azidostyrene)

The azide polymers have been found to respond quite effectively to sensitization, much in the manner of that found earlier for poly(vinyl cinnamate) [37].

Examples of photosensitive functional groups which have been attached to or incorporated within polymer chains are shown in Fig 1.

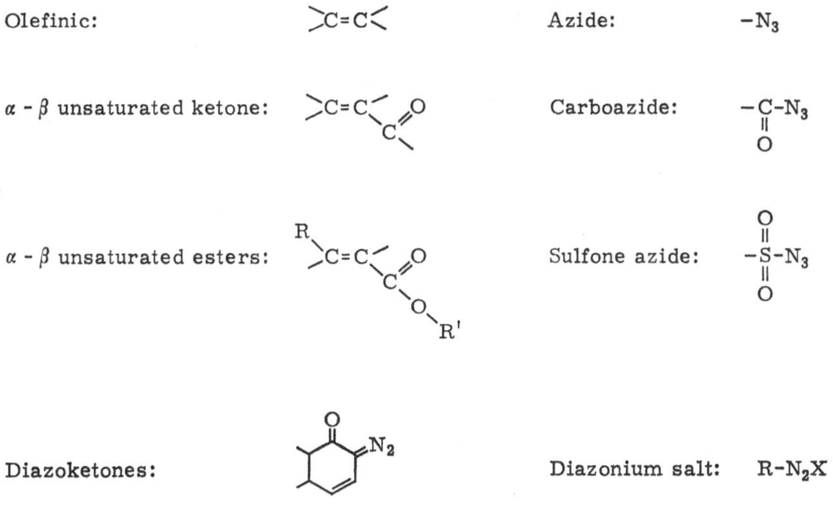

Fig. 1. Examples of photosensitive functional groups

Units used for attaching or condensing photosensitive groups to polymer chains can be drawn from esters, carbonates, phosphonates, urethanes, ethers, amides and sulfonamides. A number of such variations for the incorporation of cinnamate structures are shown in Fig. 2.

2. Sensitometry of Photocrosslinkable Polymers

A sensitometric method for the evaluation and comparison of light-sensitive polymers was developed by *Minsk* and co-workers [6,33]. Coatings of polymers of various compositions or containing different sensitizers were exposed through a step tablet to controlled quantities of light. The unexposed or partially exposed soluble polymer areas were dissolved away by solvents in a developing step. The resulting image was made visible by inking of the residual polymer. By comparison of the minimum amount of light required to produce the weakest appearing step, the comparative speed value of the polymer in question is assessed. Poly-(vinyl cinnamate) is taken as a value of 1.0.

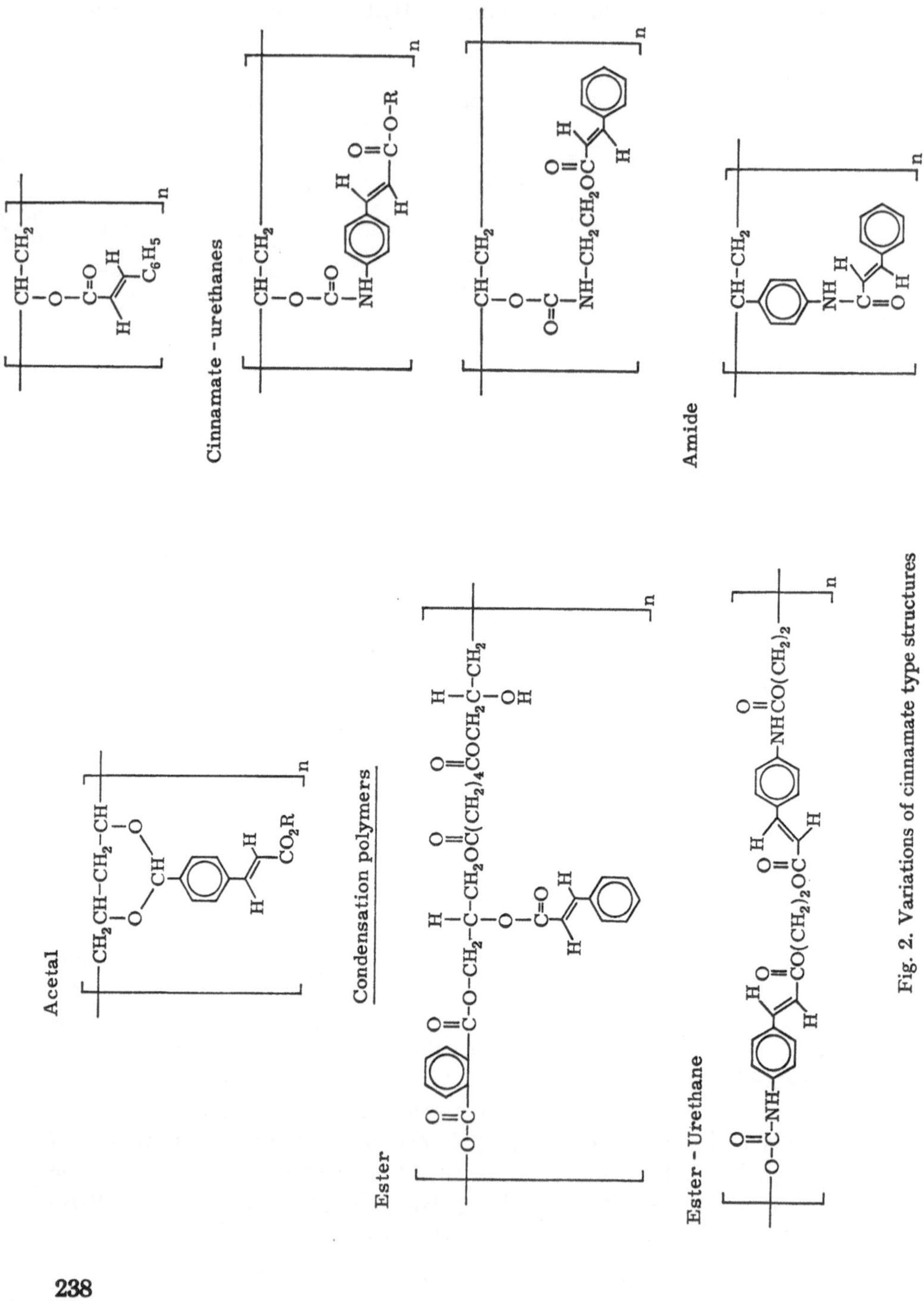

Fig. 2. Variations of cinnamate type structures

Others have used interference microscopic [36] and gravimetric methods [37] to compare the density of the residual polymer on the various steps.

3. Spectral Response

Robertson, VanDeusen and *Minsk* [33] described the use of a wedge spectrograph for the measurement of spectral response. Such a wedge spectrograph consists of a modified Bausch and Lomb monochromator, as shown in Fig. 3. Comparative spectral response patterns for polymers can be obtained on a comparative basis by exposure-development of coatings of various light-sensitive polymers. Wedge spectrograms also

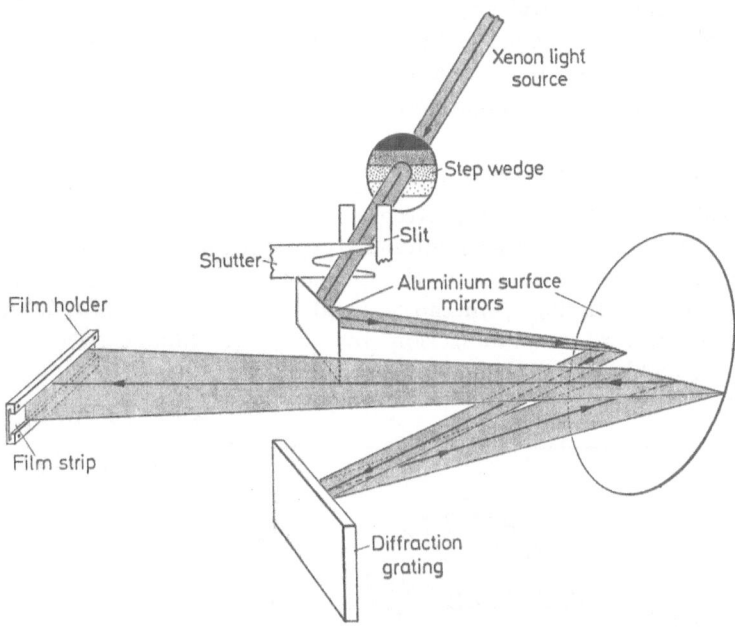

Fig. 3. A wedge spectrograph

provide a very useful and facile method of comparing the effect of the inclusion of other materials, such as sensitizers, in polymer coatings. By this means we can determine whether an addendum produces sensitization or desensitization effects. Typical wedge spectrograms for poly(vinyl acetate cinnamate) and poly(vinyl acetate cinnamate)-picramide are compared in Fig. 4.

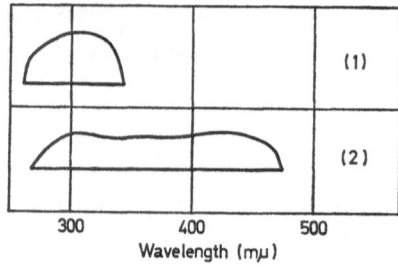

Fig. 4. Wedge spectrograms of unsensitized (1) and picramide sensitized (2) poly (vinyl cinnamate)

4. Extension of the Spectral Sensitivity of Photocrosslinkable Polymers

The range of the spectral sensitivity of light-absorbing photocrosslinkable polymers can be extended in two ways: (a) by modification of the range of light absorption of the crosslinking chromophore in the polymer, or (b) by inclusion of spectral sensitizers.

a) Chromophore Modification

Fig. 5 illustrates the wedge spectrogram (1) corresponding to the spectral sensitivity of poly(2-vinyl-1-methyl-5-styryl-pyridinium methosulfate)[38]. Spectrograms 2 and 3 indicate the increased range of spectral response of the corresponding *p*-methoxy- and *p*-dimethylamino-substituted polymers toward insolubilization by photocrosslinking. The action spectra roughly follow the absorption spectra of the polymers as shown in Fig. 5.

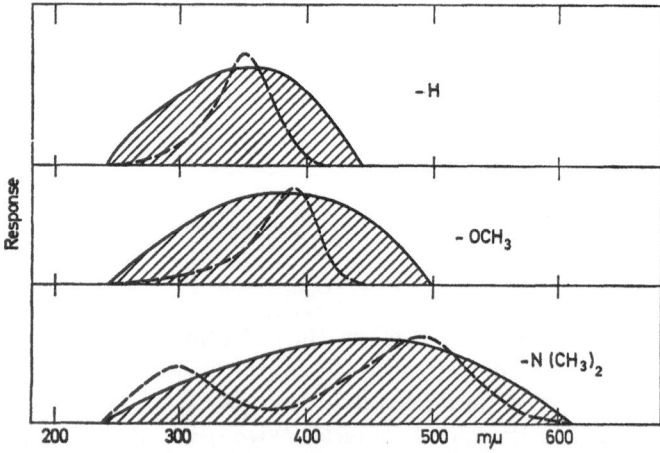

Fig. 5. Wedge spectrograms of poly(2-vinyl-1-methyl-5-(*p*-substituted styryl)pyry dinium) methosulfate; where R is H, OCH₃ and N(CH₃)₂

Another type of chromophore modification can be accomplished by extension of the unsaturated chain length. The spectral responses of poly(vinyl acetate cinnamate) and poly(vinyl acetate cinnamylidene acetate) (Fig. 6) typify this method [39].

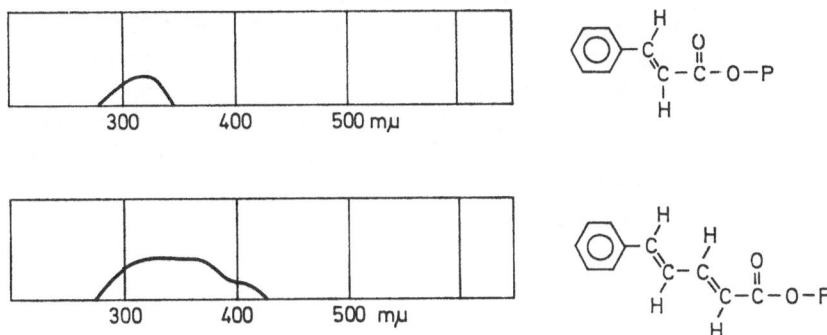

Fig. 6. Wedge spectrograms of poly(vinyl cinnamate) and poly(vinyl cinnamylidene-acetate)

b) Spectral Sensitization

Minsk, et al. [6,33], discovered the feasibility of extension of the spectral response of insolubilization of poly(vinyl acetate cinnamate). Several other investigations have extended the group of addenda which provide spectral extension of the photocrosslinking of poly(vinyl acetate cinnamate) [40,41,42,43,44]. Most striking is the spectral extension obtained when poly(vinyl acetate benzoate cinnamylidenacetate) [45] is sensitized with various pyrylium salts as shown in Fig. 7.

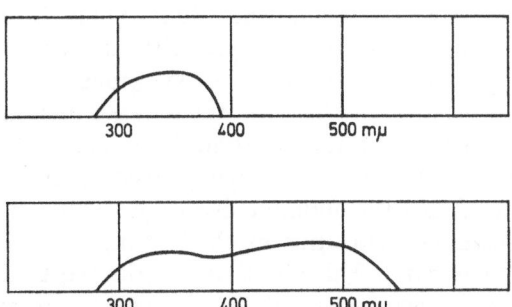

Fig. 7. Wedge spectrograms of poly(vinyl acetate benzoate cinnamylideneacetate) unsensitized and (lower) sensitized with 4-(*p*-butoxyphenyl)-2,6-bis(*p*-ethylphenyl)-thiapyrylium perchlorate

241

5. Studies of the Mechanism of Photocrosslinking

a) Poly(vinyl cinnamate)

The photocrosslinking of poly(vinyl cinnamate) was assumed by *Minsk* to involve reactions which led to the disappearance of the double bond. A number of studies have confirmed that ultraviolet irradiation of poly(vinyl cinnamate) causes diminution of the value of absorbance in the region characteristic of the cinnamate ester group. Typical experiments involved the irradiation of coatings of poly(vinyl cinnamate) on quartz plates. It was then possible to follow the disappearance of the cinnamate ester absorption spectra with irradiation time. The photoreaction taking place has been presumed by *Tsuda* [41] to be analogous to dimerization of low molecular weight cinnamate esters.

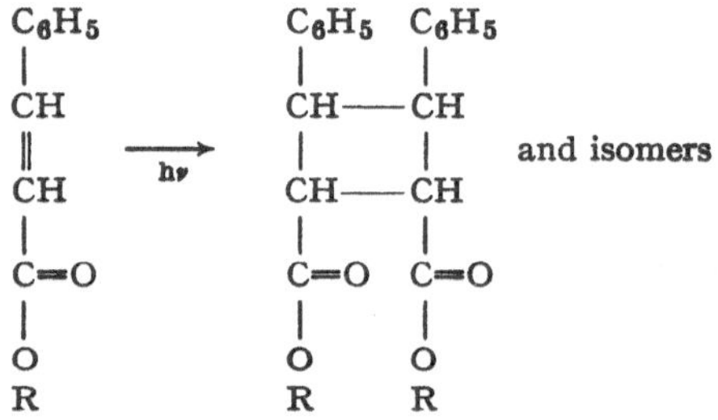

$$
\begin{array}{ccc}
C_6H_5 & C_6H_5 & C_6H_5 \\
| & | & | \\
CH & CH\!\!-\!\!-\!\!-\!\!CH & \\
\| \xrightarrow{h\nu} & | \quad\quad | & \text{and isomers}\\
CH & CH\!\!-\!\!-\!\!-\!\!CH & \\
| & | \quad\quad | & \\
C\!\!=\!\!O & C\!\!=\!\!O \quad C\!\!=\!\!O & \\
| & | \quad\quad | & \\
O & O \quad\quad O & \\
R & R \quad\quad R &
\end{array}
$$

Tsuda [41] described a relationship between the degree of esterification of poly(vinyl alcohol) with cinnamate and the amount of light necessary to produce crosslinked images. Unfortunately, the experimental verification could not be carried out since the solubility, and hence the development characteristics, change markedly with the percentage value of esterification or amount of residual hydroxyl groups.

A relationship was found between the average molecular weight and the relative sensitivity of the poly(vinyl cinnamates) prepared from parent poly(vinyl alcohols) of varying average molecular weights. *Tsuda* [41] reported that the optimum degree of sensitization was reached when the sensitizer concentration reached 10 percent by weight of that of poly(vinyl cinnamate). This effect was attributed to the necessity of intimacy between sensitizer molecules and cinnamate groups in order that the energy absorbed by the sensitizer be transferred efficiently to cinnamate. In a later study, *Tsuda* [43] investigated the role of a number of sensitizers (S) and concluded that the triplet state of the sensitizer (S^T)

transferred energy to the triplet of the cinnamate esters group which then underwent crosslinking reactions.

$$S^0 \xrightarrow{\text{h}\nu} S^1$$

$$S^1 \longrightarrow S^T$$

$$S^T + (\text{cinnamate})^0 \longrightarrow S^0 + (\text{cinnamate})^T$$

$$(\text{cinnamate})^T \longrightarrow \text{crosslinking chemistry, several steps}$$

By variation of the sensitizer (S) the energy of S^T at which crosslinking of poly(vinyl cinnamate) fails was estimated to be 17,300 cm^{-1}. This value agreed well with a phosphoroscopic value of 17,600 cm^{-1} made at 77° K of cinnamic acid in EPA glass. Methyl cinnamate failed to give any observable phosphorescence at 77° K. This was attributed to the great tendency of the ester to undergo dimerlike reactions, rather than to release the energy as phosphorescence.

A comparison of the photosensitivity of poly(vinyl cinnamate) with the photodimerization of cinnamic acid in poly(vinyl alcohol) coatings was carried out by *Nakamura* and *Kikuchi* [44]. Cinnamic acid in poly(vinyl alcohol) coatings was found to undergo photoisomerization as well as photodimerization. The ultraviolet absorption spectra used to follow the course of the reaction showed one isosbestic point together with other nonisosbestic curve intersections. Such complex curve changes indicated that photoisomerization and at least one other reaction, photodimerization, was occurring.

Corresponding ultraviolet absorption curves for the irradiation of poly(vinyl cinnamate) showed no isosbestic point, indicating that little or no photoisomerization took place. In fact, the value of the wavelength of λ_{max} shifted to longer wavelengths as the total area under the curves decreased. Photodimerization-type reactions would have been expected to produce such a decrease of area under the curves, but with no bathochromic wavelength shift. The same type of decrease in absorbance accompanied by a shift of the value of the wavelength of λ_{max} to longer wavelengths occurred when 2-nitrofluorene was used as a sensitizer in poly(vinyl cinnamate). *Nakamura* and *Kikuchi* [44] found that the photosensitivity or quantum yield of photocrosslinking decreased in proportion to the degree of disappearance of the double bonds of the cinnamate groups. A relationship between the photosensitivity, the minimum exposure time for crosslinking, and the quantum yield of crosslinking was derived therefrom.

Kikuchi and *Nakamura* [42] studied the ESR spectrum of poly(vinyl cinnamate) irradiated with light of wavelengths of 240—250 mμ at 77° K. The ESR spectrum consisted of two components, a broad singlet and a quartet. Temperature studies of the stability of the two components indicated differences which were attributed to the presence of two types of radicals. It was proposed that the broad singlet was due to crosslinked cinnamoyl radicals and the quartet to radicals on the main chain which were produced by hydrogen abstraction reactions.

Parallel phosphorescence studies indicated that the O—O band for cinnamic acid occurred at 20,000 cm^{-1}. *Tsuda* [43] reported a value of 17,600 cm^{-1}. In the same study *Kikuchi* and *Nakamura* [42] measured the quenching of the phosphorescence of cinnamic acid with a group of typical sensitizers, among which were 2-nitro-fluorene and p-nitroaniline. The critical separation distance between cinnamic acid and sensitizers for effective quenching was found to be 10 A.

A more direct comparison of monomeric and polymeric systems was made by *Curme, Natale* and Kelly [46], who studied the relationship between the photochemical behaviors of ethyl cinnamate and poly(vinyl cinnamate). In their study ethyl cinnamate containing sensitizers such as p-nitroaniline, picric acid or Michler's ketone was irradiated such that the energy was absorbed only by the sensitizer. The disappearance of cinnamate ester ultraviolet absorption was similar to the changes taking place in poly(vinyl cinnamate). As in the experiments of *Nakamura* and *Kikuchi* [44] irradiation of coatings of Michler's ketone-sensitized poly(vinyl cinnamate) exhibited the unexpected shift of the value of the wavelength, λ_{max}, of ultraviolet absorption to longer wavelengths. In the presence of oxygen the sensitizers underwent photolysis.

As a result of luminescence-quenching experiments, *Curme, Natale* and *Kelley* [46] concluded that cinnamate did not accept singlet energy from Michler's ketone or picric acid at 77° K. Again it was suggested that triplet energy was donated by the sensitizer to cinnamate ester, which then underwent dimerization-type crosslinking.

It has been tempting for researchers to assume that the crosslinking takes place when an excited double bond adds to one of its neighbors to form cyclobutane-type structures. Parallels from the poly(vinyl cinnamate) system have been drawn to the well-known dimerization of cinnamic acid to truxillic and truxinic acids. A study of a number of substituted cinnamic acids revealed that in order for efficient dimerization to take place the double bonds of two neighboring cinnamate structures must lie within 3.6—4.1 A of each other [25]. The regularity of structure in the crystalline state is not preserved in solution or in the polymeric medium. It has been difficult, therefore, to conclude that cyclobutane-type structures are formed predominantly during the photocrosslinking of poly(vinyl cinnamate).

However, *Sonntag* and *Srinivasan* [47] irradiated poly(vinyl cinnamate) and conducted paper chromatographic analysis of the hydrolysis products of the crosslinked polymer. The presence of α-truxillic acid was indicated on a qualitative basis. The presence of α-truxillate ester in the crosslinked poly(vinyl cinnamate) requires that there be some interchain stereoregularity.

It was proposed that the following type of ordering of the pendant cinnamate groups would produce a means of formation of α-truxillic ester units:

$$
\begin{array}{ccc}
& & P \\
& & | \\
P & & O \\
| & & | \\
O & & C{=}O \quad C_6H_5 \\
| & & | \qquad | \\
C{=}O \quad C_6H_5 & & CH{-\!-\!-}CH \\
| \qquad\quad | & \xrightarrow{h\nu} & | \qquad\quad | \\
CH \qquad CH & & CH{-\!-\!-}CH \quad \xrightarrow[\text{hydrolysis}]{} \quad \alpha\text{-truxillic acid} \\
\| \qquad\quad \| & & | \qquad\quad | \\
CH \qquad CH & & C_6H_5 \quad CO \\
| \qquad\quad | & & | \\
C_6H_5 \quad C{=}O & & O \\
\qquad\quad | & & | \\
\qquad\quad O & & P \\
\qquad\quad | & & \\
\qquad\quad P & &
\end{array}
$$

where P represents the backbone polymer chain.

Nakamura, Sakata and *Kichuchi* [48] have compared the thermal and photochemical crosslinking of poly(vinyl cinnamate). The critical hardening value required is approximately 1.3 crosslinks per macromolecule possessing a degree of polymerization of 1400. The minute amount of crosslinking required to insolubilize poly(vinyl cinnamate) makes the study of the process difficult; insight can be obtained only by inference. Furthermore, *Nakamura* and *Kikuchi* [44] reported that the quantum yield of the disappearance of the double bonds in poly(vinyl cinnamate) decreased rapidly with irradiation time. Extrapolation to zero time of exposure indicated a quantum yield of 0,34. It therefore seems hazardous to base conclusions on observations of products which result from long irradiation times.

Marvel and *McCain* [49] have shown that ethyl cinnamate can be homopolymerized by use of typical radical-type thermal initiators, such as benzoyl peroxide or 2,2'-azobis(isobutyronitrile). Ethyl cinnamate and poly(vinyl cinnamate) can, therefore, take part in radical addition reactions.

If crosslinking of poly(vinyl cinnamate) occurs at levels of crosslinks as low as 1.3 per macromolecule [47], and if at least two types of radicals are present [41], then the possibility of vinyl-polymerization-type photocrosslinking at the onset cannot be excluded. The problem of making studies at the onset of crosslinking has yet to be solved. The competitive role of oxygen, an efficient triplet quencher, is still unexplained.

Buettner [50] has shown that oxygen quenching of triplet states of molecules depends on the nature of the polymeric medium in which the parent molecule is dispersed. In polymers which are highly hydrogen-bonded the quenching is minimal. However, in acrylates and polymers similar to poly(vinyl cinnamate), oxygen quenching is marked. Furthermore, the sensitizers used in studying the photolysis of ethyl cinnamate undergo slow photolysis [46]. During the same study it was reported that singlet energy was not drained from Michler's ketone or picric acid by cinnamate ester. Considering this fact, the presence of oxygen, and the triplet tendencies of the sensitizers, it appears that the photocrosslinking of poly(vinyl cinnamate) could involve the triplet states as a chemical species rather than in a role involving energy transfer to cinnamate. The triplet state of the sensitizer would be involved only to the extent that the triplet state or a species derived therefrom adds to or complexes with a cinnamate unit. The result of this interaction could produce a radical species which could then initiate an inefficient polymerization-like reaction producing only the few crosslinks necessary to cause insolubilization of poly(vinyl cinnamate). Alternatively, the triplet of the sensitizer could react with triplet oxygen to form a semiquinone-like intermediate which would react further with oxygen to form radical initiators. The fact that maximum sensitization is gained at the level of 10 percent of sensitizer [41,44] indicates a critical interaction between the sensitizer and the polymer. In the polymer phase the intimacy of sensitizer and cinnamate unit is at a maximum.

b) Azide Polymers

The azide polymers can be divided into the two subclasses: the aromatic azides and the acid azides, such as carboazides and sulfoazides.

$$\text{—}\bigcirc\text{—N}_3 \qquad \text{Aromatic azide}$$

$$\text{—}\bigcirc\text{—CON}_3 \qquad \text{Carboazides}$$

$$\text{—}\bigcirc\text{—SO}_2\text{N}_3 \qquad \text{Sulfone azides}$$

In all cases it is believed that upon absorption of energy a nitrene-type intermediate and nitrogen are formed:

$$R\text{—N}_3 \longrightarrow R\ddot{\underset{\cdot}{N}} + N_2$$

The nitrene constitutes a very reactive intermediate which can take part in coupling, dimerization and hydrogen abstraction reactions.

Merrill and *Unruh* [35)] described the preparation of azidophthalate esters of poly(vinyl alcohol). They found that the reaction of a nitrene with poly(vinyl toluene) was parallel to that of ethyl benzene in that side-chain hydrogen atom abstraction took place. *p*-Carboxyphenylazide gave *p,p'*-dicarboxyazobenzene and nitrogen on irradiation in the crystalline form [35)].

$$ \text{HOC} \overset{\text{hv}}{\longrightarrow} \text{HOC} \quad \text{N=N} \quad \text{COH} $$

On the other hand, phenylazide gave complex mixtures of products. As in the case of poly(vinyl cinnamate), the onset of insolubilization masked interpretation of the chemical reactions taking place. *Merrill* and *Unruh* [35] did, however, divide the process into the photochemical reaction and the subsequent coupling reactions which follow. *Rieser* and *Marley* [51)] have studied in detail the photolysis of aromatic azides at room temperature and at 77° K. Results from the photochemical decompositon of phenylazide indicated that the primary photolytic step was indeed the formation of a nitrene and that the process was neither bimolecular nor concerted in nature. *Merrill* and *Unruh* [35)] pointed out that photolysis of aromatic azides caused the loss of nitrogen as the rate-determining step and that reactions of the resulting nitrene depended on the nature of the environment. It is not known whether the azide decomposition proceeds through a triplet or a singlet excited state. Typical triplet sensitizers have, however, been useful as sensitizers for azide-bearing polymers.

c) Chalcone-Type Polymers

Poly(4'-vinyl-*trans*-benzalacetophenone) was prepared by *Unruh* [52)] by condensation of poly(*p*-vinylacetophenone) with benzaldehyde. In his studies *Unruh* [52)] compared the photochemical behavior of poly(4'-vinyl-*trans*-benzalacetophenone) with that of the model compound, 4-ethylbenzalacetophenone. Both the polymer and its model showed typical photoisomerization behavior when irradiated in dilute solution. This behavior of the two species was consistent with the results reported for benzalacetophenone by *Lutz*, et al. [53,54)]. *Unruh* found, however, that when the *trans* polymer was photoisomerized to the *cis* form, the expected complete return to *trans* form upon addition of acid failed to result. This fatigue or loss was attributed to small amounts of intrachain dimerization or polymerization taking place in the coiled polymer chains. When the polymer was progressively irradiated as a film, a smooth series of curves following through two isosbestic points resulted. A hypsochromic shift of λ_{max} 50 mμ of ultraviolet absorption occurred. These observations indicated a very rapid, clean conversion of *cis* to *trans* isomer. The rapid shift indicated that isomerization preceded polymerization or dimer-type reactions which led to intra- and interchain crosslinking. The film became insoluble in the coating solvent at levels of less than 1 percent conversion of *trans* to *cis* polymer. Thus photoisomerization and photocrosslinking reactions compete when such films are irradiated.

IV. Conclusion

It appears that at least three processes can take place when systems containing photopolymerizable monomers or photocrosslinkable polymers are exposed to light. They are photopolymerization, photocrosslinking and photografting. These three processes lead to the formation of networks of crosslinks which provide the rigidity and stability of structure that permit photopolymerization and photocrosslinkable polymers to be of practical importance in image formation. The differences between photopolymerization and the photocrosslinking of polymers for practical applications are not great and lie only in the details of the photocrosslinking mechanism. Photopolymerization could more correctly be termed photoinitiation since the primary act of initiation is the only step which is light dependent. During long irradiation times photografting could occur during the propagation and after the termination steps leading to crosslinking or branching. The presence of two radical species appearing during photocrosslinking of poly(vinyl cinnamate) lends credence to the idea that two processes could be occurring, one due to a radical at the cinnamate unit and the other due to a radical on the polymer chain. The two radical sites would result in photodimerizing-type crosslinking and radical reactions of the backbone chain.

V. References

[1] *Hoffman, A.:* Ann. Chem. *115*, 271 (1860).
[2] *Murray, A.:* U. S. Patent 1,965,710, July 10, 1934.
[3] *Delzenne, G.:* Ind. Chim. Belge. *24*, 739, (1959).
[4] *Oster, G.*, and *N. L. Yang:* Chem. Rev. *68*, 125 (1968).
[5] *Gersheim, H.:* History of Photography, Chapter 5. London: Oxford Press 1966.
[6] *Minsk, L. M., J. G. Smith, W. P. VanDeusen,* and *J. F. Wright:* J. Appl. Polymer Sci. 2, 302 (1959).

7) *Hepher, M.:* J. Phot. Sci. *12*, 181 (1964).
8) *Beebe, M. E., A. Murray,* and *H. V. Herlinger:* U. S. Patent 1,587,264, June 1, 1926; U. S. Patent 1,587,273, June 1, 1926; U. S. Patent 1,658,510, Feb. 7, 1928.
9) Printing plates: *L. Plambeck, Jr.,* U. S. Patent 2,760,863, Aug. 28, 1956; U. S. Patent 2,791,504, May 7, 1957; *S. H. Munger,* U. S. Patent 2,923,673, Feb. 2, 1960; *R. H. Beck,* and *C. W. Smith,* U. S. Patent 3,218,170, Nov. 16, 1965; *G. A. Thommes,* Can. Patent 760,388, June 6, 1967; *H. Wilhelm, H. Henkler,* and *R. Gehm,* Belgian Patent 685,315, Aug. 10, 1966.
Image transfer: *R. W. Baxendale, G. W. Luckey,* and *H. C. Yutzy,* U. S. Patent 3,346,383, Oct. 10, 1967; *A. Thommes, P. Walker* and *G. W. Michel,* Belgian Patent 682,052, June 3, 1966; *S. Bauer,* Can. Patent 791,655, Aug. 6, 1968; *J. R. Celeste,* Belgian Patent 685,011, Aug. 3, 1966.
10) *Berkower, L. J., J. Beutel,* and *P. Walker:* Phot. Sci. Eng. *12*, 283 (1968).
11) *Smets, G., W. DeWinter.* and *G. Delzenne:* J. Polymer Sci. *55*, 767 (1961).
12) *Oster, G.,* and *O. Shibata:* J. Polymer Sci. *26*, 233 (1957).
13) *Cooper, W.,* and *M. Fielden:* J. Polymer Sci. *28*, 442 (1958).
14) *Oster, G.:* Brit. Patent 856,884, Dec. 21, 1960.
15) *Tsunoda, M., M. Tanaka,* and *N. Murata:* Kobunshi Kagaku *22*, 107 (1965).
16) *Cooper, W., G. Vaughn, S. Miller,* and *M. Fielden:* J. Polymer Sci. *34*, 651 (1959).
17) *Geacintov, N., V. Stannett, E. W. Abramson,* and *J. J. Hermans:* J. Appl. Polymer Sci. *3*, 54 (1960).
18) — — — Makromol. Chem. *36*, 52 (1959).
19) *Needles, H. L.:* Polymer Letters *5*, 595 (1967).
20) — J. Appl. Polymer Sci. *12*, 1557 (1968).
21) *Dhandraj, J.,* and *J. E. Guillet:* J. Polymer Sci., C 23, 433 (1968).
22) *Schonberg, A., G. O. Schenk,* and *H. O. Neumuller:* Preparative Organic Photochemistry. New York: Springer 1968.
23) *Hasegawa, M.,* and *Y. Suzuki:* Polymer Letters *5*, 815 (1967).
24) *Iguchi, M., H. Nakamishi,* and *M. Hasegawa:* J. Polymer Sci., Pt. A—1 *6*, 1055 (1968).
25) *Cohen, M. D.,* and *G. M. J. Schmidt:* J. Chem. Soc. 1996 (1964).
26) *Hasegawa, M., F. Suzuki, H. Nakamishi,* and *Y. Suzuki:* Polymer Letters *6*, 293 (1968).
27) *Miura, M., T. Kitami,* and *K. Nagakubo:* Polymer Letters *6*, 463 (1968).
28) *Tazuke, S.,* and *S. Okamura:* Regional Technical Conference, Society of Plastic Engineers, p. 98. Nov. 6, 1967, postprints.
29) — — Polymer Letters *6*, 173 (1968).
30) —, *M. Asai,* and *S. Okamura:* J. Polymer Sci. Pt. A—1, *6*, 1809 (1968).
31) *Reynolds, G. A., J. A. VanAllan,* and *D. G. Borden:* U. S. Patent 3,143,423, Aug. 4, 1964.
32) *Allen, C. F. H.,* and *J. A. VanAllan:* U. S. Patent 2,566,302, Sept. 4, 1951.
33) *Robertson, E. M., W. P. VanDeusen,* and *L. M. Minsk:* J. Appl. Polymer Sci. 2, 308 (1959).
34) *Unruh, C. C.,* and *A. C. Smith, Jr.:* J. Appl. Polymer Sci. *3*, 310 (1960).
35) *Merrill, S. H.,* and *C. C. Unruh:* J. Appl. Polymer Sci. 7, 273 (1963).
36) *Lyalikov, K. S.,* et al.: Zh. Nauchn. i Prikl. Fotogr. i Kinematogr. *10*, 200 (1965).
37) *Yashinago, T., N. Kan,* and *S. Kikuchi:* J. Chem. Soc. Japan (Pure Chem. Sec.) *66*, 665 (1963).
38) *Leubner, G. W., J. L. R. Williams,* and *C. C. Unruh:* U. S. Patent 2,811,510, Oct. 29, 1957.
39) —, and *C. C. Unruh:* U. S. Patent 3,257,664, June 21, 1966.
40) *Robertson, E. M.,* and *W. West:* U. S. Patent 2,732,301, Jan. 24, 1956.

[41] *Tsuda, M.:* J. Polymer Sci., Part A 2, 2907 (1964).
[42] *Kichuchi, S.,* and *K. Nakamura:* Tech. Reg. Conf. Soc. Plastics Eng., p. 175. Nov. 6, 1967.
[43] *Tsuda, M.:* J. Soc. Sci. Phot. Japan 28, 7 (1965).
[44] *Nakamura, K.,* and *S. Kichuchi:* J. Chem. Soc. Japan 87, 930 (1966).
[45] *Leubner, G. W.,* and *C. C. Unruh:* U. S. Patent 3,257,664, June 21, 1966.
[46] *Curme, H. G., C. C. Natale,* and *D. J. Kelley:* J. Phys. Chem. 71, 767 (1957).
[47] *Sonntag, F.,* and *R. Srinivasan:* Tech. Reg. Conf. Soc. Plastics Eng., p. 163. Nov. 6, 1957, postprints.
[48] *Nakamura, K., T. Sakata,* and *S. Kikuchi:* Bull. Chem. Soc. Japan 41, 1765 (1968).
[49] *Marvel, C. S.,* and *G. H. McCain:* J. Am. Chem. Soc. 75, 3272 (1953).
[50] *Buettner, A. V.:* J. Phys. Chem. 68, 3253 (1964).
[51] *Rieser, A.,* and *R. Marley:* Trans. Raraday Soc. 64, 1806 (1968).
[52] *Unruh, C. C.:* J. Appl. Polymer Sci. 2, 358 (1959).
[53] *Lutz, R. E.,* and *R. H. Jordan:* J. Am. Chem. Soc. 72, 4090 (1950).
[54] *Kuhn, L. P., R. E. Lutz,* and *C. R. Bauer:* J. Am. Chem. Soc. 72, 5058 (1950).

Received April 3, 1969

Photochemistry of *o*-Quinones and α-Diketones

Prof. Mordecai B. Rubin

Department of Chemistry, Technion-Israel Institute of Technology, Haifa, Israel

Contents

Introduction

The photochemistry of *o*-quinones and non-enolic α-diketones has a venerable history dating at least to 1886 when *Klinger* [82] reported sunlight irradiations of benzil and 9,10-phenanthrenequinone. The latter compound and biacetyl have been extensively investigated for many years and interest has broadened to embrace a wide variety of substances. The basic processes involved depend in large measure on the presence of the vicinal dicarbonyl system rather than on classification as diketone or *o*-quinone; it appears both justified and desirable to consider their photochemistry jointly. α-Diketones which exist in the enolic form behave for the most part as substituted α,β-unsaturated ketones and will not be considered. For convenience, the term dione will be used when reference is intended to both α-diketones and *o*-quinones.

A review [149] of sunlight reactions of non-enolizable α-diketones (including *o*-quinones) appeared in 1947 and *Bruce* [34] has reviewed the photochemistry of quinones recently. The reader is referred to recent books [1–4] on photochemistry for general background material particularly with reference to the related photochemistry of monoketones.

251

Except for biacetyl, hexafluorobiacetyl, and benzil, photoreactions of o-quinones and α-diketones have been investigated in solution. Some of these reactions are elegantly clean and quantitative as illustrated in Fig. 1 for the irradiation (4358 Å) of a degassed solution of phenanthrenequinone in dioxane where four isosbestic points are observed (a fifth isosbestic point appears at 316 nm).

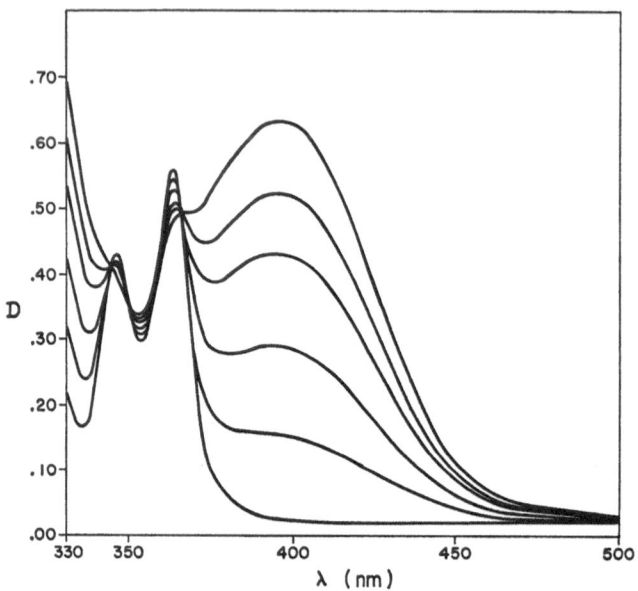

Fig. 1. Spectral changes upon irradiation of a degassed solution of phenanthren-equinone in dioxane at 4358 Å

Since diones have appreciable absorption at longer wavelengths than most organic compounds, including their products of reaction, it is usually feasible to perform photoreactions using simple filters so that light is absorbed only by the dione. The initial excited species, at least, can then be clearly defined. Proper choice of wavelength can also ensure irreversibility of reactions and eliminate problems of overirradiation.

As investigations of their solution photochemistry have proliferated, it has become increasingly clear that it is nearly impossible to find a completely unreactive solvent for some diones. Thus, for example, the quantum yield for disappearance of phenanthrenequinone in degassed benzene solution at 4358 Å is 0.25 [131] (perfluoro compounds, which might indeed be unreactive, are extremely poor solvents for many diones).

Nevertheless, differences in reactivity between solvent and substrate are often sufficiently great to allow use of "relatively" inert solvents such as benzene but their possible participation in reaction cannot be safely ignored. A special complication is introduced in hydroxylic media where addition of solvent to the dione to give monohydrate or mono-hemiketal *(1)* can occur. Such behaviour, readily detectable spectroscopically, is

$$\text{(diketone)} + \text{ROH} \rightleftharpoons \text{(hemiketal, OH, OR)}$$

1

frequently observed to an appreciable extent with α-diketones [11,14,62, 81,85,138,174] but generally not with *o*-quinones. In some cases, the equilibrium constant for reaction with solvent is not large and it is still possible to observe diketone reactions by irradiation at sufficiently long wavelengths. However, if shorter wavelengths of light are used, the observed results may reflect dione photochemistry only in part or not at all. For example, the lower quantum yield of phosphorence of aqueous solutions of biacetyl compared with hydrocarbon solutions [7] is probably due to hydration.

Another complicating factor is the role played by oxygen. In some cases oxygen appears to retard reaction and in others to accelerate. In many cases the overall course of reaction is dependent on the presence or absence of oxygen. This is illustrated in Fig. 2 for the reaction of phenanthrenequinone with ethyl acetate. In degassed solution, the spectra determined after various periods of irradiation presented an appearance very similar to Fig. 1 and the quantum yield was unity [132]. When degassing was omitted, results such as illustrated in Fig. 2 were obtained and quantum yields were variable and less than unity.

In the absence of the complications discussed above, the observed photoreactions of diones can be conveniently classified into three general types:

1) Cleavage
2) Cycloaddition
3) Hydrogen abstraction

These reactions are analogous to processes observed with monoketones but their variety is considerably enhanced by the added possibilities for reaction introduced by the second carbonyl group. Reactions in the presence of oxygen will be discussed separately. Before turning to these reactions, a brief summary of spectroscopic properties of diones is desirable.

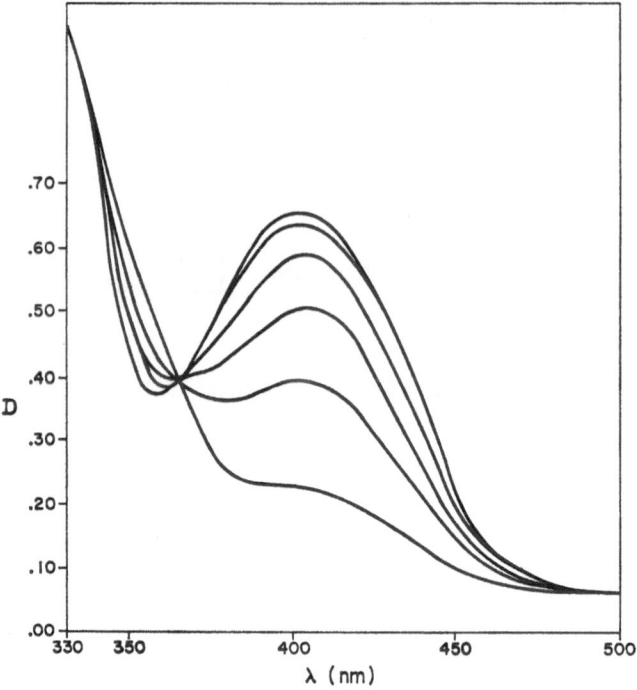

Fig. 2. Spectral changes upon irradiation of an air-saturated solution of phenan-threnequinone in ethyl acetate

Absorption Spectra

Absorption spectra of diones exhibit intense maxima due to π,π^* transitions and weak, longer wavelength maxima due to "forbidden" n,π^* transitions. The n,π^* transitions are of particular interest since (a) they often show a considerable dependence on fine details of structure, (b) their relatively long wavelengths usually allow selective irradiation in the presence of other substances and (c) the n,π^* state is believed to be the reactive state in most photoreactions of diones. The spectrum of biacetyl has been analyzed by *Sidman* and *McClure* [155] and that of camphorquinone by *Ford* and *Parry* [55].

Long wavelength maxima of representative diones are presented in Table 1. The data have been chosen to give examples of various types of compounds and provide leading references. These low intensity maxima sometimes exhibit fine structure but are often broad and poorly defined and show a considerable degree of solvent dependence [55,56,88,109]. *Forster* [56] has measured the spectrum of biacetyl in 25 solvents. Weak

Table 1. *Long wavelength absorption maxima of representative diones*

Compound	Sol-vent[a]	λ_{max}(log ε)	Lit.
Biacetyl	CH	276, 421, 439, 447 nm[b]	6,56,155)
Cyclohexane-1,2-dione	—[c]	412 (~ 1)	11)
Camphorquinone	H	270[d](1.3), 470(1.57)	55)
Bicyclo[2,2,2]octane-2,3-dione	CH	293(1.18), 447, 457, 465, 478[b]	6)
Dibenzobicyclo[2,2,2]octane-2,3-dione	CH	339(2.32), 357(2.43), 377(2.62), 458(2.99)	25)
4,4,2-Propella-11,12-dione *(2)*	CH	461(1.86)	29)
4,4,2-Propella-3,8-diene-11,12-dione *(11)*	CH	537.5(1.86)	29)
2,2,5,5-Tetramethyl-3,4-tetrahydro-furandione *(72)*	CH	280(2.56), 540	139)
3,3,5,5-Tetramethylcyclopentane-1,2-dione	CH	280(1.18), 505(1.54)	139)
3,3,6,6-Tetramethylcyclohexane-1,2-dione	A	297.5(1.46), 380(1.05)	92)
3,3,7,7-Tetramethylcycloheptane-1,2-dione	A	299(1.54), 337(1.53)	92)
3,3,8,8-Tetramethylcyclooctane-1,2-dione[e]	A	295.5(>1.64), 343(1.34)	92)
3,3,16,16-Tetramethylcyclohexadecane-1,2-dione	A	286.5(1.77), 384(1.34)	92)
Benzil	A	259(4.31), 370(1.89)	91)
2,4,6,2′,4′-Pentamethylbenzil	A	275(4.24), 400(1.94)	91)
2,4,6,2′,4′,6′-Hexamethylbenzil	A	255—290(3.41), 476(1.69) 493(1.71)	91)
2-Furil	EPA	400	50)
2-Pyridil	EPA	360	50)
Diphenylcyclobutenedione	A	322(4.21), 410(2.21)	26)
Benzocyclobutenedione *(13)*	A	301(3.72), 427(2.44)	38)
o-Benzoquinone	Bz	390(3.5), 610(1.3)	76,109)
o-Naphthoquinone	CCl$_4$	390(3.48), 498(1.59), 538(1.73)	154)
9,10-Phenanthrenequinone	Bz	411(3.24), 500(1.45)	88,154)
Acenaphthenequinone	A	337(3.29), 480(0.97)	88)

[a] A = 95% ethanol; CH = cyclohexane; EPA = ether-isopentane-ethanol; H = hexane.
[b] Log ε ~ 1.5.
[c] Various solvents.
[d] Most intense of seven bands in this region.
[e] Impure material.

charger transfer interactions of biacetyl with benzene and *p*-xylene have been suggested by *Almgren* [7].

Interactions between the two vicinal carbonyl groups allow two n,π^* transitions in diones. Both of these are observed in unconjugated diones, the shorter wavelength absorption lying in the range 270—300 nm and the longer wavelength maximum at 330—540 nm. The wide range observed for the long wavelength maximum has been attributed [92] in part to steric factors. The lowest energy conformation for open-chain diones is a *trans*-coplanar (*s-trans*) one which minimizes unfavorable dipole interactions and allows maximum overlap of π orbitals. In the presence of constraining factors, such as incorporation of the dione system in a

s-trans   s-cis

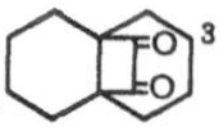

2

ring, the two carbonyl groups may be forced to assume other conformations. It has been suggested [139] that the rigid *s-cis* conformation has λ_{max} 500 nm in the absence of perturbing factors (such as ring strain). *Leonard* and *Mader* [92] have reported spectra of a series of cyclic diones in which the dihedral angle between the carbonyl groups varied with ring size. As the angle increased from 0° the long wavelength maximum shifted to shorter wavelength reaching a minimum at 90° ($\sim$ 340 nm) and then shifting back to longer wavelength as the dihedral angle approached 180° (*s-trans*, 400—450 nm). The short wavelength n,π^* absorption exhibited much small shifts in the opposite sense. Conformations of diones may also be affected by steric interactions as illustrated [91] in the comparison between benzil (370 nm) and 2,4,6,2′,4′,6′-hexamethylbenzil (476, 493 nm).

In addition, ring strain [139] and intramolecular interactions [25,29,139] may be important. Spectra of biacetyl and benzil showed marked changes when determined [175] in silica gel-cyclohexane slurry. The shift of the n,π^* maximum to shorter wavelength in the series *o*-benzoquinone, *o*-naphthoquinone, 9,10-phenanthrenequinone has been discussed by *Murrell* [103].

The short wavelength n,π^* absorption may be buried in the intense π,π^* absorption of conjugated diones. The extensive conjugation present in *o*-quinones leads to a number of intense π,π^* bands which tail into the visible. Conjugated diones such as benzil show π,π^* bands usually in the range 250—300 nm. Only the tail of the π,π^* absorption is observed in spectra of unconjugated diones determined with the usual equipment. Low temperature spectra of a variety of diones have been described [180] recently.

Emission Spectra

Diones are characterized by weak fluorescence and relatively intense phosphorescence. For biacetyl, which has been investigated by many workers under a wide variety of conditions, the phosphorescence quantum yield (ϕ_P) in the gas phase [49] (27°) at 405 or 436 nm was 0.145 and the fluorescence quantum yield (ϕ_F) was 0.0023. Relative intensities ($\phi_P : \phi_F$) determined in another laboratory [114] at 3650 or 4358 Å (26°) were 58 ($\pm$8) : 1. This ratio decreased with increasing temperature; phosphorescence was completely quenched by added oxygen [114].

α-Diketones exhibit the unusual and very useful (*vide infra*) property of fluorescence and phosphorescence in fluid solution. The intensity of phosphorescence was not appreciably affected [58] by variations in viscosity in a series of hydrocarbon solvents but appears to be dependent on a variety of factors including solvent [7,125,141], light intensity, temperature, and the presence of impurities. In a series of measurements made in one laboratory [7] with biacetyl in benzene solution, ϕ_P varied from 0.0228–0.0800 and phosphorescence lifetime (τ) from 183–644 μsec. Nonetheless, radiative lifetimes (τ/ϕ_P) determined in various laboratories were approximately 10 msec for biacetyl, benzil, and other open chain diones.

Phosphorescence spectra in low temperature glasses have been determined for a number of diones. The triplet energies calculated from the O—O bands in such spectra are collected in Table 2. The values obtained from room temperature spectra show only small deviations from values obtained at 77° K. It is interesting to note that the entire range of triplet energies is 47—59 kcal/mole for all compounds investigated including unconjugated and conjugated diketones as well as *o*-quinones.

Intersystem crossing efficiencies ($S_1 \rightarrow T_1$) have been calculated for biacetyl [7,10,122,126], 2,3-pentanedione [126], 2,3-heptanedione [126], 2,3-octanedione [126], benzil [89,115], and 9,10-phenanthrenequinone [30]. In all cases the values were close to unity supporting the assumption that the triplet state is the reactive excited state in dione photochemistry.

Parker [115] has recently reported delayed fluorescence (quantum yield of 0.0015) from benzil. The activation energy corresponded to 0.20 $\pm$ 0.02 μm^{-1} for this $T_1 \rightarrow S_1$ transition.

From the photochemical point of view, it would be useful to know the geometry of the reactive excited state of diones. For example, a system having the *s-trans* conformation would be unlikely to undergo a concerted 1,4-addition across the dione moiety. There is practically no information presently available on this point. On the basis of the large separation between absorption and fluorescence bands and the lack of mirror image symmetry between absorption and emission spectra of pivalil *(3)*, *Evans* and *Leermakers* [50] concluded that there is a significant

Table 2. *Triplet-state energies of diones* [a])

Compound	E_T(kcal/mole)			Lit.
	Hydrocarbon solvent	Polar solvent	Other	
Biacetyl	54.5 [b])	56	58 [c])	50,58,73,126)
2,3-Pentanedione	54[b])	56		50,73,126)
2,3-Heptanedione	54[b])	56		126)
2,3-Octanedione	54[b])	56		126)
Pivalil (3)		52		50)
Camphorquinone		52	48 [c])	50,55)
Cycloheptanedione		57.5		50)
Cyclodecanedione		55.5		50)
1-Phenyl-1,2-propanedione		54		50)
Benzil	54	53, 57 [d])		50,72,73,74)
α-Pyridil		59		50)
α-Furil		55		50)
β-Naphthil (4)	54, 58 [d])	54, 58 [d]) 54		74)
Acenaphthenequinone	50			154)
1,2-Anthraquinone	47			154)
9,10-Phenanthrenequinone	49			154)

[a]) From O—O bands of phosphorescence spectra at 77° K.
[b]) Identical values obtained in fluid solution at room temperature.
[c]) Crystal.
[d]) Two Bands of different phosphorescent lifetime.

(unspecified) change in geometry between ground state and emitting state. *Almgren* [7]) has suggested that the excited state geometries of biacetyl and benzil are more similar than the ground state geometries.

Existence of s-cis and s-trans triplet states of β-naphthil (4) has been inferred [74]) from the observation of two states (E_T 58 and 54 kcal/mole)

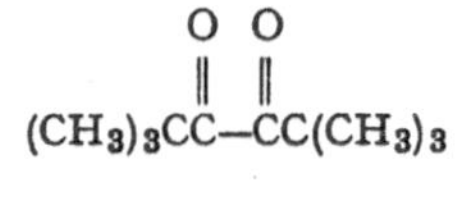

3

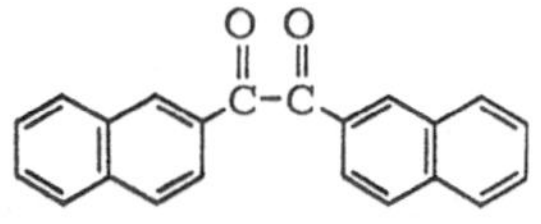

4

of different phosphorescence lifetimes in EPA or MCIP glasses. The higher energy, longer-lived triplet was considered to have the less stable s-cis conformation since it was not observed in an isopentane, 3-methyl-pentane glass whose setting temperature was very close to 77° K thereby allowing greater possibility of conformational equilibration. A similar situation may occur with benzil [74,154]. Variations in emission spectra of biacetyl, benzil, o- and p-anisyl due to variations in medium have been atributed [181], in part, to conformational factors.

Energy Transfer

Biacetyl has been employed extensively in studies of energy transfer processes due to the fortunate combination of weak light absorption ($\varepsilon_{max} < 30$) and the property of fluorescing and phosphorescing in fluid solution at ordinary temperatures. It is thus possible to irradiate a wide variety of compounds under conditions such that little, if any, light is absorbed by biacetyl and, in cases where energy transfer occurs, to observe its characteristic fluorescence and/or phosphorescence. Alternatively, biacetyl may be irradiated in the long wavelength region where many compounds do not absorb and quenching of its fluorescence and/or phosphorescence observed. Quantitative information concerning singlet-singlet or triplet-triplet energy transfer may be obtained from measurements of absolute or relative intensities of emission.

These methods have been used for characterization of excited states and for investigations of intimate details of energy transfer processes. Compounds studied include simple ketones [46,70,71,98,168], conjugated ketones [10,44,102,125,126,170], monoolefins [9,123], benzene [47,48,78,93,116, 170], and a wide variety of aromatic compounds [9,46,47,59,80,140,142, 161,170,182]. A note of caution should be sounded concerning quantitative conclusions based on emission intensities of biacetyl or other diones. As mentioned earlier, these intensities are subject to considerable deviation as a result of experimental variables including concentration, solvent, temperature and presence of trace amounts of impurities. Recently it has been shown [102] that the efficiency of benzophenone sensitized biacetyl phosphorescence decreases with increasing intensity of exciting light.

The benzil-biacetyl system provides a particularly interesting example of energy transfer studies. The absorption spectra of these two diones allow selective irradiation and the emission spectra differ sufficiently to allow calculation of the contribution of each to the total emission. These facts were exploited by *Richtol* and *Belorit* [124] to observe behaviour of both partners in a photosensitized process.

Turning to photochemical reactions, triplet energies of diones lie in an intermediate position relative to those of organic compounds in general. Diones may, therefore, undergo sensitized reactions; these will be discussed as appropriate in the sequel. Diones may also function as sensitizers and a number of examples have been recorded, mainly by

Hammond and collaborators as part of their extensive studies on photo-sensitized reactions. Examples include (a) *cis-trans* isomerization of olefins sensitized by biacetyl [42,64,65,176], 2,3-pentanedione [64,65], benzil [64,65,89], α- and β-naphil [64], and phenanthrenequinone [30,65] (b) diene and anthracene dimerizations sensitized by biacetyl [9,45,94], 2,3-pentanedione [94,162], camphorquinone [94], benzil [94,162], and β-naphthil [94], and (c) the quadricyclene-norbornadiene interconversion sensitized [66] by benzil.

Bohning and *Weiss* [30] have suggested that *cis-trans* isomerization of stilbenes may proceed via an intermediate addition product (*vide infra*) and not by energy transfer. The results obtained in the biacetyl-anthracene system are not completely clear; *Backstrom* and *Sandros* [9] reported isolation of anthracene dimer while *Dubois* and *Behrens* [45] suggested that a 1:1 adduct of unspecified structure is formed.

Cleavage Reactions

Homolytic cleavage of a bond alpha to the carbonyl group is a common reaction of monoketones in the gas phase [1-4] and is observed in solution [158] in cases involving special factors such as stabilization of intermediate radicals. A similar generalization can be applied to diones. Three types of primary cleavage might be envisioned with diones:

$$\underset{R-C-C-R'}{\overset{O\quad O}{\overset{\|\quad\|}{}}} \longrightarrow \underset{R-C\cdot + R'-C\cdot}{\overset{O\qquad O}{\overset{\|\qquad\|}{}}} \tag{1}$$

$$\longrightarrow \underset{R-C-C\cdot + R'\cdot}{\overset{O\ O}{\overset{\|\ \|}{}}} \tag{2}$$

$$\longrightarrow R\cdot + R'\cdot + 2\ CO \tag{3}$$

(1) Cleavage of the bond joining the two carbonyl groups (reaction 1 above) giving rise to two acyl radicals. (2) Cleavage between a carbonyl group and the adjacent alkyl group (reaction 2) to form an alkyl radical and an α-keto-acyl radical. (3) Simultaneous cleavage of two bonds (reaction 3) with formation of two alkyl radicals and two molecules[a] of carbon monoxide. Subsequent reactions of the various radicals suggested

---

[a] It has recently been suggested [173] that the oxalyl diradical or a species $(CO)_2$ may be extruded directly from diones and be capable of finite existence.

would be unexceptional. An electrocyclic process in which a bis-ketene is formed directly from an *o*-quinone is also conceivable (*vide infra*).

Cleavage of cyclic diones by reactions 1 or 2 might be a fast, reversible process (after necessary spin inversion if triplet states are involved) and,

$$(CH_2)_n \underset{\substack{C=O \\ \\ C=O}}{\big|} \underset{\text{dark}}{\overset{h\nu}{\rightleftharpoons}} (CH_2)_n \underset{\substack{\dot{C}=O \\ \\ \dot{C}=O}}{}$$

therefore, not be observable except under special conditions. In fact, almost all cleavages involving cyclic diones do involve some special feature such as the relief of strain upon cleavage of a four-membered ring or the stabilization gained by generation of an aromatic system. Discussion of individual cases follows.

Biacetyl. Photolysis of biacetyl was first reported in 1923 by *Porter, Ramsperger*, and *Steel* [121] who irratiated the vapor at 100° and identified the products as ethane and carbon monoxide. Subsequently methane, acetone, ketene, and 2,3-pentanedione were also recognized as products. Considerable effort has been expended to achieve a full understanding of the details of the primary processes and progress was reviewed by *Noyes, Porter* and *Jolley* [113] in 1956. Irradiations have been performed over a wide range of pressure, temperature and wavelength (including the far ultraviolet [68] and mercury photosensitization [69]). Overall and individual quantum yields varied widely.

All of the reaction products can be rationalized in terms of primary cleavage to acetyl radicals and analogy to the photolysis of acetone. However, the fact that appreciable carbon monoxide was observed [113] in iodine trapping experiments, in contrast to the results with acetone, suggests that some other primary dissociation also occurs.

Noyes, Mulac and *Matheson* [112] concluded that interaction of *two excited biacetyl molecules* occurs in gas phase reactions at 4358 Å and moderate temperature. At higher temperatures, they suggested that one of the reaction products strongly inhibited both phosphorescence and cleavage. Recently, *Lemaire* [90] proposed that this inhibitor is the enolic form of biacetyl. Irradiation of biacetyl at 120° and 4358 Å or at room temperature and shorter wavelengths resulted in intensification of absorption at 275 nm. Titration with ICl allowed calculation of an extinction coefficient of about 10,000. Although a mechanism for its formation was not proposed, the enol ($CH_2=C(OH)COCH_3$) seems to be a reasonable hypothesis. Quantum yields for its formation in the gas phase in the wavelength range 2537—3130 Å were 0.10—0.13; quantum yields in solution where ketonization might be much faster were lower.

The reactions of biacetyl in fluorocarbons and mineral oil parallel the gas phase except that quantum yields were appreciably reduced [61]. Hydrogen abstraction is the major process in less inert solvents but products containing an acetyl group (e. g. acetylcyclohexane) were observed [20] to an appreciable extent from irradiations in cyclohexane, ether, and dioxane; cyclohexene, ethylbenzene, and 2-propanol reacted only by H-abstraction. Photoirradiation of biacetyl and phenylacetic or phenoxyacetic acid in acid medium produced [12] benzyl methyl ketone

$$C_6H_5CH_2COOH + CH_3COCOCH_3 \xrightarrow[H_2O]{H^+} CO_2 + C_6H_5CH_2COCH_3$$
$$\lambda > 300$$

or phenoxyacetone; a complex mechanism not involving primary dissociation of biacetyl was proposed.

The formation of N,N-difluoroacetamide in 80% yield from irradiation of biacetyl in the presence of tetrafluorohydrazine has been reported [117].

$$CH_3COCOCH_3 + N_2F_4 \longrightarrow CH_3CONF_2$$

Ethyl radicals formed by photolysis [63] of biacetyl and triethylborine were assumed to arise *via* displacement by acetyl radicals.

Hexafluorobiacetyl. In the gas phase, hexafluorobiacetyl behaved [169] in the same manner as biacetyl, affording carbon monoxide and hexafluoroethane in stoichiometric ratio. However, in the presence of hydrocarbon vapor or in hydrocarbon solution, the disappearance of dione was not accompanied by formation of cleavage products. For example, photolysis ($\lambda > 3000$ Å, 65°) of hexafluorobiacetyl in 150 molar excess of 2,3-dimethylbutane until 84% of starting material had disappeared resulted in only 0.5% of carbon monoxide, an equivalent amount of trifluoromethane, and a complex mixture of other products. On the basis of carbonyl and hydroxyl absorption in the infrared, it was assumed that the excited dione abstracted hydrogen from hydrocarbons to give products of photoreduction, photoaddition, etc. Occurrence of H-abstraction in the gas phase is unique.

Benzil. Photolysis of benzil vapor at 200° to give carbon monoxide and benzophenone was reported [121] in 1923. The gas phase reaction has not been investigated since then.

$$\underset{\displaystyle C_6H_5C\ C\ C_6H_5}{\overset{\displaystyle O\ \ O}{\| \ \|}} \xrightarrow[h\nu]{200°} \underset{\displaystyle C_6H_5C\ C_6H_5}{\overset{\displaystyle O}{\|}} + CO$$

The quantum yield for disappearance of benzil at 3660 Å in cyclohexane solution has been estimated to be not greater than 0.25. This reaction produced [37] a complex mixture of products including benzoic acid, benzaldehyde and phenyl cyclohexyl ketone which suggest that the dione underwent cleavage to benzoyl radicals. Hydrogen abstraction appears to be the major process in other solvents.

Cleavage of benzil to benzoyl radicals has also been suggested [32] to account for the observation that irradiation of a polystyrene film containing benzil produced a film which initiated polymerization of acrylonitrile. This initiation was observed even several hours after termination of irradiation.

Benzil reacted [117] with tetrafluorohydrazine to give N,N-difluorobenzanide in unspecified yield.

Formation of aromatic systems. In the months immediately preceding completion of this review, two groups [36,173] reported a new bis-decarbonylation reaction of bicyclo [2,2,2]octadienediones to give benzene derivatives. The compounds reported to react in this way are shown below. These same compounds *(6—9)* were stable to temperatures of 180° or more and gave complex product mixtures upon pyrolysis at higher temperatures.

The formation of the aromatic system provides a convenient rationale for the photolytic reactions. However, compound *10* also eliminated carbon monoxide to form the appropriate cyclohexadiene derivative. It has been suggested [173] that an electrocyclic reaction governed by orbital symmetry considerations is involved rather than cleavage to acyl radicals which then decarbonylate.

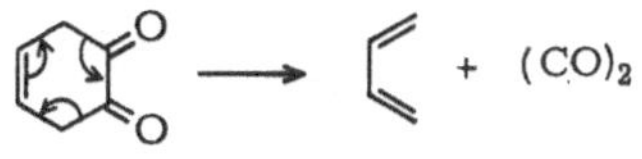

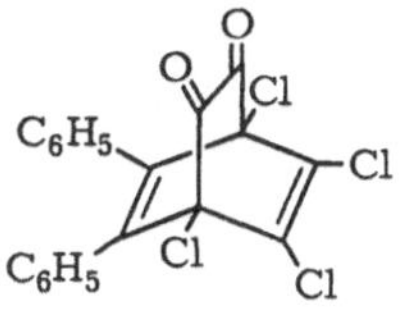

5

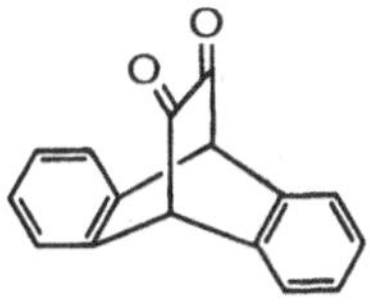

6

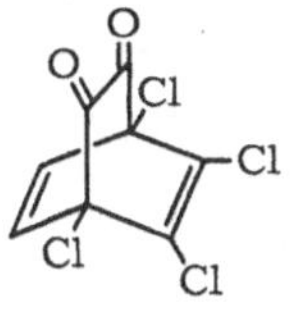

7

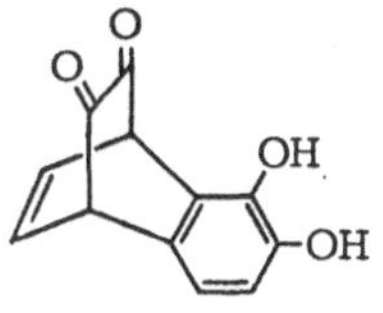

8

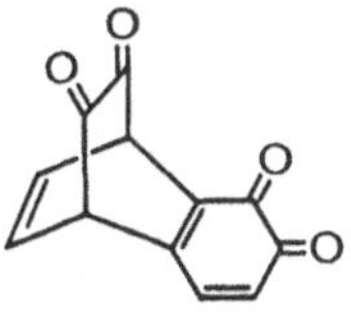

9

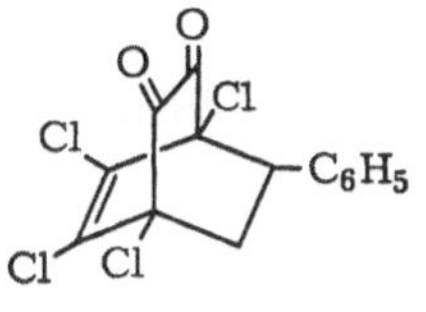

10

Cleavage of 4-membered ring diones. Cleavage has been observed in all cases where the dione system is incorporated in a four-membered ring, undoubtedly reflecting the strain present in such systems. The only derivative of cyclobutanedione which has been investigated [28] is the [4,4,2]propellane *(11)* which underwent quantitative bis-decarbonylation

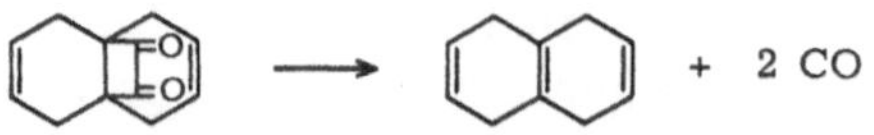

to isotetralin *(12)* upon sunlight irradiation in methanol, carbon tetrachloride, pentane, and benzene solutions.

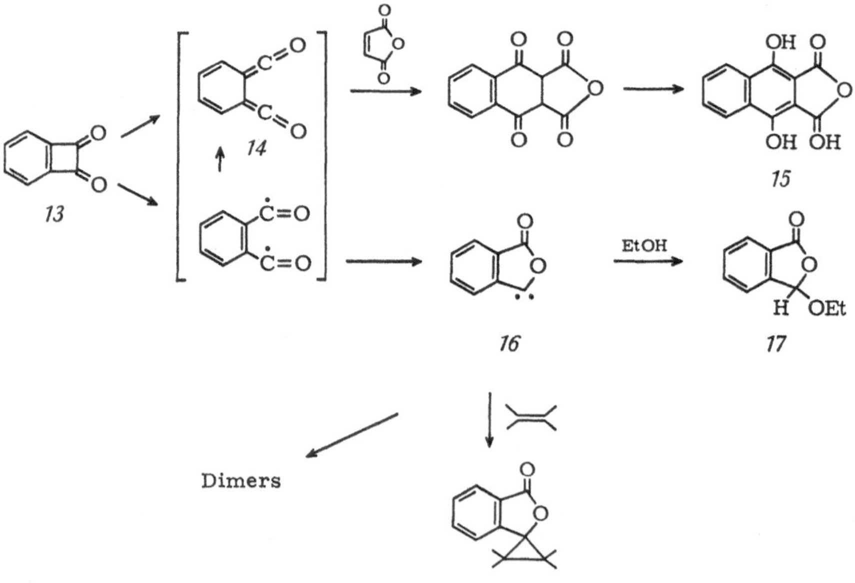

Cleavage without decarbonylation has been observed with derivatives of cyclobutenedione ("cyclobutadienequinone"), the initial product being either a bis-ketene or a diacyl radical depending on whether the reacting excited state is singlet or triplet. Thus, the product of photolysis [95] of phenylcyclobutenedione in methanol was dimethyl phenylsuccinate plus an unidentified substance. The bis ketene intermediate *(14)* formed from benzocyclobutenedione *(13)* has been trapped [157] as the Diels-Alder adduct *(15)* with maleic anhydride (and 1,4-naphthoquinone). Interestingly, *14* was not trapped by ethanol but rearranged to the presumed carbene *(16)* which reacted with ethanol to give *17*. The carbene could also be trapped as the cyclopropane derivative *18* when the irradiation was performed in the presence of olefins. Dimers of *16* were obtained [33,157] when unreactive solvents were used.

Dimethylcyclobutenedione was reported [27] to give a dimeric product upon irradiation. Diphenylcyclobutenedione has been stated [24] to be stable to light in benzene or acetonitrile solution in the *absence of oxygen*.

Cleavage of Pyracycloquinone. An additional example of dione cleavage to a ketene intermediate appeared at the time of completion of this review. Photolysis [23] of pyracycloquinone *(19)* in methanol solution afforded 30% of the diester *21a* with a quantum yield of 0.3. Inter-

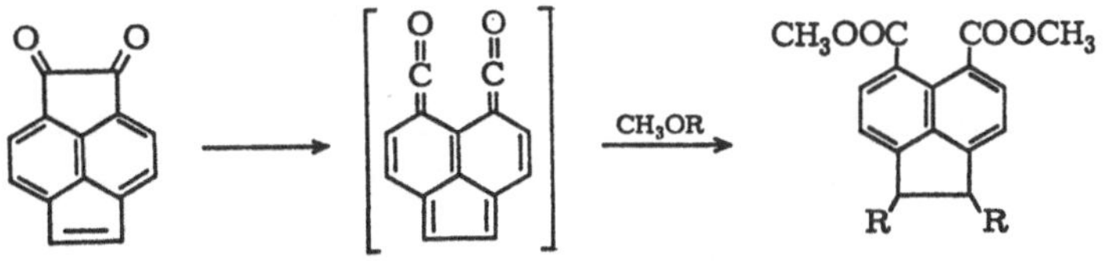

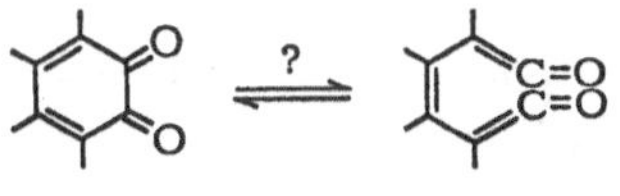

mediacy of the bis-ketene *20*, was supported by incorporation of two atoms of deuterium *(21b)* when the reaction was performed in CH_3OD. No reaction was observed in cyclohexane where *20*, if formed, would not react further but could revert to *19*. Special structural features are required for this reaction since the dihydro compound *22* was stable to light; nor has such a cleavage been reported with acenaphthenequinone. It might be noted that any o-quinone is formally capable of photolysis to a bis-ketene as illustrated but no evidence for such cleavage has been reported in the *absence of oxygen*.

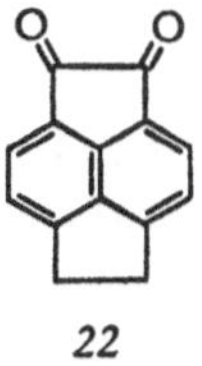

Cleavage of Tetramethyltetralindione. Photolysis of 1,1,4,4-tetramethyl-2,3-tetralindione *(23)* has been investigated [60] in a variety of solvents.

One of the byproducts observed in some cases was tetramethylindanone *(24)*; in acetic acid it constituted 6% and in benzene 14% of the total product. The most reasonable mechanism for formation of *24* is *via* cleavage and decarbonylation to give the tertiary-benzilic diradical *25* which can cyclize to *24*. The process may be facilitated by the considerable stabilization present in *25*.

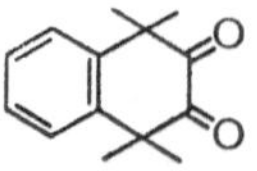

23 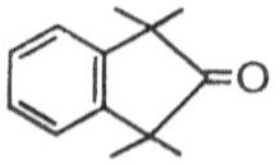   24 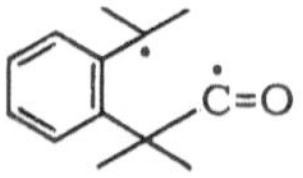   25

Cycloaddition Reactions

1. Sulfur dioxide

The photoaddition of sulfur dioxide to *o*-quinones in benzene solution to produce cyclic sulfates *(26)* in fair to high yields was reported by *Schenck* and *Schmidt-Thomée* [143)] in 1953. Products were obtained from tetrachloro-*o*-benzoquinone, 1,2-naphthoquinone, 3-nitro-1,2-naphtho-

26

quinone, phenanthrenequinone and its 2-, 3-, and 4-mononitroderivatives but not with tetrabromo-*o*-quinone, acenaphthenequinone, furil, or benzil. With benzil the pinacolic product, benzil-benzoin (*vide infra*), was obtained in unspecified yield. No further investigations of this reaction have appeared.

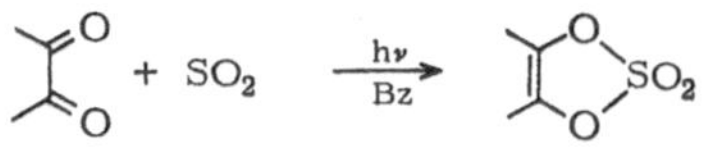

2. Triphenylphosphine

Phosphines undergo cycloaddition with diones in the dark [120)]. Recently, a report appeared [77)] describing irradiation of phenanthrenequinone in wet benzene containing triphenylphosphine. The product was considered to be a hydrate of a species such as *27*; intermediacy of *28* was also considered possible. No reaction was observed in the absence of water.

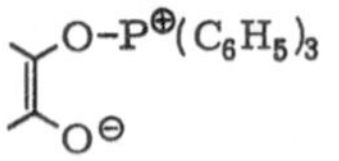

27 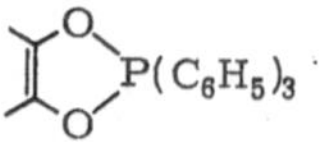   28

3. Olefins

The cycloaddition of olefins to *o*-quinones was first reported by *Schonberg* [147] in 1944 and has been reviewed recently by *Pfundt* [120]. This review, covering the literature to 1966, records more than a dozen *o*-quinones (and benzil) and innumerable olefins which participate in photochemical reaction. The only olefinic compounds which failed to react were those in which particularly efficient triplet energy transfer from quinone to olefin might be expected. In the earlier work, benzene solutions of the reactants were exposed to Cairo sunlight and the precipitated product(s) isolated, generally in fair to good yield. It was assumed in all cases that 1,4-addition of olefin across the dione moiety occurred to give 1,4-dioxenes *(30)*; questions of stereoisomerism were ignored.

Reexamination of these reactions beginning in 1965, mainly using 9,10-phenanthrenequinone *(PQ)*, has shown that 1,4-dioxenes are only one of several possible types of products and are not obtained at all in certain cases. All of the results reported before 1965 thus serve to delineate the scope of the reaction but do not establish structures of products and certainly provide no mechanistic insights. The more recent results are collected in Table 3. The following scheme serves as a reasonable working hypothesis in the light of these results. The essential features are excitation of the dione (direct excitation of olefin

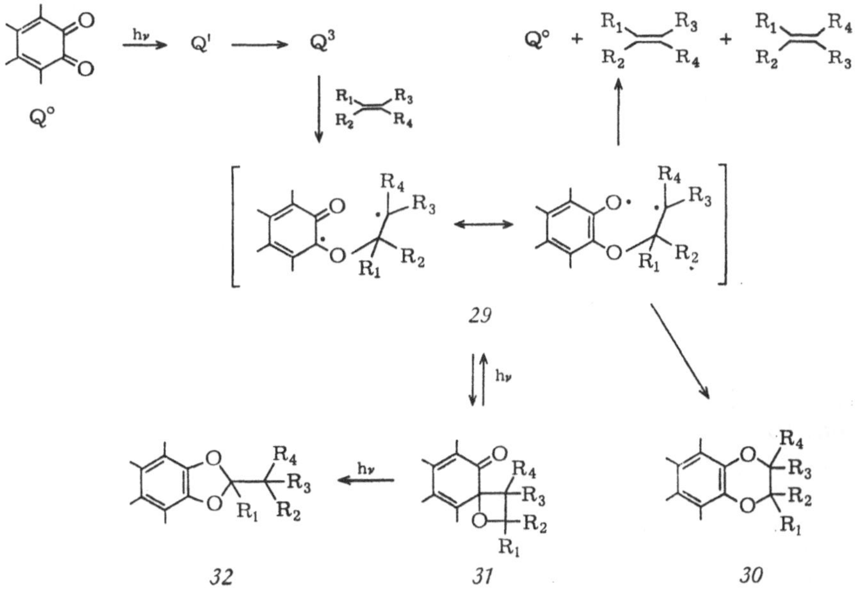

Table 3. *Photocycloadditions of diones and olefins since 1965*[a]

A. Phenanthrenequinone

Olefin	Solvent	λ (nm)	Φ	Dioxene (%)	Oxetane (%)	Other (%)	Lit.
Isobutylene	Bz	>405	0.21 [b]	75	10	10 [c]	54)
cis-2-Butene	Bz	405	0.34 [b]	~75 [d]	~5	9 [c]	39,54)
trans-2-Butene	Bz	405	0.36 [b]	~80 [d]	~5	9 [c]	39,54)
Trimethylethylene	Bz	405	0.45 [b]	~73	~5	14 [c]	54)
Tetramethylethylene	Bz	405	0.79 [b]	63	2	29 [c]	54)
trans-4-Octene	Bz	>300		$\sqrt{}$ [d]		$\sqrt{}$ [c]	39)
di-*t*-Butylethylene						10 [e]	53)
Phenyl vinyl ether	Bz	>375		48			87)
Ethyl vinyl ether	—	>375		62			87)
Ethyl vinyl sulfide	Bz	375		69		21 [c]	87)
Vinyl chloride	Bz	375		68 [f]	23 [f]		87)
cis-Stilbene	Bz	405	0.14 [g]	$\sqrt{}$ [d]			30,51,119)
trans-Stilbene	Bz	405	0.066 [g]	$\sqrt{}$ [d]			30,51,119)
α,α'-*cis*-Stilbene-d$_2$	Bz	405	0.17 [g]	$\sqrt{}$ [d]			30)
α,α'-*trans*-Stilbene-d$_2$	Bz	405	0.069 [g]	$\sqrt{}$ [d]			30)
1,1-Diphenylethylene	Bz	405	0.035 [g]				30)
Triphenylethylene	Bz	405	0.11 [g]				30)
α-Chlorostilbene						10 [e]	53)
Indene	Bz	>375		13	26		87)
N,N-Diphenyl-imidazoline						10 [e]	53)
Vinylene carbonate *(39)*	Bz	>375		—	80		52)
2,3-Dihydrofuran	Bz	375		11	—		87)
Furan	—	375		48	—		87)
2-Methylfuran	—	375		26 [d]	—		87)
2,5-Diphenylfuran	Bz	375		34	—		87)
Benzofuran	Bz	375		—	80		87)
2-Phenylbenzofuran	Bz	375		37	36		87)
Furocoumarins *(40, 41, 42)*	Bz	375		—	60—75		87)
Furochromones *(43, 44)*	Bz	375		—	70		87)
Benzothiophene-dioxide	Bz	375		—	8		87)
N,N-Diacetyl-imidazolinone *(45)*	Bz	375		6	58		159)

Table 3 (continued)

Olefin	Solvent	λ (nm)	Φ	Dioxene (%)	Oxetane (%)	Other (%)	Lit.
4-Methyl-N,N-diacetyl-imidazolinone *(46)*	Bz	>375		95	—		159)
4-Phenyl-N,N-diacetyl-imidazolinone *(47)*	Bz	375		95	—		159)
1,3-Diphenyl-Δ⁴-imidazolinone	Bz			60		6 e)	160)
1,4-Dioxene	—	375		55	40		86)
2,3,5,6-Tetraphenyl-1,4-Dioxin *(48)*	Bz	375		92	—		86)
Isocoumarin *(49)*	Bz	375		—	74		87)
2-Phenylisocoumarin *(50)*	Bz	375		71	16		87)
2,2-Dimethyl-3-Chromene *(51)*	Bz	375		25	38		87)

B. Other Diones

Dione	Olefin	Solvent	λ (nm)	Dioxene (%)	Oxetane (%)	Lit.
Benzil	Vinylene carbonate *(39)*	Bz	>375	—	37	52)
Benzil	Xanthotoxin *(42)*	Bz	375	—	94	87)
Benzil	Visnagin *(43)*	Bz	375	—	56	87)
Tetrachloro-*o*-benzoquinone	*cis*-stilbene	Acetoneⁱ⁾ >400		√ d)		33)
Tetrachloro-*o*-benzoquinone	*trans*-stilbene	Bz	400	√ d)		34)
Acenaphthene-quinone	Vinylene carbonate *(39)*	Bz	>375	—	81	52)
o-Naphtho-quinones	Dioxene	Bz	>300	√		179)

a) For earlier work see ref. 120)
b) Quantum yield for disappearance of quinone.
c) Products of H-atom abstraction.
d) Mixture of *cis*- and *trans*-isomers. Unspecified yield.
e) Dioxole.
f) Including 2:1 adducts (quinone:olefin).
g) Quantum yield for formation of adduct.
h) Reaction occurred at the unsubstituted double bond.
i) Also acetonitrile. A different reaction was observed in benzene.

269

Table 3 (continued)

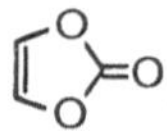

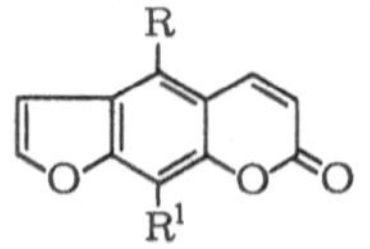

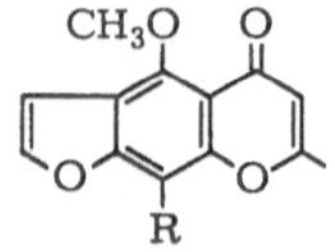

39	*40*, R = R¹ = H *41*, R = OCH₃, R¹ = H *42*, R = H, R¹ = OCH₃	*43*, R = H *44*, R = OCH₃
45, R = H *46*, R = CH₃ *47*, R = C₆H₅	*48*	*49*, R = H *50*, R = C₆H₅ *51*

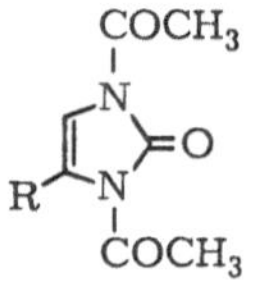

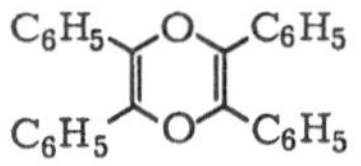

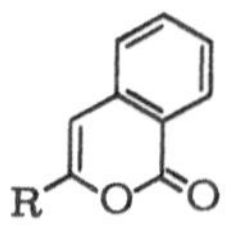

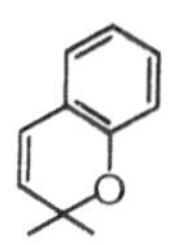

does not lead to cycloaddition) *via* a singlet state to the n,π^* triplet followed by formation of an addition product *(29)* between olefin and dione which may then collapse to 1,4-cycloadduct (dioxene, *30*) and/or 1,2-cycloadduct (keto-oxetane, *31*) and/or revert to quinone plus starting or isomerized olefin. The keto-oxetane *31* has been suggested to be photochemically unstable [87] being converted back to starting material [52] and, to a small extent [53], to dioxole *(32)*. With suitably substituted olefins, each of the above products *(30, 31, 32)* may be obtained as a mixture of stereoisomers (*cf. 34* and *35*).

The scheme above resembles the two-step mechanism for oxetane formation by cycloaddition of olefins to monoketones which is generally accepted [1–4] to proceed via the n,π^* triplet state of the ketone. It is assumed, by analogy, that the same excited state is involved in dione reactions. Additional support for intermediacy of the n,π^* triplet and concomitant stepwise formation of the new bonds derives from a kinetic study [30] of the reaction of *PQ* with *cis*- and *trans*-stilbene[b] where the

---

[b] It has been suggested [30] in the case of the stilbenes that the intermediate *29* may revert to *cis*- or *trans*-olefin and thus provide an alternate pathway for *cis* ⇄ *trans* isomerization which does not involve triplet energy transfer from quinone to olefin. It might also be noted that $\phi cis \rightarrow trans = 0.45$ and $\phi trans \rightarrow cis = 0.43$ in the *PQ*-stilbene system, values considerably larger than ϕ cycloaddition.

intersystem crossing efficiency of *PQ* was calculated to be unity, from observation of quenching by anthracene [54], and observations that the stereochemical integrity of *cis*- or *trans*-olefins is not preserved in their cycloadducts. The fact that the relative proportions of cycloadducts *30* and *31* from tetramethylethylene and *PQ* were not affected [54] by a quantity of anthracene sufficient to reduce the quantum yield by one-half suggests that 1,2- and 1,4-additions involve the same excited state.

The loss of stereochemical integrity in adducts of *PQ* has been observed with: *cis*- and *trans*-2-butene [c], both of which give a mixture of dioxenes *(30)* containing a preponderance of *cis*-isomer [39,54]; *trans*-4-octene which gave a mixture of *cis*- and *trans*-isomers while no *cis*-4-octene was detected in the unreacted olefin [39]; *cis*-ethoxystyrene [119], *trans*-1-phenylpropene [119]; and *cis*- and *trans*-stilbene [51,119].

In the latter case, the proportions of *cis*- and *trans*-diphenyldioxenes *(34* and *35)* varied with reaction temperature [51] under conditions where

34   35   36   37

little stilbene underwent *cis-trans* isomerization. At 25° the product from *cis*-stilbene contained 72% of *34* and at 70°, 60% of *34;* the corresponding values obtained with *trans*-stilbene were 19% at 25° and 35% at 70°. It was suggested that conversion of intermediate *36*, by internal rotation, to the conformer *37* competes with radical coupling and that the rotation *36 → 37* (or *vice versa*) involves a small energy barrier. Results of the kinetic study [30] mentioned above were also interpreted in terms of a short lived intermediate species. Cycloaddition involving a singlet state might be expected to proceed directly to product in a concerted fashion.

A lesser degree of loss of configuration was observed in the reactions of *cis*- [36] and *trans*-stilbene [35] with tetrachloro-*o*-benzoquinone to give dioxenes. The *cis*-stilbene reaction (in acetone or acetonitrile) afforded about 80% of *cis*-adduct while 88% of *trans*-adduct [35] was obtained from the reaction of *trans*-stilbene (both reactions at 15°, λ > 400 nm). Curiously, irradiation of the quinone and *cis*-stilbene (or diphenyl-acetylene) in benzene solution produced the mono-phenyl ether *(33)*

---

c) The probability of biacetyl photosensitized *cis⇄trans* isomerization of butenes has been estimated [42,123] to be exceedingly small.

which was reportedly formed only in trace amounts from reaction in the absence of olefin (*cf.* ref. 131).

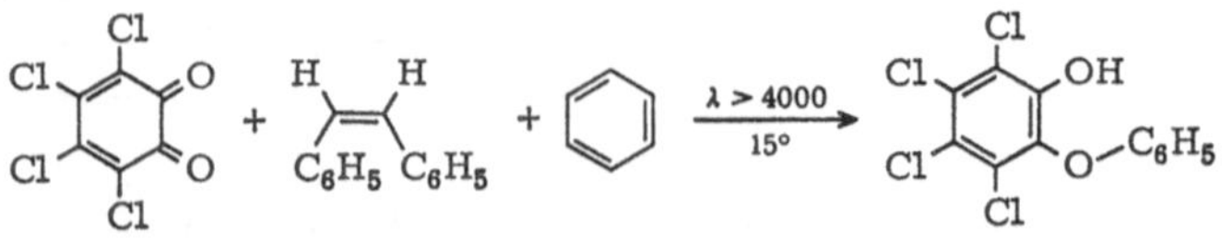

33

Examination of Table 3 shows a pattern of keto-oxetane *vs.* dioxene formation (1,2- *vs.* 1,4-addition) which is not susceptible to rational interpretation at present. Thus, for example, furan and ethyl vinyl ether gave exclusively dioxene, benzofuran gave exclusively keto-oxetane, and 2-phenylbenzofuran a 1:1 mixture; 1,4-dioxene afforded 55% dioxene and 40% keto-oxetane while vinylene carbonate gave exclusively keto-oxetane. The only generalization that seems to hold is that substitution by methyl or phenyl on the reactive double bond tends to increase the relative amount of dioxene formed but it is not clear if this effect is steric or electronic in origin. Experiments involving rational variations of olefin and dione structure would be valuable. It should be noted that the wavelengths of light used in these experiments were such that questions of photoreversibility probably need not be considered. The question of 1,2- *vs.* 1,4-addition also arises in hydrogen atom abstraction reactions (*vide infra*).

An additional complication appeared in olefin reactions when the olefin contained allylic hydrogen atoms (or other abstractable hydrogen, *cf.* ethyl vinyl sulfide). The hydrogen abstraction reaction illustrated to form 1,2-adducts *38* then competes with cycloaddition. Such

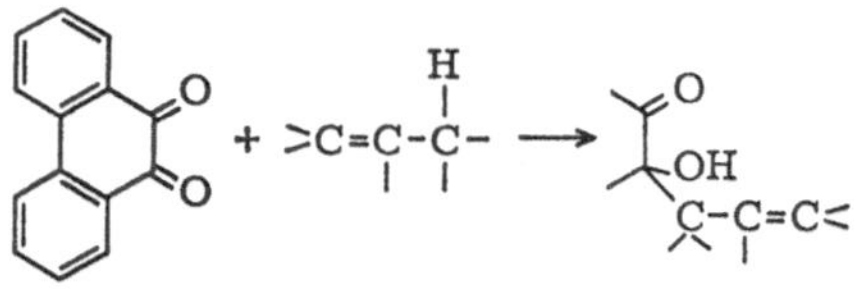

38

behaviour has been suggested in the reaction of *trans*-4-octene [39] and was investigated in some detail by *Farid* [54] in the series isobutylene, *cis*- and *trans*-2-butene, trimethylethylene and tetramethylethylene. As can be seen from the Table, both the quantum yield for disappearance of *PQ* and the extent of hydrogen atom abstraction increase with increasing methyl substitution; with isobutylene the quantum yield was 0.21, H-atom abstraction accounted for 10% of reaction while with tetramethylethylene the quantum yield was 0.79 and included 29% abstraction.

Very little information is available on reactions of unconjugated diones with olefins. The reaction of biacetyl with cyclohexene was first reported [79] in 1964 and repeated [20] in 1968; in both investigations no cycloadduct was observed. Preliminary results suggest [15] that camphorquinone reacts with cyclohexene to form, in part, a keto-oxetane.

Recently, reactions of PQ and related quinones with 1,3-disubstituted isobenzofurans have been reported [184] to involve 1,4-cycloaddition as well as cleavage of the furan ring.

Hydrogen Abstraction Reactions

Hydrogen abstraction reactions are characterized by transfer of a hydrogen atom from a suitable donor, often the solvent, to a carbonyl oxygen atom of photoexcited dione [d]. The primary products are a pair of radicals (within a solvent cage) which may proceed to stable products in a variety of ways as shown in the following general scheme:

H-abstraction:

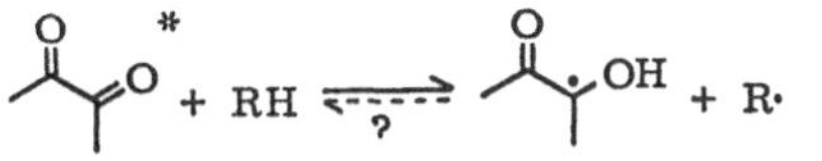

$$\text{RH} = \text{RCHO, } \mathord{>}\text{CHOH, } -\overset{\text{O}}{\underset{\|}{\text{C}}}-\text{O}-\text{CH}\mathord{<}, \mathord{>}\text{CH}-\text{O}-, \mathord{>}\text{CH}-\text{S}-, \mathord{\geqslant}\text{CH},$$

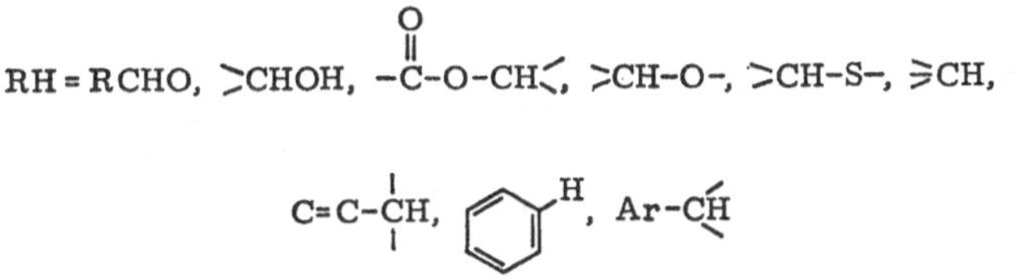

Dimerization:

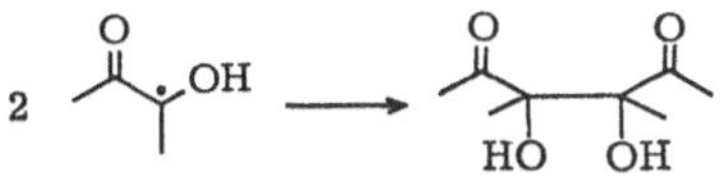

or quinhydrone   (2)

$$2\ \text{R}\cdot \longrightarrow \text{R}-\text{R} \qquad (3)$$

---

[d] This process involves reduction of the carbonyl compound and concomitant oxidation of the H-donor. The term photoreduction, referring to the carbonyl compound, is often used to designate these reactions. While this has the merit of indicating the mechanistic similarity of the initial step, it does not necessarily describe the nature of the overall process. Consistent with the suggestion that photochemical processes be described in terms generally used to classify organic reactions, it seems more appropriate to refer to photoreduction, photopinacolization, photoaddition, etc.

273

Reduction:

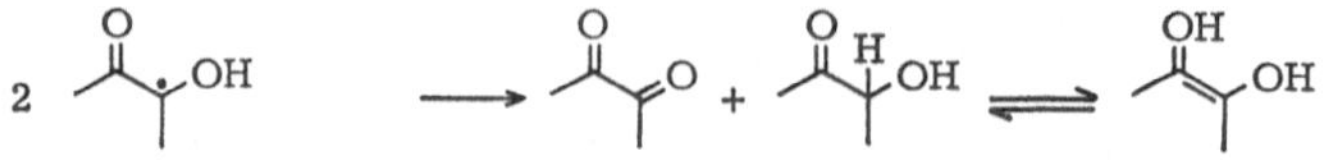

(4)

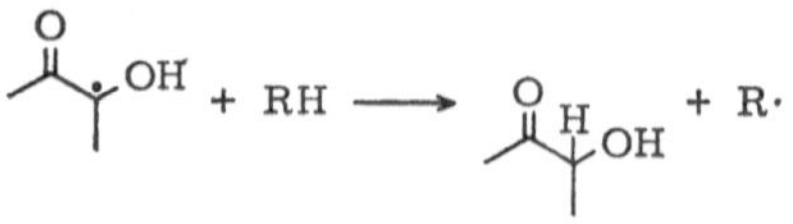

(5)

(6)

Coupling:

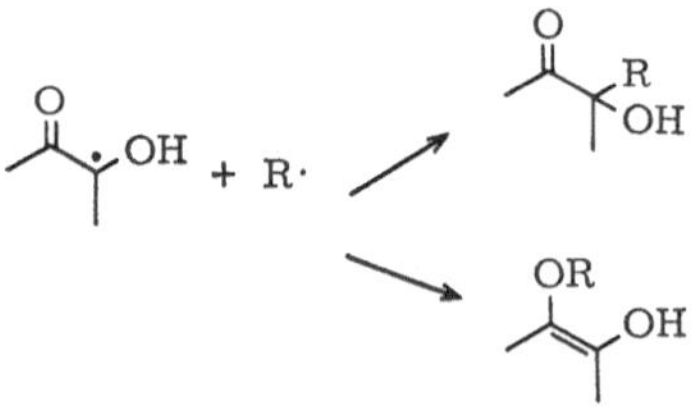

(7)

(8)

An alternative scheme was proposed by *Moore* and *Waters* [100] to account for the high yield obtained in the reaction of benzaldehyde with phenanthrenequinone. They suggested that the initially formed radical R · adds to a ground-state quinone molecule to give a new radical which then abstracts hydrogen from RH to regenerate the radical R ·

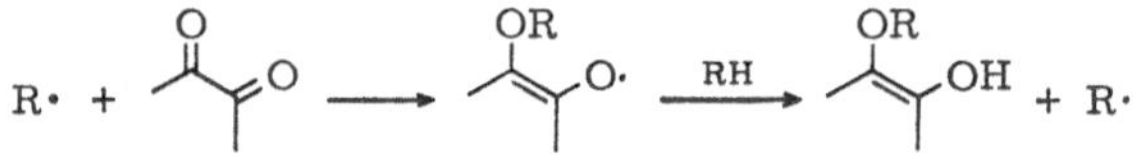

thus providing a chain mechanism for the photoaddition. With the exception of the reaction of phenanthrenequinone with ethanol [165] (*vide infra*), quantum yield data and effects of inhibitors [134] do not support such a proposal. Also, attempts to initiate reactions by thermal decomposition of peroxides have given [20,127] results which do not parallel photoreactions.

The reactive state in H-abstraction is generally accepted to be the n,π^* triplet by analogy with conclusions reached in the well-studied H-abstraction reactions of monoketones [1-4] and of p-quinones [153]. This view is supported by (a) occurrence of reaction upon irradiation at the long wavelength n,π^* region of dione absorption, (b) the high intersystem crossing efficiency of diones (*vide supra*), (c) the observation of radicals as intermediate products, and (d) sensitization by triplet sensitizers. However, participation of singlet states to some degree or in special cases cannot be excluded.

There is evidence that some of these reactions may be photochemically reversible. Thus, irradiation of the adduct of phenanthrenequinone and dioxane in benzene solution led to a mixture having approximately the same composition as that obtained from irradiation of dioxane and the quinone [128]. Irradiation [133] of the adducts of this quinone with benzaldehyde or p-xylene also resulted in regeneration of quinone.

In the following discussion, general aspects of H-abstraction will be discussed first followed by a summary of specific results classified according to H-donor.

I. Individual Steps

A. The Hydrogen Abstraction Step

1. The Dione

With the exception of those diones which undergo α-cleavage because of special structural features (*vide supra*) all diones participate in H-abstraction to some degree. As noted earlier, H-abstraction has even been observed in the gas phase (with hexafluorobiacetyl [169]). However there is little, if any, evidence which allows comparisons of the reactivities of different diones. Kinetic constants have only been determined in one series of compounds [54] and quantum yields only in the few cases summarized in Table 4.

The quantum yield provides a measure of the relative rate of chemical reaction *vs* all deactivation processes. From this point of view, it can be seen from Table 4 that the intermolecular reactions of phenanthrenequinone and the intramolecular reactions of dimethyloctanedione and decanedione are more efficient than the intermolecular reactions of camphorquinone. However, no comparisons of reactivities can be made between different reactions having quantum yields of unity. Further, there is no guarantee that the rate of H-abstraction by excited-state camphorquinone is less than the rate of abstraction by phenanthrenequinone simply because quantum yields are lower with camphorquinone. The possibility that dione may be regenerated by some process subsequent to H-abstraction (e. g. reaction 4 in the general scheme) should also be considered.

Table 4. *Quantum yields for disappearance of diones in the presence of hydrogen atom donors*

Dione	H-Donor	Solvent	λ(Å)	ϕ	Lit.
2,7-Dimethyl-4,5 octanedione	—	C_2H_5OH	4358	1.13	[164]
5,6-Decanedione	—	C_2H_5OH	4358	1.01	[164]
Camphorquinone	benzaldehyde	—	4358	0.3	[127]
Camphorquinone	dioxane	—	4358	0.2	[127]
Camphorquinone	p-xylene	—	4358	0.07	[129]
Camphorquinone	methanol	—	2537	0.02	[99]
Camphorquinone	2-propanol	—	2537	0.06	[99]
Camphorquinone	2-propanol	—	3660	0.06[a]	[99]
Camphorquinone	2-propanol	—	3660	0.26—0.90[b]	[99]
Camphorquinone	2-propanol	—	3130	0.10—0.39[b]	[99]
Phenanthrenequinone	benzaldehyde	—	4358	1	[127,134]
Phenanthrenequinone	dioxane	—	4358	1	[127,134]
Phenanthrenequinone	ethyl acetate	—	4358	1	[132]
Phenanthrenequinone	p-xylene	—	4358	0.5	[127,134]
Phenanthrenequinone	ethanol	benzene	4358	0.75—4.95[c]	[165]
Phenanthrenequinone	benzene	—	4358	0.25	[131]
Phenanthrenequinone	olefins		>405 nm [d]		[54]

[a] Sensitized by m-methoxyacetophenone.
[b] Sensitized by benzophenone. Quantum yield varied with dione conc.
[c] Quantum yield varied with quinone and ethanol concentrations.
[d] Overall quantum yields for cycloaddition and H-abstraction were given, *cf.* Table 3.

2. The Hydrogen Donor

The types of compounds presently known to donate a hydrogen atom to photoexcited quinones include aldehydes, primary and secondary alcohols, esters and lactones, ethers and thioethers, olefins having allylic hydrogen atoms, alkylbenzenes, benzene, and saturated hydrocarbons. This list is undoubtedly incomplete. Acetone, methyl ethyl ketone, acetic acid, and t-butyl alcohol react extremely slowly.

Conversion of reactive compounds to the corresponding radicals can occasionally be inferred from products of reaction. In most esr studies of dione photolyses these radicals have not been observed, presumably because of short lifetimes, but recently *Zeldes* and *Livingston*[171] detected the dioxanyl radical from photolysis of biacetyl in dioxane solution. The wide variety of compounds which are converted to the corresponding radicals is of considerable theoretical interest since formation of many different types of radicals can occur under comparable conditions allowing qualitative and quantitative comparisons of their behaviour.

Little information is available on relative reactivity of H-donors. From the quantum yield data in Table 4 it can be seen that quantum yields for reaction with camphorquinone decrease in the series benzaldehyde, dioxane, p-xylene (4358 Å) and 2-propanol, methanol (2537 Å). With phenanthrenequinone at 4358 Å, ethanol, benzaldehyde, ethyl acetate, and dioxane are more efficient donors than p-xylene or benzene. Relative rates of reaction have been determined by competition experiments in only one series. With phenanthrenequinone (not under irreversible conditions) the relative rates for benzaldehyde, dioxane, and p-xylene were [133] 50:3:1.

Competition of a different kind occurs in reactions of esters and lactones since these compounds contain hydrogen atoms alpha to a carbonyl group and alpha to an ether function. A priori, abstraction of either type of hydrogen might be expected. In fact, reaction of phenanthrenequinone at 4358 Å led in quantitative yield to products of reaction at the carbon atom alpha to the ether function (52, 53) in ethyl acetate [132], ethyl propionate [132], and γ-butyrolactone[e] [110].

52   R = H, CH₃   53

It is difficult at present to establish the factors determining ease of H-donation by various species. It would seem reasonable to assume that the stability of the radical being generated would be reflected in the transition state for abstraction. This leads to the conclusion that α-keto radicals are appreciably less stable than ether radicals in agreement with the results of *Russell* and *Lokensgard* [135] who concluded from esr spectra that the odd electron in α-keto radicals is largely localized on the α-carbon atom. However, polar factors may also play a significant role. *Walling* and *Gibian* [167] have shown that rates of photochemical H-abstraction by benzophenone from a series of substituted toluenes fit a *Hammett* relationship and *Zwitkowits* [172] has observed similar trends with phenanthrenequinone. If the transition state for H-abstraction partakes of the supposed electron deficient character of the oxygen atom in the n,π^* state of ketones, the ability of the carbon atom from which

---

[e] It is interesting to note that alkylation [177] of γ-butyrolactone in a radical chain process occurred at the carbon atom alpha to the carbonyl group.

abstraction is occurring to stabilize positive charge might help to reduce the energy of activation. Indeed, acylonium ions, oxonium ions, benzylic

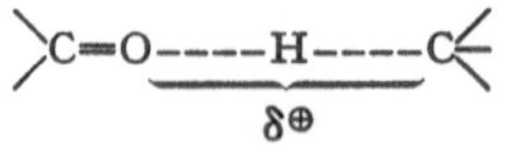

$$\overset{}{\underset{}{>}}C{=}O\text{----}H\text{----}C\overset{}{\underset{}{<}}$$
$$\delta\oplus$$

and allylic carbonium ions are energetically much more favorable than positive charge adjacent to a carbonyl or nitrile group in qualitative agreement with the overall trends and the substituent effects observed.

3. The Semidione Radical

The inference that semidione radicals are intermediates in diketone photolysis has been widely accepted for many years since it provides a satisfactory rationale for the results observed in many reactions. Recently, direct evidence for the presence of these radicals has been obtained from flash photolysis [13] of benzil in alcoholic solvents and esr studie of irradiated solutions of biacetyl [12,171,183], camphorquinone [99,183], and other diones [183]. The related radical anions, generated by chemical means, have been studied extensively [5,31,136] and reports of the radical cations have also appeared [111,137].

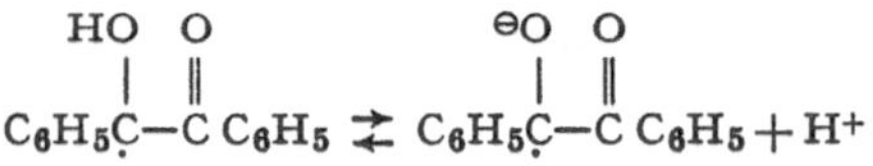

$$\underset{C_6H_5\overset{|}{C}{-}\overset{\parallel}{C}\,C_6H_5}{\overset{HO\quad O}{}} \rightleftarrows \underset{C_6H_5\overset{|}{C}{-}\overset{\parallel}{C}\,C_6H_5}{\overset{\ominus O\quad O}{}} + H^+$$

Beckett, Osborne, and *Porter* [13] calculated a *pK* of 5.9 for the dissociation of the semidione radical obtained from benzil. *Norman* and *Pritchett* [111] have suggested that radical cations are strong acids.

The semidione radical derived from open-chain diones may, like the diones themselves, assume various conformations. It has been assumed that the preferred conformation is "*s-trans*" but results appear to vary with the method used for generating radicals. The conformational factor might play a role in determining the relative importance of subsequent reactions.

Two resonance structures, in which the odd electron is localized on carbon *(54)* or on oxygen *(55)*, may be written for a semidione radical. Based on the conclusion [135], mentioned earlier, that the odd electron of α-keto radicals is mainly localized on carbon, the canonical form *55*

      54   *55*

may be relatively unimportant. Writing the two forms emphasizes the fact that radical coupling involving the semidione radical occurs both at carbon and at oxygen, but is not a necessary condition for such ambident character. In the case of quinones, additional resonance forms in which the odd electron is localized on carbon atoms of the quinone skeleton may also be written and one case of nuclear substitution (4-cyano-1,2-naph-thoquinone) has been reported [8,144].

Two different semidione radicals might be formed from an unsymmetrical dione and these could then undergo further reaction to form two isomeric products. Thus, for example, photolysis [129] of camphorquinone in *p*-xylene solution afforded the isomeric 1,2-adducts *56* and *57* in approximately 2:1 ratio. At first glance it might seem that the prepon-

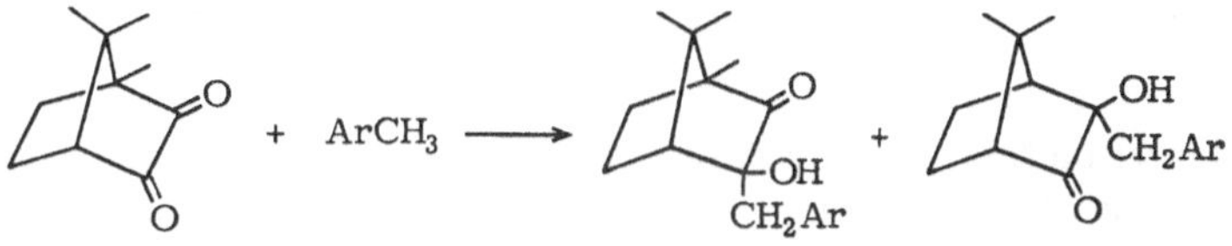

56   *57*

derance of *56* over *57* reflects the greater ease of formation of semidione *58 vs. 59* due to steric factors, and that the product composition is determined in the H-abstraction step in which *58* and *59* are formed. However, *Monroe, Weiner*, and *Hammond* [99] concluded, on the basis of the simplicity of its esr spectrum, that the semidione radical derived from camphorquinone has the symmetrical structure *60* or involves rapid (on the esr time scale) tautomerization between *58* and *59*, as illustrated.

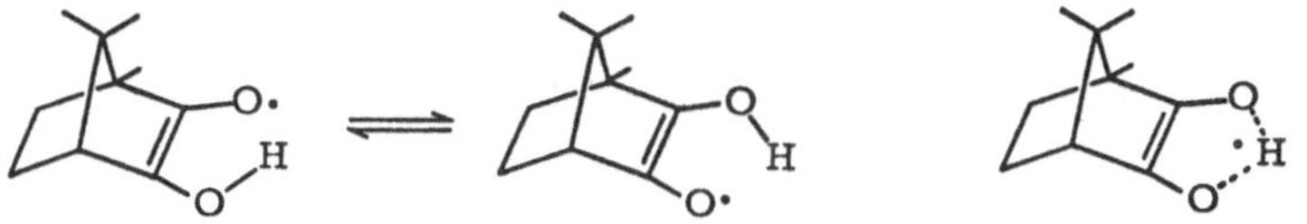

58   *59*   *60*

(*Norman* and *Pritchett* [111] concluded that rapid tautomerization is involved in chemically generated biacetyl semidione radical; the results *Zeldes* and *Livingstone* [171] observed with the photochemically generated radical differ). If this view is correct, product composition is determined in a step subsequent to H-abstraction. This is in agreement with the results obtained [130] with camphorquinone and aromatic aldehydes in benzene solution where the two isomeric products *61* and *62* were obtained in equal amounts in contrast to the results with *p*-xylene. This result cannot be explained by the assumption that initial H-abstraction determines product ratios since there is no reason to believe that there should be any significant difference between the two cases. The result is explicable in terms of a symmetrical specie or rapidly equilibrating tau-tomers where steric hindrance to coupling at carbon favors formation of *56* over *57* but no steric effect is felt at the more remote oxygen atoms. The problem of coupling at oxygen *vs.* coupling at carbon will be discussed in the section on radical coupling.

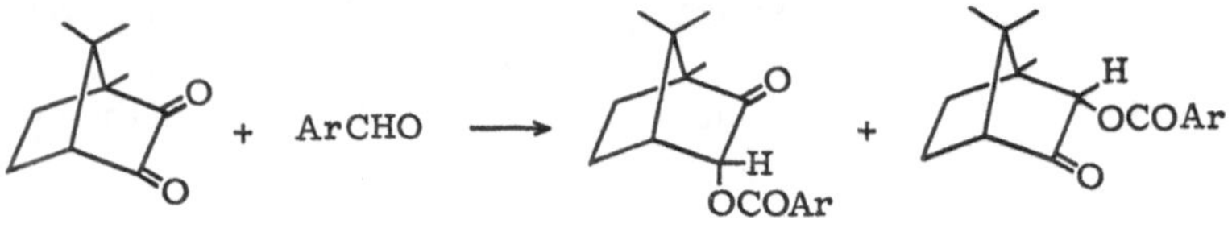

61   *62*

A novel method for generating semidione radicals has recently been reported by *Monroe, Weiner,* and *Hammond* [178] who found that the quantum yield for photoreduction of camphorquinone in 2-propanol was markedly enhanced when benzophenone was added and the solution irradiated at wavelength 3660 Å where most of the light was absorbed by benzophenone. Instead of benzpinacol formation, the dione underwent photoreduction. Similar enhancement was not observed with *m*-methoxy-acetophenone which does not abstract hydrogen from 2-propanol. The conclusion was that the ketyl radical, formed in the efficient H-abstraction reaction of benzophenone, transferred a hydrogen atom to camphor-quinone to generate the semidione radical. It was suggested that this phenomenon be called "chemical sensitization".

B. Dimerization of Intermediate Radicals

The inital H-abstraction step results in formation of a pair of radicals in a solvent cage. To the extent that one of the partners escapes from the cage, encounters between like radicals will be possible and dimeric products may be formed. Thus dimers of the semidione radical, of the radical derived from the donor, or both are sometimes observed.

In fact, dimerization of semidione radicals appears to be important only with the open chain diones biacetyl and benzil (and possibly hexa-fluorobiacetyl). The diketopinacol (*63*, $R = C_6H_5$) was first obtained by

Klinger [82] in 1886 and called benzilbenzoin since it decomposed thermally to benzil and benzoin; the accepted structure was proposed by *Cohen* [41] in 1916. This compound was the only product reported from reactions of benzil in ether [82], alcohols [40,41], aldehydes [17], and alkyl benzenes [17] (except acenapththene [96]); some of these results bear

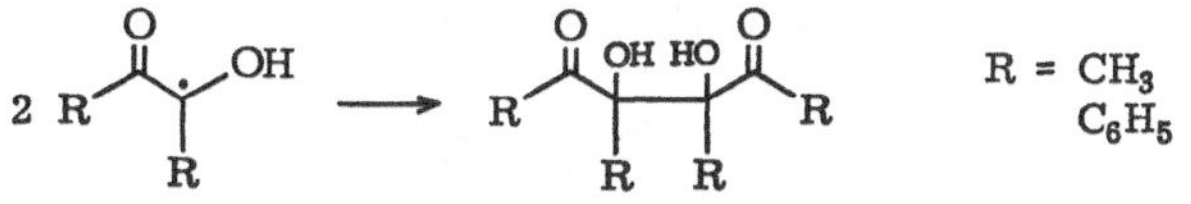

63

repetition. Biacetyl gave exclusively diketopinacol (*63*, R=CH₃) in alcohol solutions [21,41,163] and appreciable yields [20] of *63* in other solvents. Other semdione radicals show little tendency to dimerize. 2,3-Pentanedione and higher aliphatic diones undergo [163,164] an efficient intramolecular reaction; all the cyclic α-diketones whose photochemistry has been investigated (camphorquinone, α,α'-tetramethyldiones) present considerable steric hindrance to dimerization; phenanthrenequinone afforded quinhydrone in some cases.

The hydrogen donors for which dimers of the corresponding radical have been reported are hydrocarbons. Thus biacetyl [20,79] and camphorquinone [15] reactions with cyclohexene afforded, in part, bicyclohexenyl. Dimers have also been reported in reactions of alkylbenzenes [15,20,101, 129,133], diphenylmethane [148], and thioxanthene [148] (*64*).

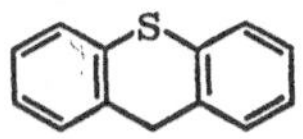

64

Biphenyl was observed [131] in the photoreaction of phenanthrenequinone with benzene. A four percent yield of a dimer of diethyl ether has been reported [20]. Minor amounts of dimeric products in other cases may have been overlooked.

Rate constants for dimerization of a number of semidione radicals have been determined [183] recently and lie in the range $2—4 \times 10^8$ M⁻¹ sec⁻¹.

C. Reduction

Transfer of a hydrogen atom to a semidione radical results in overall reduction of diones to α-hydroxyketones or enediols (dihydric phenols). Sinces these are in tautomeric equilibrium no distinction has been possible between the two possible routes to a given product, as illustrated below. Dihydric phenols undergo ready oxidation to quinones upon exposure to air but their presence in deoxygenated solutions has been indicated [39,131].

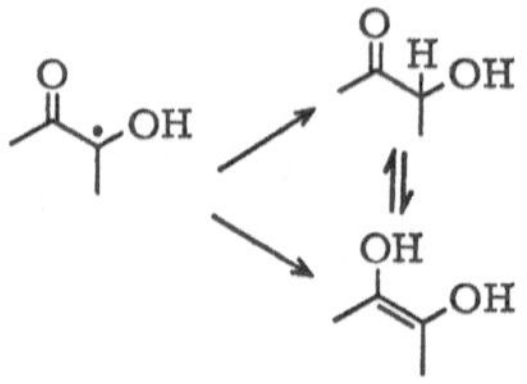

Reduction is observed principally with those α-diketones which do not form diketopinacols *(63)* including camphorquinone [15,97,99], tetramethyltetralindione [60], and 2,2,5,5-tetramethyltetrahydrofuran-3,4-dione [15]. A low yield of benzoin was reported [37] in the reaction of cyclohexane solutions of benzil. The isomeric diols obtained from camphorquinone had the *endo* configuration [15,97,99].

The best condition for reduction is use of a primary or secondary alcohol as hydrogen donor since the resulting carbinyl radicals show little tendency to couple and are readily converted to aldehydes or ketones by transfer of a second H-atom. Quantitative yields of reduction products have been obtained with 2-propanol. It has been suggested [178] that two semidione radicals may disproportionate affording one molecule of reduction product and regenerating a dione molecule. The rate constants for a number of such disproportionation have been determined [183] recently and lie in the range $1-4 \times 10^7$ M^{-1} sec^{-1}.

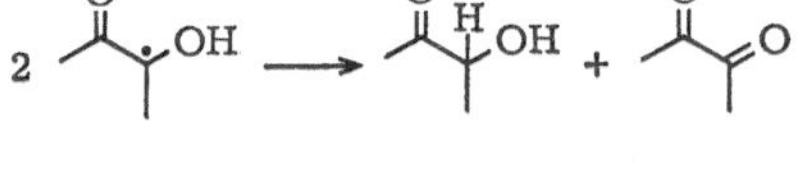

Polarographic reduction of diones was influenced by irradiation with light [21,22].

D. Radical Coupling

With the exception of reactions involving alcohols as hydrogen donors, the major process observed upon irradiation of diones in the presence of H-donors is formation of 1:1 adducts derived from coupling of the pair of radicals generated in the H-abstraction step. The ambident character of the semidione radical allows such coupling to occur at either of two positions, coupling at carbon resulting in overall 1,2-addition to the dione and coupling at oxygen in overall 1,4-addition. (As noted earlier, one case of nuclear coupling has also been reported [8,144].) Formation of adducts could be a concerted process, as illustrated, in reactions proceed-

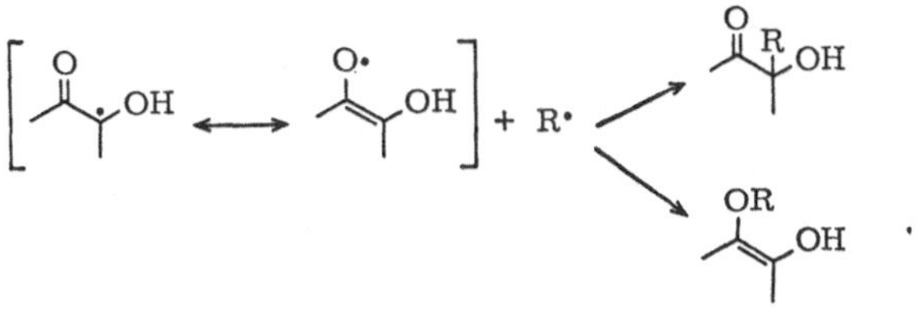

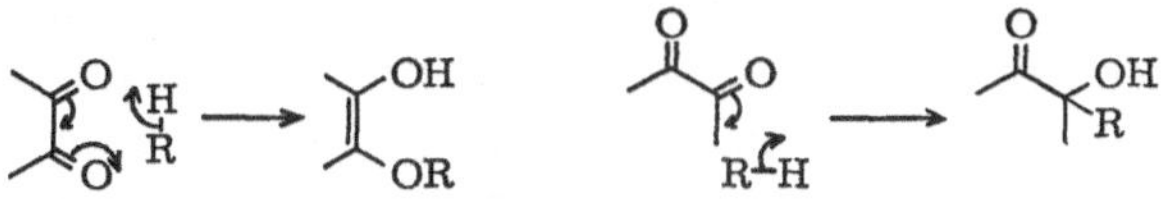

ing via a singlet excited state; as has been discussed earlier, it seems much more likely that triplet states are involved, in which case a concerted process is not possible.

In those cases where a hydrogen atom is located on a carbon gamma to a carbonyl group, intramolecular H-abstraction can occur and radical coupling then results in formation of a four membered ring. This process occured with high efficiency with certain aliphatic diones [163,164], the only compounds investigated to date in which such intramolecular H-

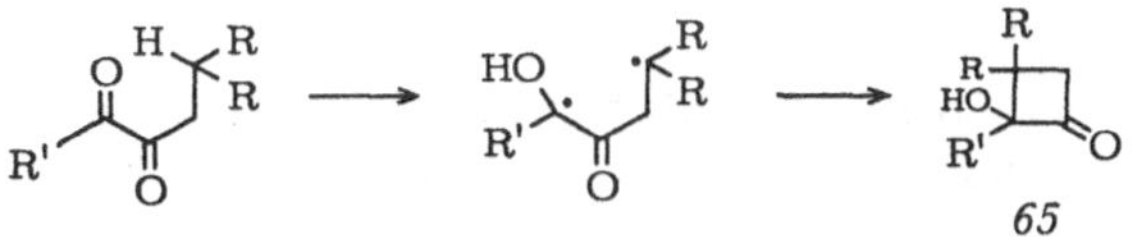

abstraction is feasible. The exclusive products are cyclobutanolones *(65)*; this specificity was attributed to favorable conformational factors. The analogous reactions of monoketones are much more complex.

Coupling at both carbon (1,2-addition) and at oxygen (1,4-addition) of the semidione is observed in other cases. In general, *o*-quinones exhibit a greater tendency to couple at oxygen than do cyclic diketones or biacetyl. The position of coupling also varies with the hydrogen donor. Aromatic acyl radicals combine almost exclusively at oxygen, aliphatic acyl radicals and ether radicals exhibit mixed behaviour but with a greater tendency to combine at oxygen than benzyl radicals.

Both polar and steric factors play a role in determining the orientation of coupling with the semidione radical. The steric factor is illustrated in the reactions of camphorquinone with aldehydes. Aromatic aldehydes, butyraldehyde and pivalaldehyde reacted [15,130)] exclusively to give 1,4-adducts but acetaldehyde gave a mixture of both 1,2- and 1,4-adducts.

The importance of polar factors was shown [134)] in a series of reactions of substituted toluenes with phenanthrenequinone under irreversible conditions (4358 Å). These reactions produced mixtures of 1,2- and 1,4-adducts *(66, 67)*; the ratio *66:67* decreased in the sequence *p*-methoxytoluene, *p*-t-butyltoluene, *p*-methyltoluene, toluene, *p*-nitrotoluene. In other words, greater electron-supplying ability of a *p*-substituent resulted in a higher proportion of coupling at carbon. It was suggested [134)] that the semidione radical is polarized by the inductive effect of the oxygen atoms so that there is slightly decreased electron density at carbon; increase of electron density at the reactive position of the donor radical should then enhance the rate of coupling at carbon, and *vice versa*. This suggestion is predicated on the assumption that radical coupling is an exothermic process of low activation energy and, therefore, the transition state closely resembles the reactants. This factor may also contribute to the greater tendency of *o*-quinone semidiones to couple at oxygen.

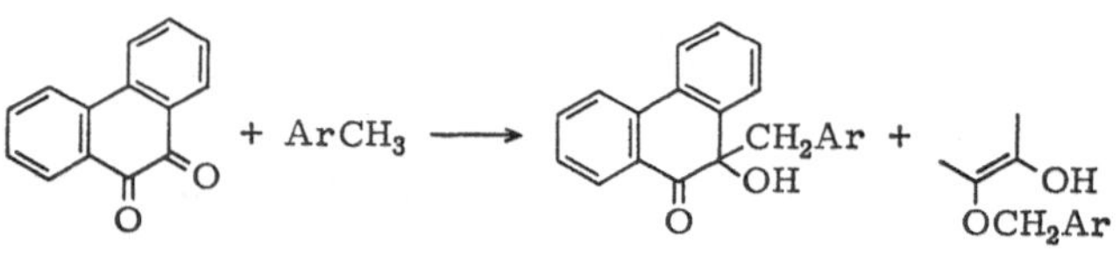

66   *67*

An additional factor, electron transfer from H-donor radical to semidione radical, may play a role in the transition state for radical coupling. As illustrated below, this results in contribution from two stabilized

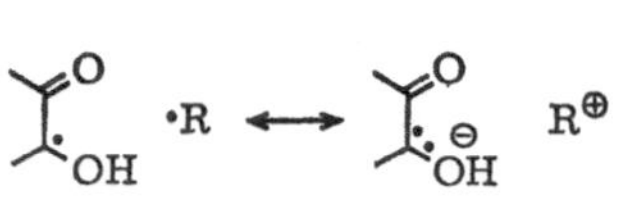

species, the enolate ion of α-hydroxyketone and a carbonium ion (e. g. acylonium, oxonium, or benzyl carbonium ion). This effect, which has been suggested [166)] in radical additions to neutral molecules, could be reflected in relative rates of radical coupling and conceivably could be involved in orientation effects.

Products of photoaddition, particularly of *o*-quinones, may exhibit appreciable light absorption and be subject to further photoreaction.

Thus, irradiation [172] of phenanthrenequinone in anisole at 4358 Å produced a 1,4-adduct which rearranged to 1,2-adduct upon irradiation at shorter wavelength. Similar photorearrangement [133] of the 1,4-adduct (*67*, Ar = *p*-CH$_3$C$_6$H$_4$) of *p*-xylene and phenanthrenequinone was shown [134,172] not to involve reversal to quinone and *p*-xylene. Presumably, the adduct is cleaved to the same pair of radicals from which it was formed, followed by recombination and so on until the more light stable product is formed. Irradiations of quinones at short wavelengths should be interpreted with care.

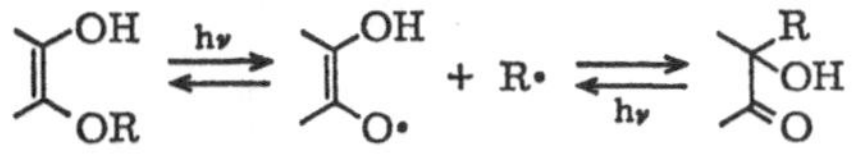

II. Reactions of Diones and Hydrogen Donors

A. Diones with Aldehydes

The diones and aldehydes whose photoreactions have been reported are summarized in Table 5 where it can be seen that this system has been investigated almost to excess. Most of the quinone reactions were performed in Cairo sunlight using benzene solutions of quinone containing a moderate excess of aldehyde. The adducts crystallized from solution usually in good to excellent yield; no attempts were made to analyze the total reaction product. With one exception, all quinone adducts are products of 1,4-addition; both cyclic [146] and open-chain [83] tautomeric structures have been proposed for these enediol monoesters. Infrared

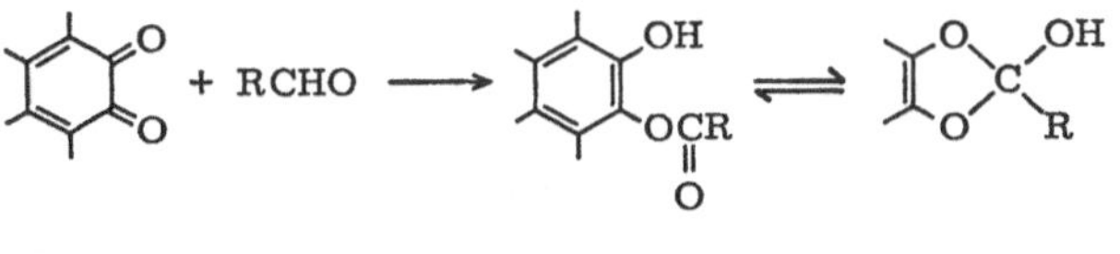

spectra of the phenanthrenequinone-benzaldehyde adduct show that the open-chain form is predominant (if not exclusive) in the solid state [100,128] and in methylene chloride solution[128]; spectra of *p*-tolualdehyde, *m*- and *p*-nitrobenzaldehyde, and *p*-cyanobenzaldehyde adducts in the solid state were similar to that of the benzaldehyde adduct. Some of the quinones investigated are unsymmetrical and two open-chain forms, interconvertible through the common cyclic form, could be formed. No information is available on this point.

Table 5. *Dione-aldehyde photoreactions*

Dione	Aldehydes
3,4-Hexanedione	Propionaldehyde [163]
4,5-Octanedione	n-Butyraldehyde [163]
Benzil	Isobutyraldehyde [16], benzaldehyde [16,84], Salicylaldehyde [16]
Tetramethyl-3,4-tetrahydrofurandione *(72)*	Acetaldehyde [15], propionaldehyde [15], n-butyraldehyde [15], benzaldehyde [15], anisaldehyde [15]
Camphorquinone	Acetaldehyde [130], propionaldehyde [130], n-butyraldehyde [130], pivalaldehyde [130], phenylacetaldehyde [130], benzaldehyde [130], *p*-tolualdehyde [130], anisaldehyde [130], *p*-chlorobenzaldehyde [130], *p*-nitrobenzaldehyde [130]
1,1,4,4-Tetramethyl-tetralindione *(73)*	*n*-Butyraldehyde [60], benzaldehyde [15], anisaldehyde [15], *p*-chlorobenzaldehyde [15]
Tetrachloro-*o*-quinone	Acetaldehyde [145], Benzaldehyde [145], *p*-tolualdehyde [145], anisaldehyde [145], cinnamaldehyde [145]
Tetrabromo-*o*-quinone	*p*-Tolualdehyde [145], anisaldehyde [145]
o-Naphthoquinone	Acetaldehyde [8]
4-Chloro-1,2-Naphthoquinone	Acetaldehyde [8], *p*-nitrobenzaldehyde [8]
6-Bromo-1,2-Naphthoquinone	Benzaldehyde [106], *m*-tolualdehyde [106], *p*-tolualdehyde [106], *o*-methoxybenzaldehyde [106], anisaldehyde [106], *o*-chlorobenzaldehyde [106]
3,4-Dichloro-1,2-naphthoquinone	Anisaldehyde [145]
4-Cyano-1,2-naphthoquinone	Acetaldehyde [8,144], propionaldehyde [144], anisaldehyde [144], *p*-nitrobenzaldehyde [144], cinnamaldehyde [144]
4-Cyanomethyl-1,2-naphthoquinone	Acetaldehyde [8]
4-(2′,5′-Dimethylphenoxy)-1,2-naphthoquinone	Acetaldehyde [8]
4-(3′,5′-Dimethylphenoxy)-1,2-naphthoquinone	Acetaldehyde [8], anisaldehyde [8]
4-(β-Naphthoxy)-1,2-naphthoquinone	Anisaldehyde [8]
Acenaphthenequinone	Benzaldehyde [156], anisaldehyde [156], salicylaldehyde [156], cinnamaldehyde [156]

Table 5 (continued)

Dione	Aldehydes
9,10-Phenanthrenequinone	Acetaldehyde [83], chloral [84], isovaleraldehyde [83], benzaldehyde [83,100,128,146], *p*-isopropylbenzaldehyde [150], anisaldehyde [84,100], 2,4-dimethoxybenzaldehyde [150], salicylaldehyde [84], furfural [84], *p*-chlorobenzaldehyde [146], phthalaldehyde [106], isophthalaldehyde [106], terephthalaldehyde [106], cinnamaldehyde [84], 2-methoxy-1-naphthaldehyde [150], 2-formylquinoline [105], 9-anthraldehyde [105], 3-pyrenealdehyde [106] *p*-tolualdehyde [172], *p*-nitrobenzaldehyde [172], *m*-nitrobenzaldehyde [172], *p*-cyanobenzaldehyde [172]
3-Phenylbenzo(f)quinoxaline-5,6-quinone *(74)*	Benzaldehyde [106], *m*-tolualdehyde [106], *p*-tolualdehyde [106], *o*-methoxybenzaldehyde [106], anisaldehyde [106], *o*-chlorobenzaldehyde [106]
Chrysenequinone	Benzaldehyde [104], anisaldehyde [104], 9-anthraldehyde [105]
1,2-Benzophenazine-3,4-quinone *(75)*	Benzaldehyde [151], anisaldehyde [151]
Retenequinone	Anisaldehyde [104], 9-anthraldehyde [105]
Benzo(h) quinoline-5,6-quinone *(76)*	Benzaldehyde [107], *p*-tolualdehyde [107] anisaldehyde [107], *o*-chlorobenzaldehyde [107]
2-N-Phenyl-α,β-naphtho-1,2,3-triazolequinone *(77)*	Anisaldehyde [108], 3,4-diethoxybenzaldehyde [108]

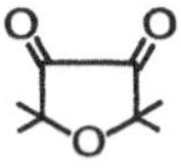

72

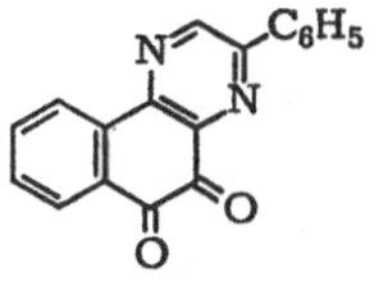

74

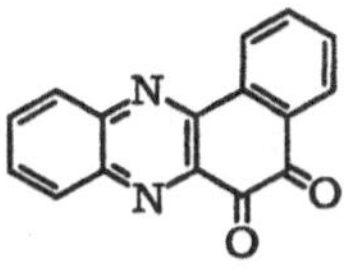

75

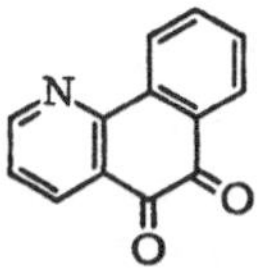

76

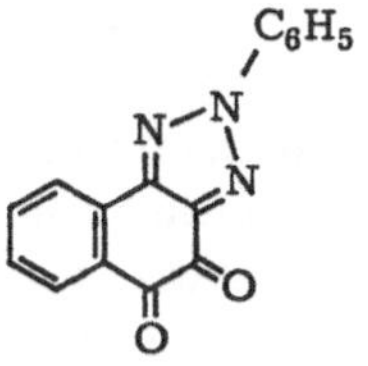

77

The single exception to 1,4-addition is 4-cyano-1,2-naphthoquinone which reacted normally with aromatic aldehydes but underwent [8,144] nuclear substitution to give 1,2-dihydroxy-3-acyl-4-cyanonaphthalenes *(69)* with acetaldehyde (unspecified yield) and propionaldehyde (17% yield). The structural assignments were based on chemical and spectro-

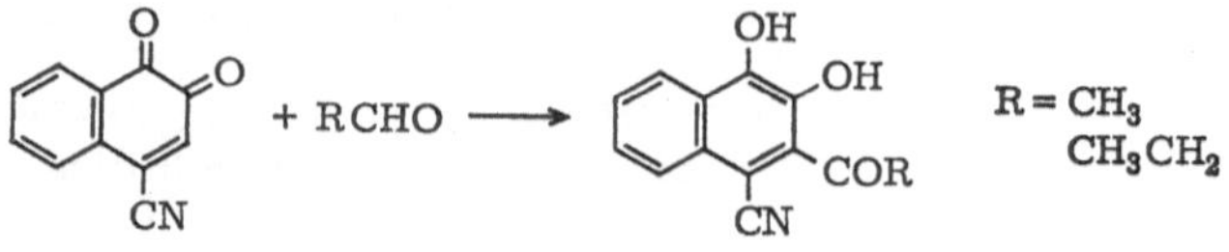

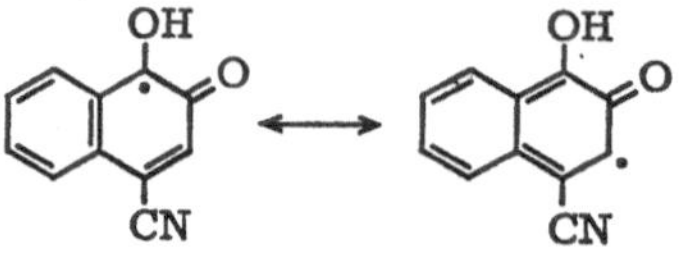

scopic evidence. The result can be rationalized in terms of the usual semidione radical intermediate *(70)* since one of its canonical forms *(71)* has the odd electron localized at C-3. Analogous products did not crystallize from irradiations of other 1,2-naphthoquinones. Further investigation would appear to be desirable.

The enediol-monoesters initally formed by 1,4-addition of aldehydes to α-diketones tautomerize to the more stable keto-form so that the observed products are α-acyloxyketones. In reaction of camphorquinone [15,130], the acyloxy group had the more stable, *endo* configuration

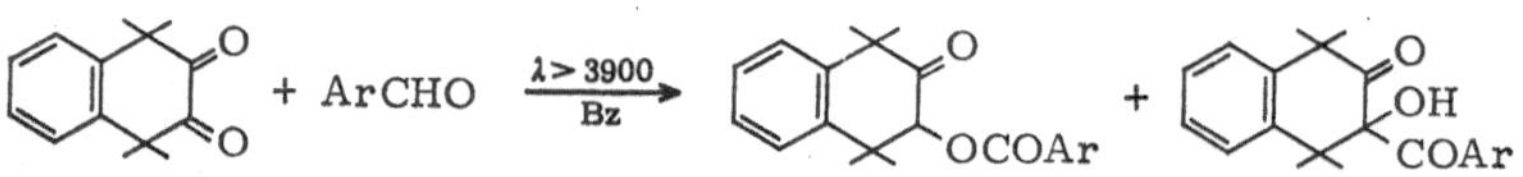

in both isomers *(61, 62,* formed in 1:1 ratio). Esterification of photo-reduction products of camphorquinone produced [15] the same esters indicating that stereochemistry of both processes is identical.

1,4-Addition has been the exclusive result in all reactions of aromatic aldehydes reported to date with the exception [15] of reactions of aromatic aldehydes with 1,1,4,4-tetramethyl-2,3-tetralindione *(73)*. These reactions afforded high yields of adduct mixtures containing 1,4-adduct *(78)* and a small fraction of 1,2-adduct *(79)*. This observation should allow evalution of polar factors in aldehyde addition reactions. Reactions of 4,5-octanedione in butyraldehyde and 3,4-hexanedione in propion-aldehyde produced [163] 1,4-adducts in addition to the cyclobutanolones resulting from intramolecular reaction.

Cyclic diketones showed a greater tendency to undergo 1,2-addition with aliphatic aldehydes. As noted earlier, camphorquinone gave [130] approximately equal amounts of 1,2- and 1,4-adducts with acetaldehyde but only 1,4-adducts with butyraldehyde or pivalaldehyde. Tetramethyl-tetrahydrofurandione gave [15] about 30% of 1,2-adduct and *73* gave 50% with butyraldehyde.

The reaction of phenylacetaldehyde with camphorquinone produced [130] about 10% of 1,4-aldehyde adduct *(61, 62)* but the major product was a mixture of isomers *(56, 57)* identical with those obtained by reaction of the dione with toluene. Clearly, the intermediate phenylacetyl radicals underwent decarbonylation to form benzyl radicals.

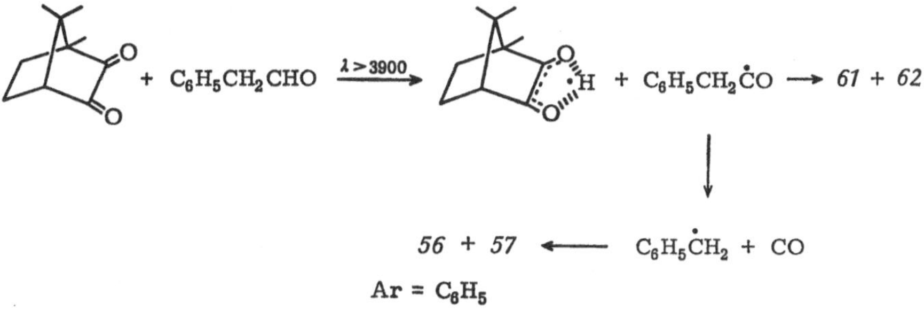

$$\text{(camphorquinone)} + C_6H_5CH_2CHO \xrightarrow{\lambda > 3900} \text{(adduct)} + C_6H_5CH_2\overset{\bullet}{C}O \longrightarrow 61 + 62$$

$$56 + 57 \longleftarrow C_6H_5\overset{\bullet}{C}H_2 + CO$$

$$Ar = C_6H_5$$

As usual, the only product reported [16,84] from reactions of benzil in the presence of aldehydes was benzilbenzoin *(63)*. The reported failure [16] of benzil to react with valeraldehyde, furfural and cinnamaldehyde seems questionable.

The crystalline product obtained from reaction of chloral with phenanthrenequinone was reported [84] to be chlorine free.

B. Diones with Alcohols

The well-known fact that primary and secondary alcohols are efficient donors of H-atoms is emphasized by the fact that the strongest esr signals for semidione radicals were obtained [99,171] when diones were irradiated in 2-propanol solution. The subsequent fate of the semidione radicals depended on their structure; open chain species giving diketopinacols and cyclic species undergoing reduction.

Thus, biacetyl was converted to *63* (R=CH₃) in high yield in ethanol [41] and 2-propanol [20,163] solution and benzil to benzilbenzoin (*63*, R=C₆H₅) in ethanol [40] and 2-propanol [37]).

Reduction to α-hydroxyketones has been observed with tetramethyl-tetrahydrofurandione (*72*, 2-propanol [15]), camphorquinone (methanol [97,99], 2-propanol [15,99]), 3,3-dimethylindanedione (*80*, methanol, ethanol, 2-propanol [85]), tetramethyltetralindione (*73*, methanol, 2-propanol [60]). The only quinone which has been reacted with alcohols is phenanthrenequinone (ethanol [165], cholestanol [128]). These reactions were unexceptional except for the following points.

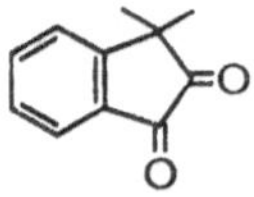

80 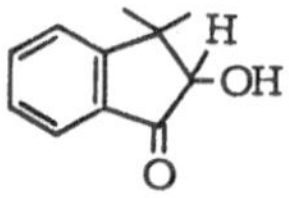   81

The two carbonyl groups in the indanedione *80* differ in that one is conjugated with the aromatic system and the other is not. Reaction was stated [65] to lead exclusively to reduction of the unconjugated carbonyl group (compound *81*).

The reduction of phenanthrenequinone was mentioned earlier as the only example of a hydrogen abstraction reaction having a quantum yield greater than unity. A chain process was suggested [165] in which an hydroxyethyl radical transfers an H-atom to ground state quinone and the resulting semidione radical abstracts hydrogen from ethanol to give reduced quinone and a new hydroxyethyl radical.

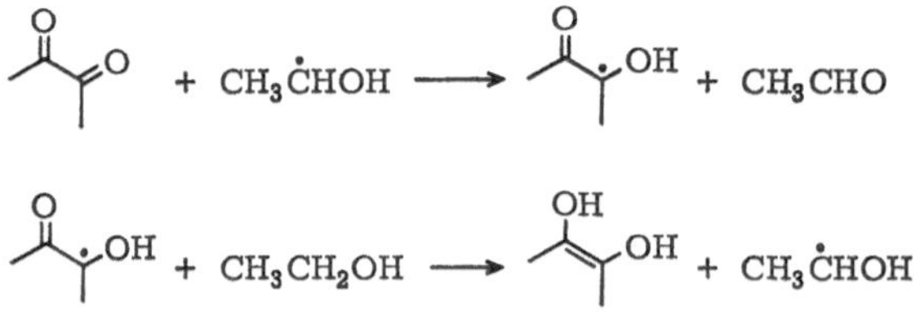

The *endo*-configuration of the hydroxyketones *82* and *83* formed from camphorquinone has been unequivocally established [15] and shown to be related to configurations of the aldehyde adducts. This stereochemical result requires attack from the normally more hindered side of the molecule. As discussed earlier (*cf.* Table 4) *m*-methoxyacetophenone sensitized the reduction of camphorquinone in 2-propanol, and benzophenone effected a new type of sensitization by transfer of hydrogen from benzophenone ketyl to camphorquinone.

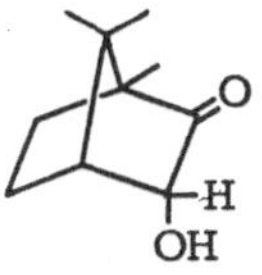

82 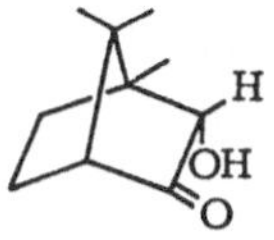   83 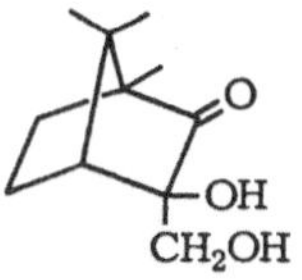   84 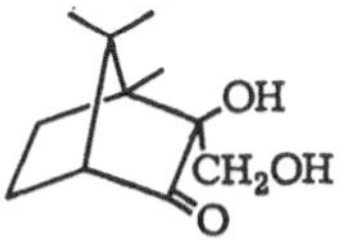   85

Alcohols are usually oxidized to aldehydes or ketones in these reactions; cholestanone has been isolated [128] from reaction of cholestanol and phenanthrenequinone in benzene solution. Two cases have been reported, both involving reactions in methanol, where the intermediate hydroxymethyl radical coupled (in part) with semidione radical. Thus, the 1,2-adducts (34%) 84 and 85 were obtained [99] with camphorquinone at 2537 Å in addition to 82 and 83 (66%). 1,2-Adduct (35%) predominated [60] over reduction product (18%) in reaction of 73 in methanol. This reaction led to a very complex mixture of products, some or all of which may reflect reactions of the monohemiketal since light filtered through Pyrex was used and decolorization of the dione was observed in methanol solution.

Tertiary alcohols are fairly resistant to reaction with diones.

C. Diones with Ethers and Thioethers

Photochemistry of diones began with Klinger's [82] report of the sunlight reactions of benzil and phenanthrenequinone with diethyl ether but reactions of diones with ethers have been investigated to a very limited extent since then. The reported reactions include biacetyl (dioxane [20]), benzil (diethyl ether [82], dioxane [128]), tetramethyltetralindione (73, dioxane [60]), tetrachloro-o-benzoquinone (dioxane [128]), acenaphthenequinone (dioxane [128]) and phenanthrenequinone (diethyl ether [82], di-isopropyl ether [127], di-n-butyl ether [127], tetrahydrofuran [128], dioxane [128], anisole [128], methoxycholestane (86) [128], methoxy-5-cholestene [128]).

The only reactions which proceeded with good yields were those of phenanthrenequinone. With the exceptions of the steroid ethers, the ether was used as solvent and 1,4-adducts (87) could be isolated in high yield when irradiations were performed at λ > 3900 Å. At shorter wavelengths the dioxane and anisole adducts rearranged to the corresponding 1,2-adducts (88), rearrangement of the anisole adduct being quite facile. The 1,4-adducts of aliphatic ethers were unstable in air [127].

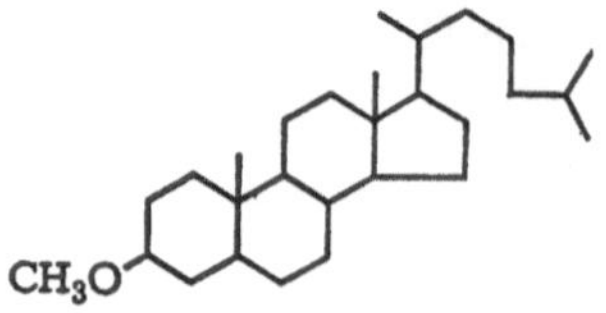

86 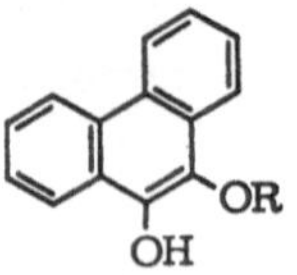   87 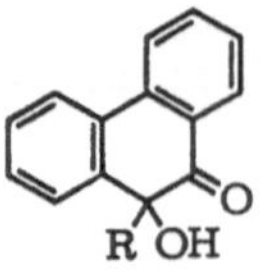   88

With *73*, only 1,2-adducts (*89*, a mixture of diastereomers) were obtained in addition to 8% of reduction product of *73* and an unexplained 19% of anhydride *90* (the reaction was reportedly run in inert atmosphere).

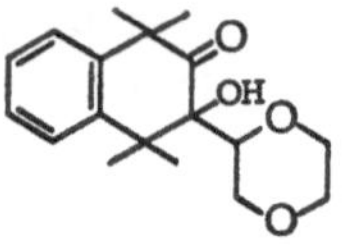

89 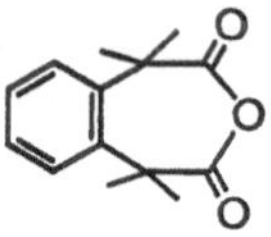   90 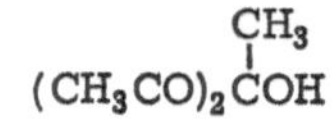   91

The reaction of biacetyl and dioxane was complex affording modest yields of 1,2-adduct, diketopinacol *63*, acetyldioxane, and compound *91* whose origin is obscure. Benzil in diethyl ether furnished benzilbenzoin *(63)*; a mixture from which no pure products were isolated was obtained with dioxane. A similar result was obtained with acenaphthenequinone and dioxane. Tetrachloro-*o*-benzoquinone and dioxane afford 20% of a 1,4-adduct analogous to *87*; the remaining products were not characterized.

In 1,4-adducts *(87)*, the normally unreactive ether function has been converted to the readily hydrolyzable acetal or ketal group. Thus, *Klinger* did not isolate the adduct of diethyl ether and phenanthrenequinone, but detected acetaldehyde upon acid hydrolysis. An attempt has been made [128] to utilize this change in functionality to provide a photochemical method for cleavage of ethers. Methoxycholestane *(86)* was used as a model compound. Acid hydrolysis of the product of irradiation of approximately equal amounts of *86* and phenanthrenequinone in benzene afforded 19% of cleavage products in addition to recovered *86*. Assuming that problems of photoequilibrium were involved, the reaction was repeated in benzene-acetic acid and 50% of cleavage product resulted. No ether cleavage was observed with 3-methoxy-5-cholestene; the products appeared to be cycloadducts resulting from reaction at the double bond. Reactions of safrole and isosafrole [127] with phenanthrenequinone also occurred at the olefinic bond and not at the methylenedioxy function.

Two reactions of thioethers with phenanthrenequinone have been investigated. As mentioned earlier (*cf.* Table 3), ethyl vinyl sulfide afforded [86] 69% of cycloaddition product; it was shown later [54] that 21% of diastereomeric 1,2-adducts (88, R=CH(CH$_3$)SCH=CH$_2$). p-Dithiane yielded [127] a 1,4-adduct (87, R=$\overset{\text{S}}{\underset{\text{S}}{\bigcirc\!\!\!\!\diagup}}$) which could only be characterized spectroscopically because of its instability.

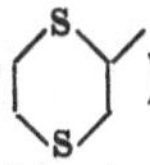

D. Diones with Esters and Lactones

The reactions of ethyl acetate, ethyl propionate, and γ-butyrolacetone with phenanthrenequinone to give exclusively products of reaction (52, 53) at the carbon atom alpha to the ether oxygen have been discussed. Methyl acetate reacted [110] similarly. Preliminary results [127] suggest that suitable substitution on the carbon atom alpha to the carbonyl group can affect the orientation of addition.

The only other reported case of a dione-ester reaction is the tetramethyltetralindione (73) — methyl formate reaction included in *Gream*, *Paice*, and *Ramsay's* [60] survey of reaction of 73. Two radicals, 92 and 93, can be formed by H-abstraction from methyl formate. The methoxycarbonyl radical, 92, accounted directly for 54% of the product

$$CH_3O\overset{\text{O}}{\overset{\|}{C}}\cdot\ \rightarrow CH_3\cdot + CO_2$$

92

$$\cdot CH_2O\overset{\text{O}}{\overset{\|}{C}}H$$

93

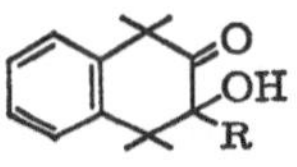

94

as the 1,2-adduct 94, (R=COOCH$_3$). An additional 10% of 1,2-adduct 94 (R=CH$_3$) may have been formed *via* coupling of methyl radicals formed from decarboxylation of 92. About 3% of the product was derived from reaction of 93 either as the ester (94, R=CH$_2$OCHO) or the free alcohol (94, R=CH$_2$OH). As in other reactions of 73 in supposedly inert atmosphere, a considerable amount (16%) of the anhydride 90 was also formed.

E. Diones with Hydrocarbons

For convenience, these reactions have been subdivided into (1) reactions with saturated hydrocarbons, (2) intramolecular H-abstraction from saturated carbon, (3) reactions with olefins, (4) reactions with alkylbenzenes, and (5) reactions with benzene.

(1) *Reactions with Saturated Hydrocarbons*. The limited number of examples of this type indicate that H-abstraction can occur with saturated hydrocarbons. The gas phase reaction of hexafluorobiacetyl with 2,3-dimethylbutane was mentioned earlier; H-abstraction also occurred [169] in solution but products were not identified. Reaction of biacetyl in cyclohexane afforded [20] 10% of 1,2-adduct *96*, 36% of diketopinacol *63*, 24% of acetylcyclohexane, and 7% of the β-diketone *91*. With tetramethyltetralindione [60] *(73)*, 1,2-adduct *94* (R = cyclohexyl) accounted

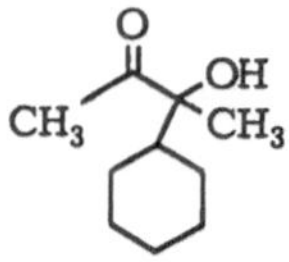

96    97

for 64% of the product; in addition 17% of reduction product and 15% of the diacid monoester *97* were obtained. The formation of *97* remains unexplained.

Reaction of benzil [27] in cyclohexane was complex, as mentioned earlier and it is not clear if any of the products were formed by H-abstraction.

(2) *Intramolecular H-abstraction from Saturated Carbon*. Urry and coworkers [163,164] have reported reactions of 2,3-pentanedione, 3,4-hexanedione, 4,5-octanedione, 2,7-dimethyl-4,5-octanedione, 5,6-decanedione, and 1,2-cyclodecanedione. Intramolecular H-abstraction from a gamma carbon atom, followed by intramolecular radical coupling produced cyclobutanolones *(65, 98)* in high yield. The cyclization product *98*

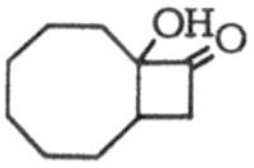

98

from cyclodecanedione was obtained in 74% yield accompanied by 9% of cyclooctanone which was presumed to be formed by elimination of ketene from *98*. The selectivity of these reactions was attributed to conformational factors as discussed earlier. The reaction of 4,5-octanedione could be sensitized by benzophenone; retardation by oxygen, naphthalene, and anthracene was observed in reactions of 2,7-dimethyloctanedione. Reactions of pentanedione and 3,4-hexanedione, which involve abstraction of primary hydrogen, were reportedly slower than those of longer chain compounds and could be partly intercepted by a good hydrogen donor (propionaldehyde).

The selectivity and high yields of these reactions should make them useful for preparation of otherwise difficulty available cyclobutane derivatives. Comparable reactions of long-chain monoketones are much more complex.

(3) *Diones and Olefins.* The first report of abstraction of allylic hydrogen involving reaction of biacetyl and cyclohexene was made by *Jolly* and de *Mayo* [79] in 1964 and confirmed recently [20]. In addition to 59% of the 1,2-adduct *99*, products of radical dimerization (*63* and bicyclohexenyl) were also obtained in appreciable yield.

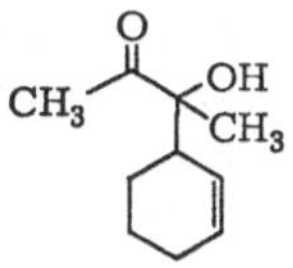

99

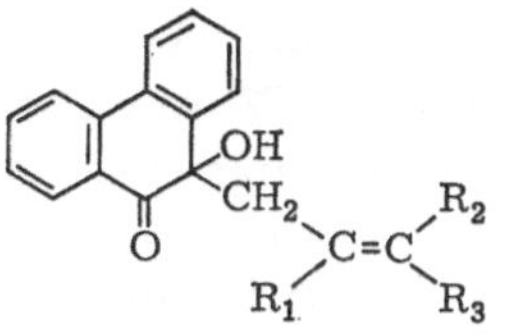

100

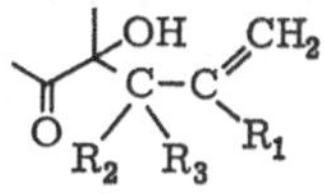

101

The reactions of isobutylene, *cis-* and *trans*-2-butene, trimethyl- and tetramethylethylene with phenanthrenequinone reported by *Farid* and *Scholz* [54] were mentioned earlier in connection with cycloaddition. While the major product was the cycloadduct in every case, appreciable amounts (9—29%) of 1,2-adducts were also obtained. They included 1,2-adducts *(100, 101)* derived from coupling at either possible position of the intermediate allylic radicals. Surprisingly, the stereochemistry of the original olefin was reportedly preserved in adducts *(100)* not involving rearrangement of the olefin *both* in the case of *cis-* and of *trans*-2-butene. Steric factors appeared to play a role in limiting the total number of possible products.

(4) *Alkylbenzenes.* Abstraction of benzylic hydrogen proceeds readily and was observed in the early years of dione photochemistry. Reactions of phenanthrenequinone have been reported [19,101,118,133,148,172] with toluene, o-, m- and p-xylene, ethylbenzene, isopropylbenzene, cumene, pseudocumene, p-cymene, t-butyltoluene, p-methoxytoluene, p-nitrotoluene, diphenylmethane, thioxanthene, tetralin, collidine, and quinaldine. Only dimers of the hydrocarbon radicals were described [148] in the reactions of diphenylmethane and thioxanthene and only phenanthrene-

quinhydrone [19] in reactions of ethylbenzene, p-cymene, and collidine. Reactions of isopropylbenzene and tetralin gave [101] dehydrogenated products (α-methylstyrene, dihydronaphthalene) in addition to quinhydrone. Adducts were obtained, generally in good yield, in the remaining cases.

After a period of some confusion regarding their structures, it was shown [118,133] that the products obtained using a broad spectrum of light were 1,2-adducts *(67)* but that both *67* and 1,4-adducts *(68)* were formed using light of wavelength longer than 3900 Å. 1,4-Adducts could also be detected [172] when reactions at shorter wavelengths were interrupted at an early stage. The rearrangement of *68* to *67* was shown [172] not to involve reversal of *68* to quinone and alkylbenzene by irradiation of the p-xylene, 1,4-adduct *(68*, Ar $= p$-CH$_3$C$_6$H$_4$) in dilute dioxane solution; the sole product was *67*. Any quinone formed would have been trapped by reaction with the large excess of more reactive dioxane. The isolation of *67* in preparative experiments thus represents a favorable photoequilibrium. The reasons for the greater photostability of *67* are unclear. As noted earlier, the ratio *67:68* depended [134] on the p-substituent in a series of p-substituted toluenes.

Reactions of alkylbenzenes with other diones produced only 1,2-adducts even upon irradiation under irreversible conditions. Biacetyl and ethylbenzene gave [20] 50% of 1,2-adduct *101* plus radical dimers ((*63*) and 2,3-diphenyl butane), tetramethyltetrahydrofurandione *(72)* gave [15] high yield of adduct *102* with p-xylene, tetramethyltetralindione *(73)*

| 102 | 101 | 103 |

and toluene afforded [60] 87% of *94* (R $=$ C$_6$H$_5$CH$_2$), and camphorquinone gave [129] about 65% of isomeric 1,2-adducts *(56, 57)* in approximately 2:1 ratio with o- or p-xylene.

Benzil afforded [96] a low yield of 1,2-adduct *(103)* with acenaphthylene. Its reactions with toluene, xylene, ethylbenzene, cumene, and pseudocumene supposedly produced [17] only the ubiquitous benzilbenzoin *(63)*, benzaldehyde, and benzoic acid.

(5) *Benzene.* Abstraction of hydrogen from benzene would not be expected to be a particularly favorable process. However, dilute benzene

solutions of phenanthrenequinone decolorize rapidly in sunlight and, as noted in the Introduction, the quantum yield for disapperance of quinone at 4358 Å in rigorously degassed benzene was 0.25. The monophenyl ether of 9,10-dihydroxyphenanthrene (*87*, R=C$_6$H$_5$), biphenyl, and phenanthrenequinhydrone were isolated [131] from preparative reactions. These results suggest the intermediacy of semiquinone and phenyl radicals formed by H-abstraction.

A similar result has been observed [36] in reactions of tetrachloro-*o*-benzoquinone in benzene solution where the phenyl ether *33* was isolated. The reaction reportedly required *cis*-stilbene or diphenylacetylene and proceeded extremely slowly in their absence.

The disappearance of biacetyl upon irradiation in benzene solution has been reported [20], no products were identified.

Photoreduction of *73* in benzene solution to the α-hydroxyketone in 54% yield was suggested to involve H-abstraction from *73* itself.

Reactions in the Presence of Oxygen

The important role which oxygen may play in dione photoreactions was mentioned in the Introduction where the phenanthrenequinone-ethyl acetate reaction was cited as an example of reduction in quantum yield and change in product composition (*cf.* Fig. 2) due to oxygen. A number of dione reactions have been investigated with oxygen deliberately present. Products, generally carboxylic acids, have been characterized but mechanistic details are obscure at present. These reactions are summarized in Table 6. The photooxidation of biacetyl has been reviewed by *Hoare* and *Pearson* [75].

The results obtained with camphorquinone in *p*-xylene are of particular interest since this reaction has been investigated [129] under otherwise identical conditions in the presence and absence of oxygen. In carefully degassed *p*-xylene solution at 4358 Å the major products were the 1,2-adducts *56* and *57*; the quantum yield for disappearance of dione was 0.07. In air-saturated *p*-xylene the quantum yield *increased* to 0.16, *56* and *57* were not formed, instead the products were camphoric acid *(104)* and camphoric anhydride *(105)*. No explanation has been offered for the different course of reaction in benzene where cyclic lactones *106* and *107* were obtained [97]. Intermediacy of the cyclic diacyl peroxide *108* has been suggested [97]. It is tempting to speculate that the diacyl radical *109* may have been trapped by oxygen.

Intermediacy of a compound such as *110*, formed by a *cis*-stilbene → dihydrophenanthrene type of cyclization, has been suggested [24] to account for formation of phenanthrene-9,10-dicarboxylic acid anhydride from diphenylcyclobutenedione.

Phenanthrenequinone [152] has been reported to sensitize the photo-oxidation of neoabietic acid.

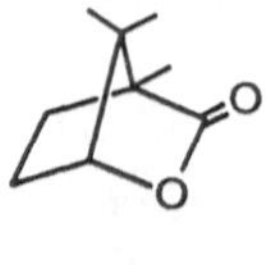

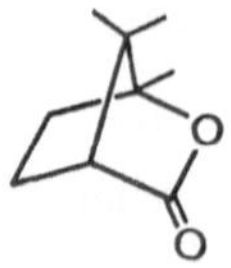

Table 6. *Dione reactions in the presence of oxygen*

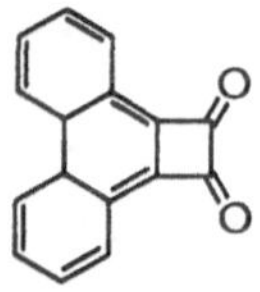

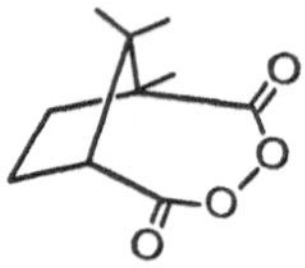

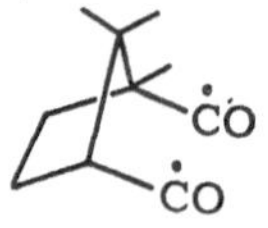

Dione	Solvent	λ (Å)	Products	Lit.
Biacetyl	Gas phase	a)	CO, CO_2, CH_3OH, CH_2O, H_2O, CH_3COOH, etc.	75)
Camphorquinone	CCl_4, t-buOH, MeOH[b]	> 3000	camphoric anhydride (105)	97)
	benzene	> 3000	(105), camphoric acid (104) (106, 107), derived hydroxy acids	97)
	p-xylene	4358	(104, 105)	129)
Diphenylcyclobutene-dione	benzene acetonitrile	sun	phenanthrene-9,10-dicarboxylic acid anhydride	24)
Benzil	benzene, acetonitrile	sun	benzoic anhydride	24)
Phenanthrenequinone	toluene, o-, m-, p-xylene, cumene ethylbenzene	sun	diphenic acid	18)
	benzene, acetonitrile	sun	diphenic anhydride	24)
Acenaphthenequinone	benzene, acetonitrile	sun	naphthalene-1,8-dicarboxylic acid anhydride	24)

a) Various wavelengths.
b) 80% Photoreduction.

Literature

1) *Calvert, J. G.,* and *J. N. Pitts, Jr.:* Photochemistry. New York: Wiley 1966.
2) *Kan, R. O.:* Organic Photochemistry. New York: McGraw-Hill 1966.
3) *Turro, N. J.:* Molecular Photochemistry. New York: Benjamin 1967.
4) *Neckers, D. C.:* Mechanistic Organic Photochemistry. New York: Reinhold 1967.
5) *Adams, M., M. S. Blois, Jr.,* and *R. H. Sands:* Paramagnetic Resonance Spectra of Some Semiquinone Free Radicals. J. Chem. Phys. *28*, 774 (1958).
6) *Alder, K., H. K. Schäfer, H. Esser, H. Krieger* u. *R. Reubke:* Über das Nortricyclen. Darstellung und spektroskopisches Verhalten einiger Homologer des Campherchinons. Ann. *593*, 23 (1955).
7) *Almgren, M.:* The Natural Phosphorescence Lifetime of Biacetyl and Benzil in Fluid Solution. Photochem. Photobiol. *6*, 829 (1967).
8) *Awad, W. I.,* and *M. S. Hafez:* Studies on 4-Substituted-β-naphthoquinones. J. Am. Chem. Soc. *80*, 6057 (1958).
9) *Bäckström, H. L. J.,* and *K. Sandros:* The Quenching of the Long-lived Fluorescence of Biacetyl in Solution. Acta Chem. Scand. *12*, 823 (1958).
10) — — Transfer of Triplet State Energy in Fluid Solutions I. Sensitized Phosphorescence and Its Application to the Determination of Triplet State Lifetimes. Acta Chem. Scand. *14*, 48 (1960).
11) *Bakule, R.,* and *F. A. Long:* Keto-Enol Transformation of 1,2-Cyclohexanedione. I. Hydration and Keto-Enol Equilibria. J. Am. Chem. Soc. *85*, 2309, 2313 (1963).
12) *Baum, E. J.,* and *R. O. C. Norman:* Electron Spin Resonance Studies. Part XVI. Photochemical Reactions of Biacetyl with Some Benzenoid Compounds. J. Chem. Soc. (B) 227 (1968).
13) *Beckett, A., A. D. Osborne,* and *G. Porter:* Primary Photochemical Processes in Aromatic Molecules. Part II. Radicals and Radical Anions Derived from Benzaldehyde, Acetophenone and Benzil. Trans. Faraday Soc. *60*, 873 (1964).
14) *Bell, R. P.,* and *A. O. McDougall:* Hydration Equilibria of Some Aldehydes and Ketones. Trans. Faraday Soc. *56*, 1281 (1960).
15) *Ben-Basat, J. M.:* Unpublished results.
16) *Benrath, A.:* Über Synthesen in Sonnenlicht. J. Prakt. Chem. *73*, 383 (1906).
17) — Über das Benzilbenzoin. J. Prakt. Chem. *87*, 416 (1913).
18) — u. *A. von Meyer:* Über die Autoxydation des Phenanthrenchinons in Gegenwart von Benzol-Kohlenwasserstoffen. Ber. *45*, 2707 (1912).
19) — — Über die Einwirkung von Benzol-Kohlenwasserstoffen auf Phenanthrenchinon im Sonnenlicht. J. Prakt. Chem. *89*, 258 (1914).
20) *Bentrude, W. G.,* and *J. R. Darnall:* Biacetyl Photochemistry: Products in Solution. Chem. Commun. 810 (1968).
21) *Berg, H.:* Naturwissenschaften *22*, 513 (1960).
22) — Kinetik und Mechanismus der polarographischen, katalytischen und photoaktivierten Reduktion von Chinonen und Ketonen. Z. Chem. *2*, 237 (1962).
23) *Beringer, F. M., R. E. K. Winter,* and *J. A. Castellano:* Photolysis of Pyracycloquinone in Methanol. An Example of Benzenoid-Quinoid Valence Tautomerism. Tetrahedron Letters 6183 (1968).
24) *Bird, C. W.:* The Photoxidation of α-Diketones. Chem. Commun. 1537 (1968)
25) *Birnbaum, H., R. C. Cookson,* and *N. Lewin:* Absorption Spectra of Ketones. Part VI. Intramolecular Charge Transfer Spectra, and Further Examples of Intensified $n \rightarrow \pi^*$ Transitions. J. Chem. Soc. 1220 (1961).
26) *Blomquist, A. T.,* and *E. A. Lalancette:* Diphenylcyclobutenedione. Synthesis and Structure. J. Am. Chem. Soc. *83*, 1387 (1961).

27) —, and *R. A. Vierling:* Dimethylcyclobutenedione. Tetrahedron Letters 655 (1961).

28) *Bloomfield, J. J., J. R. Smiley Irelan,* and *A. P. Marchand:* 1,2-Cyclobutanediones. I. The Synthesis and Reactions of Tricyclo [4,4,2,0^{1,6}]dodeca-3,8-diene-11,12-dione. [4,4,2]-Propella-3,8-diene-11,12-dione. Tetrahedron Letters 5647 (1968).

29) —, and *R. E. Moser:* 1,2-Cyclobutanediones. II. Spectral Characteristics of [4.4.2]Propella-3,8-diene-11,12-dione and Its Di- and Tetrahydroderivative. J. Am. Chem. Soc. *90,* 5625 (1968).

30) *Bohning, J. J.,* and *K. Weiss:* A Kinetic Study of the Photochemical Reaction of Phenanthrenequinone with Olefins. J. Am. Chem. Soc. *88,* 2893 (1966).

31) *Brandon, R. W.,* and *E. A. C. Lucken:* The Electron-spin Resonance Spectra and Odd-electron Distribution of a Number of Polycyclic Semiquinones and Their Derivatives. J. Chem. Soc. 4273 (1961).

32) *Breitenbach, J. W.,* u. *H. Frittum:* Freie Radikale in Fester Phase. J. Polymer Sci. *29,* 565 (1958).

33) *Brown, R. F. C.,* and *R. K. Solly:* The Photodimerization of Benzocyclobutenedione. Tetrahedron Letters 169 (1966).

34) *Bruce, J. M.:* Light-induced Reactions of Quinones. Quart. Rev. *21,* 405 (1967).

35) *Bryce-Smyth, D.,* and *A. Gilbert:* Specificity in the Thermal- and Photo-additions of *trans*-Stilbene to Tetrachloro-*o*-benzoquinone. Importance of Charge-correlation Factors in Cycloadditions which involved Donor-Acceptor Pairs. Chem. Commun. 1701 (1968).

36) — — Photochemical and Thermal Cycloadditions of *cis*-Stilbene and Tolan to Tetrachloro-*o*-benzoquinone. Photodecarbonylation of an α-Diketone. Chem. Commun. 1702 (1968).

37) *Bunbury, D. L.,* and *C. T. Wang:* The Photolysis of Benzil in Cyclohexane Solution. Can. J. Chem. *46,* 1473 (1968).

38) *Cava, M. P., D. R. Napier,* and *R. J. Pohl:* Benzocyclobutadienequinone: Synthesis and Simple Transformations. J. Am. Chem. Soc. *85,* 2076 (1963).

39) *Chow, Y. L.,* and *T. C. Joseph:* Mechanism of Phenanthrenequinone Photocycloaddition to Olefins. Chem. Commun. 604 (1968).

40) *Ciamician, G.,* u. *P. Silber:* Chemische Lichtwirkungen. Ber. *34,* 1530 (1901).

41) *Cohen, W. D.:* Photochemische Reduktion von α-Diketonen. Chem. Weekblad *13,* 590 (1916). Zentr. *1916,* II. 480.

42) *Cundall, R. B.,* and *T. F. Palmer:* The Photosensitized Isomerization of Butene-2. Trans. Faraday Soc. *56,* 1211 (1960).

43) *Dubois, J. T.:* Primary Processes in the Gaseous Photolysis of Polyatomic Molecules. I. Biacetyl. J. Chem. Phys. *33,* 229 (1960).

44) — The Selective Sensitization of Biacetyl Triplet State in the Vapour Phase. J. Am. Chem. Soc. *84,* 4041 (1962).

45) —, and *H. Behrens:* Interaction of Anthracene and Biacetyl at Different Excitation Wavelengths. J. Chem. Phys. *48,* 2647 (1968).

46) —, and *M. Cox:* Singlet-Singlet Energy Transfer in Fluid Solutions. J. Chem. Phys. *38,* 2537 (1963).

47) —, and *R. L. Hemert:* Lifetime of Excited States in Solution by the Quenching Method. J. Chem. Phys. *40,* 923 (1964) .

48) —, and *F. Wilkinson:* Triplet State of Benzene. J. Chem. Phys. *38,* 2541 (1963).

49) — — Radiative Lifetime of Triplet Biacetyl. J. Chem. Phys. *39,* 899 (1963).

50) *Evans, T. R.,* and *P. A. Leermakers:* Emission Spectra and Excited-State Geometry of α-Diketones. J. Am. Chem. Soc. *89,* 4380 (1967).

51) *Farid, S.:* Photoreactions of o-Quinones with Olefins. Mechanism of the 1:4-Cycloadditions. Chem. Commun. 1268 (1967).

52) — *D. Hess* u. *C. H. Krauch:* Strahlenchemische Bildung und Photolyse von α-Ketooxetanderivaten aus Vinylencarbonat. Chem. Ber. *100*, 3266 (1967).

53) — —, *G. Pfundt, K.-H. Scholz,* and *G. Steffan:* Photoreactions of o-Quinones with Olefins: a New Type of Reaction Leading to Dioxole Derivatives. Chem. Commun. 638 (1968).

54) —, and *K.-H. Scholz:* Photoreactions of o-Quinones with olefins: Competition between Cyclo- and R—H-addition. Chem. Commun. 412 (1968).

55) *Ford, R. A.,* and *F. Parry:* Electronic and Vibrational States of Carbonyl Compounds — I Electronic States of Camphorquinone. Spectrochim. Acta *12*, 78 (1958).

56) *Forster, L. S.:* The Effect of Solvent on the Visible Absorption Spectrum of Biacetyl. J. Am. Chem. Soc. *77*, 1417 (1955).

57) — Singlet-Triplet Absorption in Biacetyl Solutions. J. Chem. Phys. *26*, 1761 (1957).

58) —, *S. A. Greenberg, R. J. Lyon,* and *M. E. Smith:* Luminescence and Internal Conversion in Biacetyl Solutions. Spectrochim. Acta *16*, 128 (1960).

59) *Gorrell, J. H.,* and *J. T. Dubois:* Smoluchowski Equations for Systems which Do Not Obey the Stokes-Einstein Relationship. Trans. Faraday Soc. *63*, 347 (1967).

60) *Gream, G. E., J. C. Paice,* and *C. C. R. Ramsay:* Photochemistry of α-Diketones. I. Some Photochemical Reactions of 1,1,4,4-Tetramethyl-2,3-Dioxotetralin. Australian J. Chem. *20*, 1671 (1967).

61) *Greenberg. S. A.,* and *L. S. Forster:* The Photolysis of Biacetyl Solutions. J. Am. Chem. Soc. *83*, 4339 (1961).

62) *Greenzaid, P., Z. Luz,* and *D. Samuel:* A Nuclear Magnetic Resonance Study of the Reversible Hydration of Aliphatic Aldehydes and Ketones. I. Oxygen-17 and Proton Spectra and Equilibrium Constants. J. Am. Chem. Soc. *89*, 749 (1967).

63) *Grotewold, J.,* and *E. A. Lissi:* Photolysis of Biacetyl in the Presence of Triethylborine. J. Chem. Soc. (B) 264 (1968).

64) *Hammond, G. S., J. Saltiel, A. A. Lamola, N. J. Turro, J. S. Bradshaw, D. O. Cowan, R. C. Counsell, V. Vogt,* and *C. Dalton:* Photochemical cis-trans Isomerization. J. Am. Chem. Soc. *86*, 3197 (1964).

65) —, *N. J. Turro,* and *P. A. Leermakers:* Energy Transfer from the Triplet States of Aldehydes and Ketones to Unsaturated Compounds. J. Phys. Chem. *66*, 1144 (1962).

66) —, *P. Wyatt, C. D. DeBoer,* and *N. J. Turro:* Photosensitized Isomerism Involving Saturated Centers. J. Am. Chem. Soc. *86*, 2532 (1964).

67) *Hanna, R., C Sandris,* et *G. Ourisson:* Étude de Cétones Cycliques VI Comparison d'α-dicétones Polycycliques. Bull. Soc. Chim. France 1454 (1959).

68) *Harrison, A. J.,* and *J. S. Lake:* Photolysis of Low Molecular Weight Oxygen Compounds in the Far Ultraviolet Region. J. Chem. Phys. *63*, 1489 (1959).

69) —, and *F. P. Lossing:* Hg-Photosensitized Decompositions of Biacetyl, Acetylacetone and Acetonylacetone. Can. J. Chem. *37*, 1478 (1959).

70) *Heicklen, J.:* The Fluorescence and Phosphorescence of Biacetyl Vapor and Acetone Vapor. J. Am. Chem. Soc. *81*, 3863 (1959).

71) —, and *W. A. Noyes, Jr.:* The Photolysis and Fluorescence of Acetone and Acetone-Biacetyl Mixtures. J. Am. Chem. Soc. *81*, 3858 (1959).

72) *Heller, A.,* and *E. Wasserman:* Intermolecular Energy Transfer from Excited Organic Compounds to Rare-Earth Ions in Dilute Solutions. J. Chem. Phys. *42*, 949 (1965).

301

73) *Herkstroeter, W. G., A. A. Lamola,* and *G. S. Hammond:* Mechanisms of Photochemical Reactions in Solution. XXVIII. Values of Triplet Excitation Energies of Selected Sensitizers. J. Am. Chem. Soc. *86,* 4537 (1964).

74) —, *J. Saltiel,* and *G. S. Hammond:* Stereoisomeric Triplet States of an α-Diketone. J. Am. Chem. Soc. *85,* 482 (1963).

75) *Hoare, D. E.,* and *G. S. Pearson:* In Advances in Photochemistry. Vol. III, p. 83; ed. by *W. A. Noyes, Jr., G. S. Hammond,* and *J. N. Pitts, Jr.* New York: Interscience 1964.

76) *Horner, L.,* u. *H. Lang:* Zur Kenntnis der o-Chinone IX: Darstellung und Eigenschaften einiger Amino-o-chinone. Ber. *89,* 2768 (1956).

77) *Hughes, A. N.,* and *S. Uaboonkul:* Photochemical Addition of Water and Triphenylphosphine to 9,10-Phenanthrenequinone. Chem. Ind. 1253 (1967).

78) *Ishikawa, H.,* and *W. A. Noyes, Jr.:* Photosensitization by Benzene Vapor: Biacetyl. The Triplet State of Benzene. J. Chem. Phys. *37,* 583 (1962).

79) *Jolly, P. W.,* and *P. deMayo:* The Photochemical Addition of Cyclohexene to Diacetyl and Ethyl Pyruvate. Can. J. Chem. *42,* 170 (1964).

80) *Kellmann, A.:* Réactions Photochimiques de L'Acridine dans les Solvants Hydrogénés. J. Chim. Phys. *63,* 936 (1966).

81) *Kendall, E. C.,* and *Z. G. Hajos:* Tetrahydro-3,4-furandione. I. Preparation and Properties. J. Am. Chem. Soc. *82,* 3219 (1960).

82) *Klinger, H.:* Über das Isobenzil und die Einwirkung des Sonnenlichts auf einige organische Substanzen. Ber. *19,* 1862 (1886).

83) — Über die Einwirkung des Sonnenlichts auf organische Verbindungen. Ann. *249,* 137 (1888).

84) — Über Synthesen durch Sonnenlicht. Ann. *382,* 211 (1911).

85) *Koelsch, C. F.,* and *C. D. LeClaire:* The Reactions and Enolization of Cyclic Diketones V Some Carbonyl Reactions. J. Org. Chem. *6,* 516 (1941).

86) *Krauch, C. H., S. Farid* u. *D. Hess:* Photoaddition von Phenanthrenchinon-(9.10) an Vinylchlorid und 1.4-Dioxen. Chem. Ber. *99,* 1881 (1966).

87) — —, u. *G. O. Schenck:* Photoaddition von Phenanthrenchinon-(9.10) an benzocyclische Olefine zu Keto-oxetanen als neuartige Cycloadditionsreaktionen der Chinone. Chem. Ber. *98,* 3102 (1965).

88) *Kuboyama, A.:* The n-π* Transition Band of Acenaphthenequinone. Bull. Chem. Soc. Japan *33,* 1027 (1960).

89) *Lamola, A. A.,* and *G. S. Hammond:* Mechanism of Photochemical Reactions in Solution XXIII Intersystem Crossing Efficiences. J. Chem. Phys. *43,* 2129 (1965).

90) *Lemaire, J.:* Photoenolization of Biacetyl. J. Phys. Chem. *71,* 2653 (1967).

91) *Leonard, N. J.,* and *E. R. Blout:* The Ultraviolet Absorption Spectra of Hindered Benzils. J. Am. Chem. Soc. *72,* 484 (1950).

92) —, and *P. M. Mader:* The Influence of Steric Configuration on the Ultraviolet Absorption of 1,2-Diketones. J. Am. Chem. Soc. *72,* 5388 (1950).

93) *Lipsky, S.:* Evidence for Triplet-Triplet Transfer from Benzene to Biacetyl in Cyclohexane Solution. J. Chem. Phys. *38,* 2786 (1963).

94) *Liu, R. S. H., N. J. Turro, Jr.,* and *G. S. Hammond:* Activation and Deactivation of Conjugated Dienes by Energy Transfer. J. Am. Chem. Soc. *87,* 3406 (1965).

95) *Mallory, F. B.,* and *J. D. Roberts:* Reactions of Phenylcyclobutadienequinone with Methanol. J. Am. Chem. Soc. *83,* 393 (1961), footnote 6.

96) *deMayo, P.,* and *A. Stoessl:* The Irradiation of Acenaphthene with Benzil. Can. J. Chem. *40,* 57 (1962).

97) *Meinwald, J.,* and *H. O. Klingele:* Photochemical Reactions of Camphorquinone. J. Am. Chem. Soc. *88,* 2071 (1966).

98) *Michael, J. L.*, and *W. A. Noyes, Jr.*: The Photochemistry of Mixtures of 2-Pentanone and 2-Hexanone with Biacetyl. J. Am. Chem. Soc. *85*, 1027 (1963).
99) *Monroe, B. M., S. A. Weiner*, and *G. S. Hammond*: Photoreduction of Camphorquinone. J. Am. Chem. Soc. *90*, 1913 (1968).
100) *Moore, R. F.*, and *W. A. Waters*: Photochemical Addition of Benzaldehyde to Quinones. J. Chem. Soc. 238 (1953).
101) — — Some Photochemical Reactions between Quinones and Hydrocarbons. J. Chem. Soc. 3405 (1953).
102) *Murov, S., D. S. McClure*, and *N. C. Yang*: Dependence of Sensitized Biacetyl Phosphorescence Intensity on Exciting-Light Intensity. J. Chem. Phys. *45*, 2204 (1966).
103) *Murrell, J. N.*: The Electronic Spectra of Organic Molecules. Wiley, Methuen 1963.
104) *Mustafa, A.*: Reactions in Sunlight of (a) Phenanthrenequinone, Retenequinone, and Chrysenequinone with Ethylenes; (b) Retenequinone and Chrysenequinone with Aromatic Aldehydes; and (c) o-Formylbenzoic Acid with isopropyl Alcohol. J. Chem. Soc., Suppl. S 83 (1949).
105) — Reactions in Sunlight of Certain Diketones with Ethylenes and Aromatic Aldehydes. Photopolymerization of 9-Anthraldehyde. J. Chem. Soc. 1034 (1951).
106) —, *A. H. E. Harhash, A. K. E. Mansour*, and *S. M. A. E. Omran*: Photochemical Addition and Photochemical Dehydrogenation Reactions in Sunlight. J. Am. Chem. Soc. *78*, 4306 (1956).
107) —, *A. K. Mansour*, and *A. F. A. M. Shalaby*: Photoreactions in Sunlight. Experiments with Benzo (h) quinoline-5,6-quinone. J. Am. Chem. Soc. *81*, 3409 (1959).
108) — —, and *H. A. A. Zaher*: Photochemical Reactions in Sunlight. Experiments with 2-N-Phenyl-α,β-naphtho-1,2,3-triazolequinone and Monoxime Derivative in Sunlight and in Dark. J. Org. Chem. *25*, 949 (1960).
109) *Nagakura, S.*, and *A. Kuboyama*: Dipole Moments and Absorption Spectra of o-Benzoquinone and its Related Substances. J. Am. Chem. Soc. *76*, 1003 (1954).
110) *Neuwirth-Weiss, Z.*: Unpublished Results.
111) *Norman, R. O. C.*, and *R. J. Pritchett*: An Investigation of Some Aliphatic Semidione and Protonated Semidione Radicals. J. Chem. Soc. (B) 378 (1967).
112) *Noyes, W. A., Jr., W. A. Mulac*, and *Max S. Matheson*: Photochemical Primary Process in Biacetyl Vapor at 4358 Å. J. Chem. Phys. *36*, 880 (1962).
113) —, *G. B. Porter*, and *J. E. Jolley*: The Primary Photochemical Processes in Simple Ketones. Chem. Rev. *56*, 49 (1956).
114) *Okabe, H.*, and *W. A. Noyes, Jr.*: The Relative Intensities of Fluorescence and Phosphorescence in Biacetyl Vapor. J. Am. Chem. Soc. *79*, 801 (1957).
115) *Parker, C. A.*, and *T. A. Joyce*: Activation-controlled Delayed Fluorescence of Benzil. Chem. Commun. 1421 (1968).
116) *Parmenter, C. S.*, and *B. L. Ring*: Energy Transfer between Benzene and the Lifetime of Triplet Benzene in the Gas Phase. J. Chem. Phys. *46*, 1998 (1967).
117) *Petry, R. C.*, and *J. P. Freeman*: Tetrafluorohydrazine. A Versatile Intermediate for Synthesis of N-Fluoro Compounds. J. Am. Chem. Soc. *83*, 3912 (1961).
118) *Pfundt, G.*: Beitrag zur Photochemie der Chinone. Photochemische Additionsreaktionen des 9,10-Phenanthrenchinons und des Chloranils. Doctoral Dissertation, Göttingen, 1962.
119) —, u. *S. Farid*: NMR-Untersuchung zur Konformation von Derivaten des 1,4-Dioxens. Tetrahedron *22*, 2237 (1966).

M. B. Rubin

120) —, and *G. O. Schenck:* In: 1,4-Cycloaddition Reactions, ed. by *J. Hamer,* Chap. 11. New York—London: Academic Press 1967.
121) *Porter, C. W., H. C. Ramsperger,* and *C. Steel:* The Action of Ultraviolet Light upon Diketones. J. Am. Chem. Soc. 1827 (1923).
122) *Porter, G. B.:* Photooxidation of Biacetyl. J. Chem. Phys. *32,* 1587 (1960).
123) *Rebbert, R. E.,* and *P. Ausloos:* Quenching of the Triplet State of Acetone and Biacetyl by Various Unsaturated Hydrocarbons. J. Am. Chem. Soc. *87,* 5569 (1965).
124) *Richtol, H. H.,* and *A. Belorit:* Simultaneous Donor Quenching and Acceptor Sensitization in Phosphorescence Studies of Triplet Energy Transfer: The Biacetyl-Benzil System. J. Chem. Phys. *45,* 35 (1966).
125) —, and *F. H. Klappmeier:* Hydrogen Bonding Effects on Triplet Energy Transfer in Solution. J. Am. Chem. Soc. *86,* 1255 (1964).
126) — — Luminescence and Energy Transfer in Some Aliphatic Diketones. J. Chem. Phys. *44,* 1519 (1966).
127) *Rubin, M. B.:* Unpublished Results.
128) — Photochemical Reactions of Diketones. The 1,4-Addition of Ethers to 9,10-Phenanthrenequinone. J. Org. Chem. *28,* 1949 (1963).
129) —, and *R. G. LaBarge:* Photochemical Reactions of Diketones. IV. The Photoaddition of *o-* and *p-*Xylene to Camphorquinone. J. Org. Chem. *31,* 3283 (1966).
130) — —, and *J. M. Ben-Bassat:* The Photochemical Addition of Aldehydes and Camphorquinone. Israel J. Chem. (Proc.) *5,* 39. (1967).
131) —, and *Z. Neuwirth-Weiss:* Photochemical Reaction of Phenanthrenequinone and Benzene. Chem. Commun. 1607 (1968).
132) —, and *R. A. Reith:* Photochemical Addition of Esters to 9,10-Phenanthrenequinone. Preferential Formation of Ether Radicals. Chem. Commun. 431 (1966).
133) —, and *P. Zwitkowits:* Photochemical Reactions of Diketones. II. The 1,2-Addition of Substituted Toluenes to 9,10-Phenanthrenequinone. J. Org. Chem. *29,* 2362 (1964).
134) — — Polar Effects in Free Radical Reactions. The Mechanism of Photochemical Addition of Ethers and Substituted Toluenes to 9,10-Phenanthrenequinone. Tetrahedron Letters 2453 (1965).
135) *Russell, G. A.,* and *J. Lokensgard:* α-Keto Radicals. J. Am. Chem. Soc. *89,* 5059 (1967).
136) —, *E. T. Strom, E. R. Talaty, K. Y. Chang, R. D. Stephens,* and *M. C. Young:* Application of Electron Spin Resonance Spectroscopy to Problems of Structure and Conformation. Aliphatic Semidiones. Rec. Chem. Progr. *27,* 3 (1966).
137) —, *E. R. Talaty,* and *M. C. Young:* Radical Cations Derived from α-Diketones. J. Phys. Chem. *70,* 1321 (1966).
138) *Sandris, C.,* et *G. Ourisson:* Étude Spectrale de Cétones Cycliques. II L'hydratation d'α-dicétones non Enolisables. Bull. Soc. Chim. France 338 (1958).
139) — — Étude Spectrale de Cétones Cycliques. IV. Comparison d'α-dicétones de Series Homomorphes. Bull. Soc. Chim. France 350 (1958).
140) *Sandros, K.:* Transfer of Triplet State Energy in Fluid Solutions. III. Reversible Energy Transfer. Acta Chem. Scand. *18,* 2355 (1964).
141) —, and *M. Almgren:* The Relative Yields of Fluorescence and Phosphorescence of Biacetyl in Fluid Solutions. Acta Chem. Scand. *17,* 552 (1963).
142) —, and *H. L. J. Bäckström:* Transfer of Triplet State Energy in Fluid Solution. II. Further Studies of the Quenching of Biacetyl Phosphorescence in Solution. Acta Chem. Scand. *16,* 958 (1962).

143) *Schenck, G. O.* u. *G. A. Schmidt-Thomee:* Photochemische Reaktionen IV. Die Photosynthese von cyclischen Schwefelsaureestern durch Addition von SO₂ an o-Chinone. Ann. *584*, 199 (1953).

144) *Schonberg, A., W. I. Awad,* and *G. A. Mousa:* Photochemical Reactions in Sunlight. Experiments with 4-Cyano-1,2-Naphthoquinone. J. Am. Chem. Soc. *77*, 3850 (1955).

145) —, *N. Latif, R. Moubasher,* and *A. Sina:* (a) Photoreduction of Phenylglyoxylic Acid. (b) Photo-reactions between Aldehydes and o-Quinones. (c) Reactions between o- Quinones and Ethylenes in the Dark and in the Light. J. Chem. Soc. 1364 (1951).

146) —, and *R. Moubasher:* Photo-reaction between Phenanthraquinone and Aromatic Aldehydes. A New Passage from Phenanthraquinone to Fluorenone. J. Chem. Soc. 1430 (1939).

147) —, and *A. Mustafa:* Reaction of Ethylenes with Phenanthraquinone, J. Chem. Soc. 387 (1944).

148) — — (c) Dehydrogenation effected by p-Benzoquinone and Phenanthraquinone. J. Chem. Soc. 657 (1945).

149) — — Reactions of Non-Enolizable Diketones in Sunlight. Chem. Rev. *40*, 181 (1947).

150) — — Reactions with Phenanthraquinone, 9-Arylxanthenes, and Diphenyl Triketone. J. Chem. Soc. 997 (1947).

151) — —, and *S. M. A. D. Zayed:* Experiments with 1,2-Benzophenazine-3,4-quinone. J. Am. Chem. Soc. *75*, 4302 (1953).

152) *Schuller, W. H.,* and *R. V. Lawrence:* Air Oxidation of Resin Acis. III. The Photosensitized Oxidation of Neoabietic Acid and the Configurations of the Pine Gum Resin Acids. J. Am. Chem. Soc. *83*, 2563 (1961).

153) *Schulte-Frohlinde, D.,* u. *C. V. Sonntag:* Zur Photoreduktion von Chinonen in Losung. Z. Phys. Chem. *44*, 314 (1965).

154) *Shcheglova, N. A., D. N. Shigorin,* and *M. V. Gorelik:* Elektronic Spectra of Aromatic α-Diketones. Russ. J. Phys. Chem. *39*, 471 (1965); Chem. Abstr. *63*, 3782a (1965).

155) *Sidman, J. W.,* and *D. S. McClure:* Electronic and Vibrational States of Biacetyl and Biacetyl-d₆. I. Electronic States. J. Am. Chem. Soc. *77*, 6461 (1955).

156) *Sircar, A. C.,* and *S. C. Sen:* Studies in Acenaphthenequinone Series. J. Ind. Chem. Soc. *8*, 605 (1931).

157) *Staab, H. A.,* u. *J. Ipaktschi:* Photochemische Reaktionen des Benzocyclobuten-1,2-dions. Chem. Ber. *101*, 1457 (1968).

158) *Starr, J. E.,* and *R. H. Eastman:* Structural Features Facilitating the Photo decarbonylation of Cylic Ketones. J. Org. Chem. *31*, 1393 (1966).

159) *Steffan, G.,* u. *G. O. Schenck:* Lichtreaktionen der 1,3-Diacetyl-Δ⁴-imidazolinone-2. Chem. Ber. *100*, 3961 (1967).

160) — Lichtreaktionen des 1,3-Diphenyl-Δ⁴- imidazolinons-(2). Chem. Ber. *101*, 3688 (1968).

161) *Stevens, B.,* and *J. T. Dubois:* Simultaneous Quenching of Molecular Fluorescence by Oxygen and Biacetyl in Solution. Trans. Faraday Soc. *59*, 2813 (1963).

162) *Turro, N. J.,* and *G. S. Hammond:* The Photosensitized Dimerization of Cyclopentadiene. J. Am. Chem. Soc. *84*, 2841 (1962).

163) *Urry, W. H.,* and *D. J. Trecker:* Photochemical Reactions of 1,2-Diketones. J. Am. Chem. Soc. *84*, 118 (1962).

164) —, and *D. A. Winey:* Photochemistry of 1,2-Diketones. 1,2-Cyclodecanedione and 2,3-Pentanedione. Tetrahedron Letters 609 (1962).

165) *Walker, P.:* The Role of Phenanthraquinone in the Photochemical Oxidation of Ethanol. J. Chem. Soc. 5545 (1963).

166) *Walling, C.:* Free Radicals in Solution. New York: Wiley 1957.

167) —, and *M. J. Gibian:* Hydrogen Abstraction Reactions by the Triplet States of Ketones. J. Am. Chem. Soc. *87*, 3361 (1965).

168) *Weir, D. S.:* Energy Exchange in the 3-Methyl-2-butanone-biacetyl System at 3130 Å. J. Am. Chem. Soc. *84*, 4039 (1962).

169) *Whittemore, I. M.,* and *M. Swarc:* Photoreduction of Hexafluorobiacetyl in the Gaseous Phase and in Solution. J. Phys. Chem. *67*, 2492 (1963).

170) *Wilkinson, F.,* and *J. T. Dubois:* Energy Transfer Studies by Spectrophoto-fluorimetric Method. J. Chem. Phys. *39*, 377 (1963).

171) *Zeldes, H.,* and *R. Livingston:* Magnetic Resonance Studies of Liquids during Photolysis, IV Free Radicals from Acetaldehyde, Diacetyl and Acetoin. J. Chem. Phys. *47*, 1465 (1967).

172) *Zwitkowits, P.:* Photochemical Reactions of 9,10-Phenanthrenequinone. Doctoral Dissertation, Carnegie Institute of Technology, 1964.

173) *Strating, J., B. Zwanenburg, A. Wagenaar,* and *A. C. Udding:* Evidence for the expulsion of bis-CO from bridged α-diketones. Tetrahedron Letters *125* (1969).

174) *Hooper, D. L.:* NMR Measurements of Equilibria involving Hydration and Hemiacetal Formation of Some Carbonyl Compounds. J. Chem. Soc. (B) *169* (1967).

175) *Leermakers, P. A., H. T. Thomas, L. D. Weiss,* and *F. C. James:* Spectra and Photochemistry of Molecules Adsorbed on Silica Gel. J. Am. Chem. Soc. *88*, 5075 (1966).

176) *Weis, L. D., B. W. Bowen,* and *P. A. Leermakers:* The Photosensitized *cis-trans* Isomerization of Piperylene in Silica Gel-Benzene Matrices. J. Am. Chem. Soc. *88*, 3176 (1966).

177) *Elad, D.,* and *R. D. Youssefyeh:* The Light-induced Addition of γ-Butyrolactone to Olefins. Chem. Commun. 7 (1965).

178) *Monroe, B. M.,* and *S. A. Weiner:* Photoreduction of Camphorquinone. J. Am. Chem. Soc. *91*, 450 (1969).

179) *Horspool, W. M.,* and *G. D. Khandelwal:* Photochemical Addition of 1,2-Naphthoquinones to *p*-Dioxen. Chem. Commun. 1203 (1967).

180) *Kuboyama, A., R. Yamazaki, S. Yabe,* and *Y. Uehara:* The n → π* Bands of Phenyl Carbonyl Compounds, α-Diketones, and Quinones at Low Temperatures. Bull. Chem. Soc. Japan, *42*, 10 (1969).

181) *Almgren, M.:* Phosphorescence Spectra of α-Diketones in Low Temperature Glasses. Photochem. Photobiol. *9*, 1 (1969).

182) *Turro, N. J.,* and *R. Engel:* Quenching of Biacetyl Fluorescence and Phosphorescence. A New Mechanism for Quenching of Ketone Excited States. Mol. Photochem. *1*, 143 (1969).

183) *Weiner, S. A. E. J. Hamilton Jr.,* and *B. M. Monroe:* The Rates of Termination of Radicals in Solution. III. Semidione Radicals, in press.

184) *Friedrichsen, W.:* Photochemische Reaktionen von o-Chinonen mit Isobenzofuranen. Tetrahedron Letters 1219 (1969).

Received March 17, 1969

Photochemical Reactions of Cycloheptatrienes and Related Compounds

Prof. Dr. Lee B. Jones and Dr. Vera K. Jones

Department of Chemistry, University of Arizona, Tucson, Arizona, USA

Contents

1. Introduction

The cycloheptatriene ring system is of great interest as a subject for excited state reactions induced by both thermal and photochemical excitation. Major contributions to the photochemistry of the troponoid system and subsequently, to the photochemistry of the cycloheptatriene system have been made by *Chapman* [1] whose pioneering work in this field has stimulated much interest in photochemistry in general. The photochemistry of the troponoid system has been reviewed quite thoroughly [1], and mention will be made only of recent developments in this area. Dimerization reactions will not be discussed.

The troponoid system undergoes various types of interesting reactions upon irradiation such as valence isomerization [2-7], rearrangment [8-10], dimerization [11-13], and addition [6]. In most cases, irradiation leads to valence isomerization to produce bicycloheptadienones as the primary reaction. There are three principle modes of photoisomerization termed type A, B and C cyclizations [2b]. Type A represents valence isomerization to give norcaradienones, which subsequently give rise to benzene derivatives by elimination of carbon monoxide. This reaction occurs during the photolysis of tropone [1] in low yield. Type B cyclization occurs during irradiation of simple monocyclic tropolones [8, 9] and their methyl ethers [2, 3] to give bicyclo[3.2.0]heptadien-2-ones, in which formation of a new carbon-carbon bond occurs at the carbon atom bearing the hydroxy or methoxy substituents. Type C cyclizations are observed in polycyclic troponoids such as colchicine [4, 5] and isocolchicine [6, 7], and result in cyclization at carbon atoms not bearing the methoxyl group.

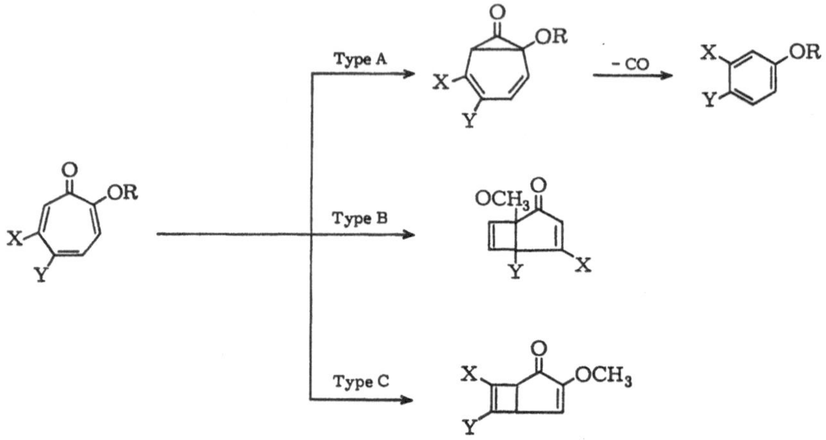

In his discussion of the valence isomerization of troponoids, *Chapman* [1] has pointed out that 2-, 3-, and 4-methoxytropones undergo different photochemical behavior. A possible function of the methoxy group may be visualized as providing stabilization for charge-separated excited states as shown in structures *1*, *2*, and *3*. In *1* and *3* (produced by excitation of 2- and 4-methoxytropone, respectively) smooth electron redistribution is available for collapse to type B cyclization products as indicated by the arrows. The corresponding intermediate *2* from 3-meth-

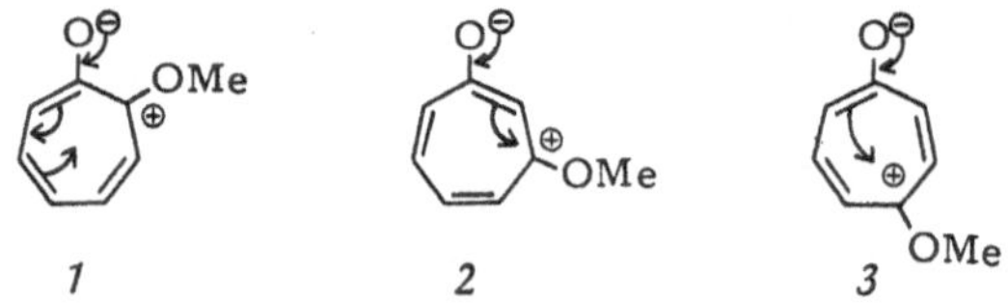

oxytropone can only return to starting material by electron redistribution without giving the bicyclic product. This generalization does not hold for the photochemistry of colchicine and isocolchicine, and has been explained by *Chapman* [5] and *Dauben* [6] as being controlled by both steric as well as styrene chromophore effects to lead to only type C cyclizations.

Cycloheptatrienes, on the other hand, normally do not contain polarizable substituents in the ring system, and thus cycloheptatriene photochemistry, in addition to being of fundamental interest, is more difficult to rationalize. As will be demonstrated with individual examples, the major photochemical transformations of cycloheptatrienes in solution are valence isomerization to the bicyclo[3.2.0]hepta-3,6-diene system or isomerization via selective 1,7-hydrogen shifts to give isomeric cycloheptatrienes. The photocyclization reaction is not unexpected, as photoisomerization of cycloheptatrienes [1] and of tropones (as shown above) is one of the most generally applicable photochemical reactions. The formation of the isomeric cycloheptatrienes are markedly affected by the presence of various substituents and as a result any general understanding of these reactions must account for the pronounced substituent effects. The formulation of the Woodward-Hoffmann rules for electrocyclization and sigmatropic migration reactions [14, 15], in addition to other molecular orbital considerations of trienes makes it possible to understand this second mode of photochemical transformations of cycloheptatrienes. These will be dealt with in greater detail in specific examples presented below.

The photochemistry of other cyclic conjugated trienes (e.g., 1,3,5-cyclooctatriene and substituted benzenes) will be discussed and their valence isomerization reactions will be compared with those of substituted cycloheptatrienes. The photochemistry of acyclic conjugated trienes is necessarily more complex than that of cyclic trienes due to the possibility of *cis-trans* isomerization. The presence of several geometric isomers in a reaction mixture opens up the possibility of other modes of reaction. These transformations will be briefly summarized. Nonconjugated acyclic and cyclic trienes will not be considered in this article.

2. Photolysis Reactions of Monosubstituted Cycloheptatrienes

Cycloheptatriene *(4)* upon irradiation in ether solution gives the bicyclic photoisomer *5* [16]. Cycloheptatriene is regenerated upon pyrolysis of *5*. Photolysis of *4* in the vapor phase [17] leads to the formation of toluene *(6)* as well as *5*. The quantum yield for toluene formation increases with decreasing pressure, the extrapolated value at zero pressure being unity within experimental error. Under the most favorable conditions, no more than five percent of the excited cycloheptatriene molecules isomerize to *5*. *Srinivasan* [17] has suggested that isomerization to toluene occurs not from an electronically excited molecule of *4*, but rather from a vibrationally excited ground state molecule formed by internal conversion from a higher electronic excited state. The formation of the bicyclic compound *5* is believed to occur from the first excited singlet state of *4*. The quantum yield for toluene formation in the gas phase decreases with increasing pressure due to collisional deactivation of vibrationally excited 1,3,5-cycloheptatriene before rearrangement. In solution only *5* is produced because rearrangement of vibrationally excited *4* is too inefficient to compete with vibrational deactivation in solution.

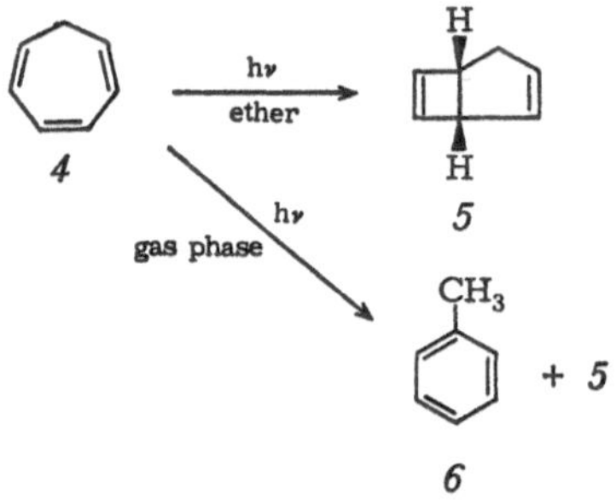

ter Borg and *Kloosterziel* [18)] studied the photolysis of neat 7-deuterio-cycloheptatriene using nmr and the results were in accordance with a series of consecutive 1,2 (or 1,7) hydrogen shifts ($7 \rightarrow \rightarrow 10, X = D$). Under their reaction conditions, the photochemically induced hydrogen migrations proceeded five hundred times as rapidly as the formation of the bicyclic product.

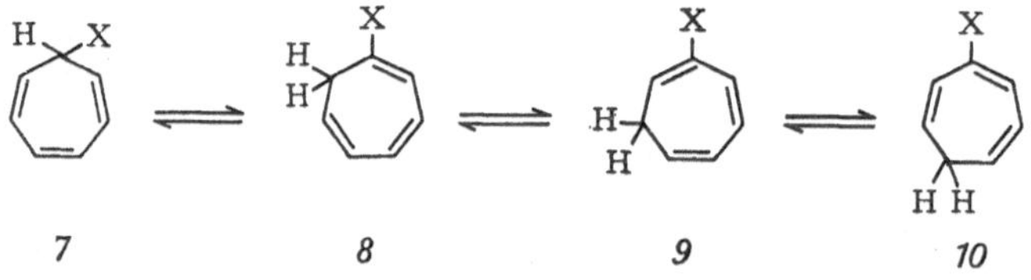

Upon irradiation of 7-phenylcycloheptatriene ($7, X = C_6H_5$), the formation of 1-phenylcycloheptatriene ($8, X = C_6H_5$) could not be formally established by nmr. Instead, only 2-phenylcycloheptatriene ($9, X = C_6H_5$) could be detected. The reaction is relatively inefficient, with *9* being the major isomer after photoequilibrium was established, at which time the formation of a bicyclic product was estimated to be present to the extent of *ca.* 15%.

Similar results on the photolysis of 7-deuteriocycloheptatriene in methanol solution were observed by *Roth* [19], who also observed that the 1,2-hydrogen shift predominates. *Doering* and *Gaspar* [20] showed that upon irradiation of 7-deuteriocycloheptatriene in benzene with added carbon tetrachloride, migration of the 7-hydrogen had occurred mainly to the adjacent position, as determined by nmr.

Photolysis of 7-methyl and 7-ethylcycloheptatriene [19] in methanol solution showed that the major photoproducts are l-methyl and l-ethyl-cycloheptatriene, as determined by nmr. Irradiation of 7-methylcyclo-heptatriene in benzene solution [21] also showed the only photoproduct to be 1-methylcycloheptatriene, as determined by gas chromatography and comparison of the nmr spectrum of the pure material with that published for 1-methylcycloheptatriene [22].

Photolysis of 7-vinylcycloheptatriene [19] upon short photolysis in methanol solution gives 1-vinylcycloheptatriene, which because of its long wavelength ultraviolet absorption (λ_{max} 298, ε 7000) is very rapidly transformed by 1,2-hydrogen shifts to the 2- and 7-vinyl isomers.

Irradiation of 7-methoxy or 7-ethoxycycloheptatriene *(11)* in ether solution [23] gives in each case the 1-alkoxybicyclo[3.2.0]hepta-3,6-diene *(12)* in yields in excess of 90%. Irradiation of the 1-alkoxycyclohepta-triene *(13)* gave *(12)*. Irradiation of *11a* at low pressures in the vapor phase gave a mixture which consisted of *11a*, *13a*, and *12a*. Formation

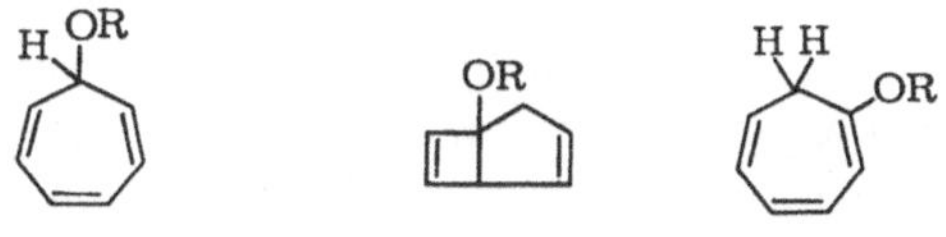

11   a, R = CH$_3$ *12*   *13*
        b, R = C$_2$H$_5$

of *13* could occur *via* a sequence of two 1,5-hydrogen migrations which are known to occur thermally [24]. If this were the case, 3-methoxycyclo-heptatriene *(14)* would be an intermediate. Irradiation of *14* yielded a complex mixture, in contrast to the clean reactions observed for *11* and *13*. The mixture from *14* consisted of at least three bicyclic products, one of which was 1-methoxybicyclo[3.2.0]hepta-3,6-diene *(15)*, and other methoxycycloheptatrienes. The bicyclic compounds *16* and/or *17* were

14   *15*   *16*   *17*

also demonstrated to be present in the bicyclic mixture. In this way it was demonstrated that *14* is most likely not an intermediate in the photolysis of *11*. When *11* is photolyzed, a photostationary state is established between the 7-isomer *(11)* and the 1-isomer *(13)*, which is displaced by conversion of *13* to the bicyclic compound *12*. Thus, the 1,7-hydrogen shift and the electrocyclization reactions are quite specific for *11* and *13*, respectively.

The conversion of triene *11* to *13*, and of *13* to *12* could not be photosensitized by acetophenone, nor quenched by oxygen. These data suggest that the reversible photochemical 1,7-hydrogen shift in *11* and *13* involves a singlet state species. Orbital symmetry arguments [14, 15] also suggest that the interconversion of *11* and *13* occurs via excited electronic states. Hence, the reactive state of *11* and *13* is most likely the lowest excited singlet state of each triene.

The high degree of selectivity observed for the transformation *11* ⇌ *13* is of particular interest. Orbital symmetry arguments [15] suggest that photochemical [1,7] suprafacial, sigmatropic migration reactions are allowed in cycloheptatrienes. A single [1,7] sigmatropic hydrogen migration in *11* leads to *13*. Triene *13*, on the other hand, could, in principle, undergo 1,7-hydrogen shifts in two directions leading to *11* and *9* ($X = OCH_3$). Experimentally, only one of these two processes is observed *(13→11)*. The transition state for conversion of *13* to *11* is asymmetric and simple orbital symmetry arguments cannot be applied. However, *Chapman* [22] has pointed out that the highest occupied molecular orbital for the electronically excited nine electron eight-atom π-system (the ether oxygen atom is included in the basic π-framework) is such that suprafacial hydrogen transfer from position seven to position one with continuous overlap is feasible, while suprafacial transfer from position seven to position six with continuous overlap is not feasible.

The above examples suggest that the ratio of valence tautomerization to 1,7-hydrogen migration might be quite sensitive to the nature of the substituent in the cycloheptatriene ring. To determine the relative efficiencies of these two processes as a function of substituent (X), a number of 1-substituted cycloheptatrienes were investigated [25]. The general reaction scheme is indicated below. In the neat liquid, the ratio of the two processes (as derived [18] from the results with 7-deuteriocycloheptatriene) is 0,002. With the electron-attracting cyano group in the 1-position of *18*, a shift of hydrogen is the only process observed. The direction of the hydrogen shift is highly specific: only *19a* is formed. When electron-donating groups are present in the 1-position of *18*, the hydrogen shift occurs in the opposite direction to give *19b* as the major product. With increasing electron-donating capacity of the substituent, valence tautomerization gains in relative importance. When X = dime-

19a 19b 18 20a 20b

thylamino, hydrogen migration products could not be detected. These data are summarized in the Table 1.

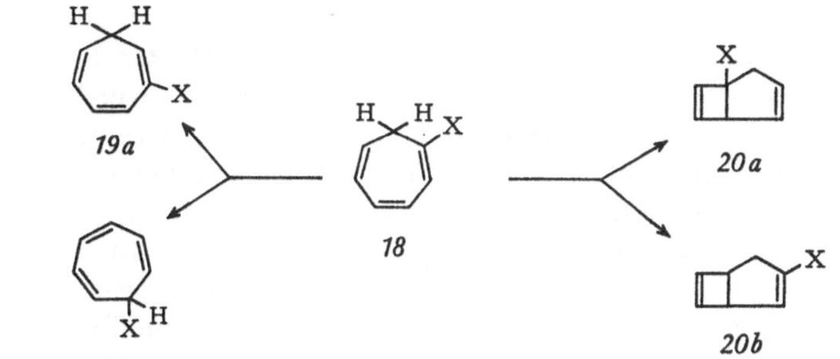

Table 1. *(Initial)products of irradiation (in %) of 1-substituted cycloheptatrienes* [25]

Substituent	Products *(19a)*	*(19b)*	*(20a)*	*(20b)*
CN	100	—	—	—
Ph	50	50	—	—
Me	2	98	—	—
SMe	—	65	35	—
OMe	—	35	65	—
NMe_2	—	—	97	3

3. Photolysis Reactions of 7,7-Disubstituted Cycloheptatrienes

7,7-Disubstituted cycloheptatrienes are of particular interest because they do not possess hydrogens available for [1,7] sigmatropic shifts as a primary reaction. As a result, other photochemical transformations should be observable. Irradiation of 3,7,7-trimethylcycloheptatriene *(21)* [26] in benzene (or cyclohexane) solution gives a mixture consisting of 2,2,6-trimethylbicyclo[3.2.0]hepta-3,6-diene *(22)*, 1,5,7-trimethylcycloheptatriene *(23)*, 1,3,7-trimethylcycloheptatriene *(24)*, and unreacted

21 22 23 24

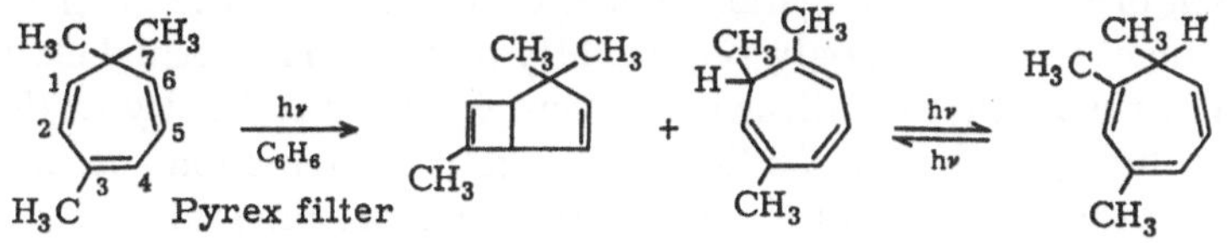

21. Compounds *22* and *23* are primary reaction products while *24* arises from a secondary photochemical reaction of *23*. Deuterium labelling established that the methyl and hydrogen migrations which result in the formation of *23* and *24*, respectively, are highly selective. In principle, a methyl group at C_7 in *21* could migrate under the influence of light to either C_1 or C_6. In fact, migration to C_1 is the only such transformation observed. Similarly, after the methyl group has shifted to yield a new triene *(23)*, only one of the possible [1,7]sigmatropic hydrogen migrations was observed *(23 ⇌ 24)*. This demonstrates that the methyl group on carbon atom three in *21* exerts an overwhelming directive influence on the course of these migration reactions.

The formation of *23* (from *21*) can be understood by applying the Woodward-Hoffmann rules [15] for sigmatropic migrations, as well as other molecular orbital considerations to these substituted trienes. If it is assumed [15] that the transition state of the methyl migration reaction is composed of a linear C_7 radical containing seven π-electrons and a methyl radical, one is led to the prediction based on the orbital symmetry of the highest filled molecular orbital of the excited state radical that migration can take place [1,7] to give either *23* or *25* (neglecting the effect of the C_3 methyl group). The two possible transition states for [1,7] methyl migration are *26* and *27*, where the $+$ and $-$ signs

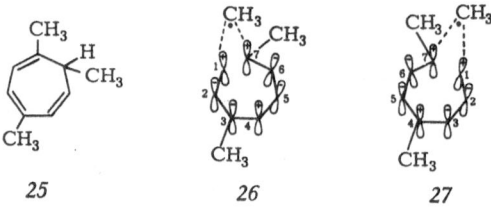

25   26   27

indicate orbital symmetry. Qualitatively, a choice between *26* and *27* may be made by considering the nonuniform charge distribution present in each of the two transition state models. HMO calculations of the excited radical (transition state) *26* indicate a relatively large positive charge on carbon atom 3. In transition state *27*, a relatively large negative charge is centered on the ring carbon 4 bearing the methyl substituent. In each of these two transition states, C_7 which bears the other methyl substituent is, at most, only very slightly positive. The normal electronic effect of the methyl substituent is to supply electron density to the ring carbon by inductive and/or hyperconjugative interaction mechanisms. This has the net effect of increasing the Coulomb integral α of the ring carbon bearing this substituent and tends to stabilize any electron deficiency at this carbon. Similarly, an increased α tends to destabilize those systems in which a negative charge density is located on carbon. As a result, 26 should represent a lower energy transition state and triene 23 should predominate. Similar considerations of charge distributions and symmetry arguments indicate that 23 should be transformed selectively into 24 upon further irradiation. Irradiation of triene 24 should result in the formation of 23.

The electrocyclization reaction of 21 to produce 22 also represents a highly selective transformation. Application of orbital symmetry arguements [14)] indicates that either 22 or 28 may be formed by a disrotatory cyclization reaction. HMO calculations do not allow for a choice between

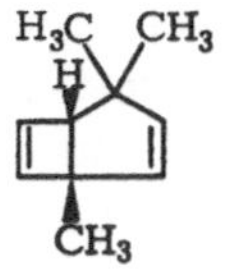

28

22 und 28. The observed selectivity can, however, be rationalized on the basis of steric considerations. The formation of 28 would result in a methyl-hydrogen eclipsing interaction at the bridgehead positions, whereas the formation of 22 avoids this interaction by placing the methyl group on one of the trigonal carbons of the cyclobutene ring yielding a more highly substituted double bond.

Irradiation of 2,7,7-trimethylcycloheptatriene (29) [27)] yields a mixture consisting of 2,2,4-trimethylbicyclo[3.2.0]hepta-3,6-diene (30), 1.3.7-trimethylcycloheptatriene (24), 1,5,7-trimethylcycloheptatriene (23), and unreacted 29. Compound 23 is a secondary photoproduct arising from further irradiation of 24. As observed in the previous system, the methyl and hydrogen migrations, as well as the cyclization reaction are highly selective. As in the case of the 3,7,7-trimethyl isomer (21),

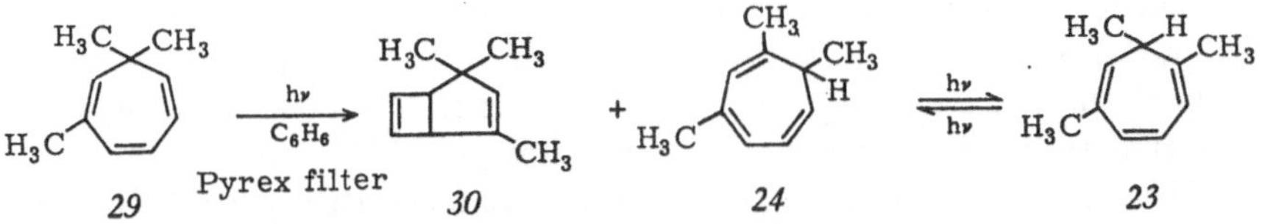

Woodward-Hoffmann orbital symmetry arguments and molecular orbital considerations allow for a distinction between the two possible paths for sigmatropic [1,7] suprafacial methyl migration (transition states 31 and 32), with 31 having the lower energy for the same reason

as those outlined above. As a result, methyl migration occurs in *29* to yield *24*. Selective hydrogen migrations observed for *23* and *24* are accounted for in a similar manner.

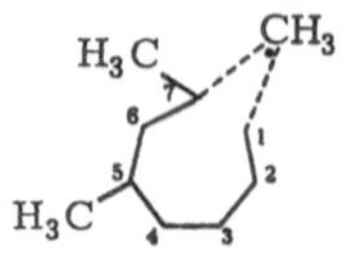

31 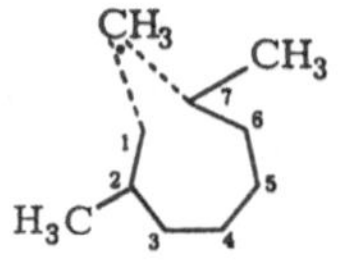   *32*

The selectivity noted in the cyclization reaction of *29* is most readily rationalized on the basis of steric considerations. The ring fusion in product *30* must be *cis*. A consideration of a Dreiding model of *30* reveals that there are no serious nonbonded interactions in this molecule. However, compound *33* (the other possible bicyclic product) has one serious interaction between the methyl group on carbon 7 and one of the gem-dimethyl groups at carbon 2. Dreiding models indicate that these two methyl groups are crowded to about the same extent in *33* as are the diaxial methyl substituents in *cis*-1,3-dimethylcyclohexane *(34)*. This interaction in *34* leads to an unfavorable energy term of about 3.7 kcal/mole [28] and should, therefore, favor the formation of *30* in the electrocyclization reaction.

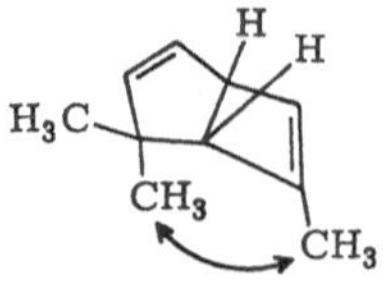

33 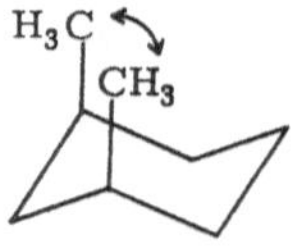   *34*

The reactions of *21* and *29* cannot be photosensitized and oxygen has no effect on either reaction. In addition, the observed cyclization and migration reactions are consistent with orbital symmetry considerations only if these are excited electronic state reactions. Consequently, it is felt that these most likely are all excited singlet state reactions.

The photolysis of 2,3,7,7-tetramethylcycloheptatriene *(35)* [29] was carried out to determine whether the methyl group at position 2 or 3 has a greater directive influence on the direction of the methyl migration reaction. The reaction was found to be complex, resulting from a concurrent electrocyclization reaction of the new trienes as they were formed. However, the major bicyclic product produced was found to be 2,2,6,7-

tetramethylbicyclo[3.2.0]hepta-3,6-diene *(36)*. Of the new cyclohepta-trienes formed, spectral evidence suggested that the major component was the cycloheptatriene *39*. This conclusion was further substantiated by the identification of the bicyclic compounds *37* and *38* (epimeric at C_2) produced by further reaction of *39*. The results indicate that the

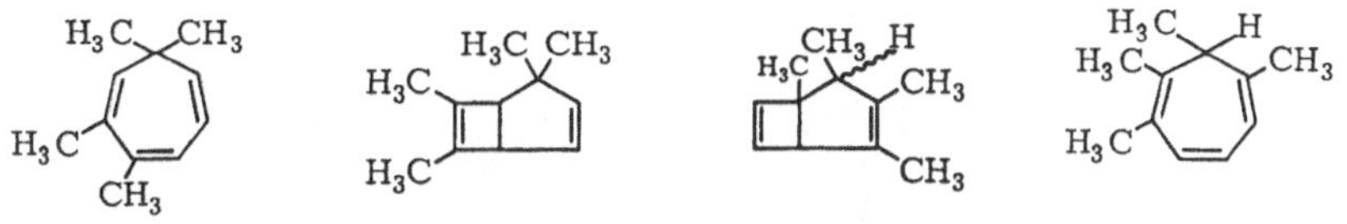

| 35 | 36 | 37, 38 | 39 |

methyl substituent at C_3 in *35* exhibits a stronger directive influence in the methyl migration reaction than the methyl substituent at C_2, and therefore the predominant mode of migration is from C_7 to C_1. This is consistent with the HMO treatment discussed for the migration reactions of *21* and *29*. The preferred mode of cyclization involves the formation of *36* which minimizes steric interactions and gives the product with the more highly substituted double bonds.

The effect of a polar substituent bonded to the cycloheptatriene system was investigated by *Chapman* and *Smith* [30], who studied the photochemistry of thujic acid *40a* and its methyl ester *40b*. Irradiation of *40* gave two products: the methyl migration product *41*, and the

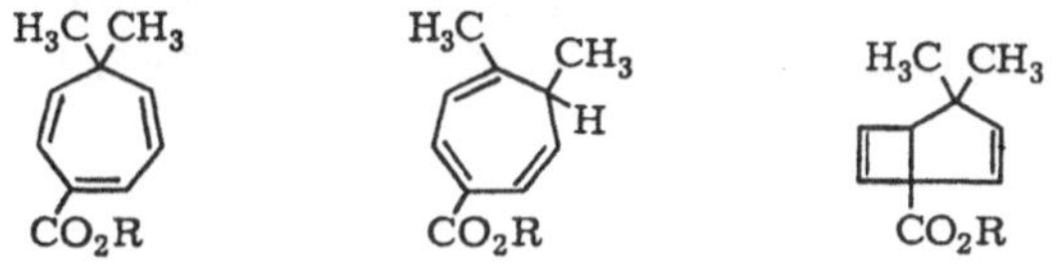

| 40 | a) R = H | 41 | 42 |
| | b) R = CH₃ | | |

electrocyclization product *42*. The formation of these products was ra-tionalized on the basis of a polar mechanism (charge separated excited state). However, the methyl migration reaction could also be rationalized using Woodward-Hoffmann rules and other HMO considerations. It is not clear in this system whether the reative state of *40* is π—π^* or n—π^*. It should be noted that a selective [1,7] sigmatropic hydrogen migration in *41* would result in an identical product. This reaction can only be detected using deuterium labelling experiments which would be rather difficult in the thujic acid system.

Photolysis of 3-acetoxy-7,7-dimethylcycloheptatriene *(43)* [31] surprisingly gave a complex mixture of products consisting of at least six major bicyclic products and a mixture of new trienes which could not be resolved by gas chromatography. Of the new bicyclic compounds only *44* was positively identified. Because of the apparent complexity of this reaction and the difficulty of separating the reaction products, this reaction was not investigated further.

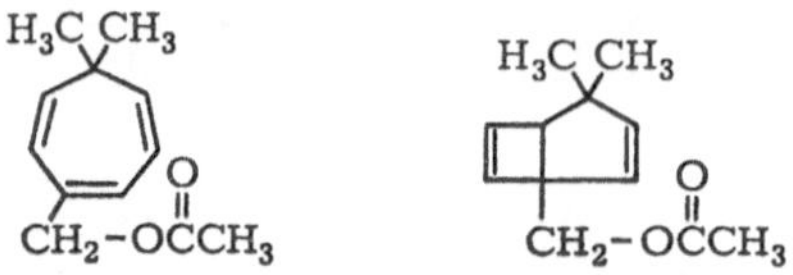

43   44

Photolysis of 7,7-*bis*(trifluoromethyl)cycloheptatriene *(45)* [32] in benzene solution yielded the bicyclic compound *46* and a small amount of the aromatic derivative *47*, in addition to unreacted *45*.

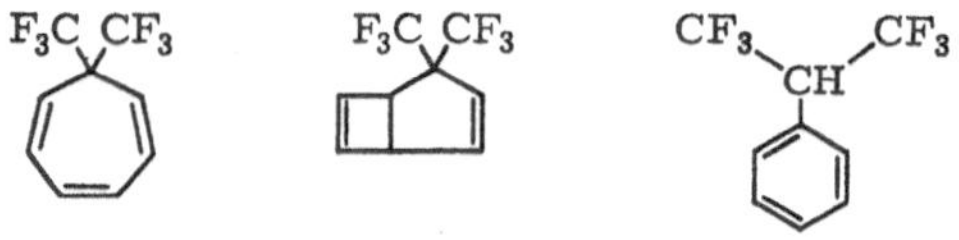

45   46   47

Photolysis of the perfluorinated analogue, perfluoro-7,7-difluoromethylcycloheptatriene *(48)*, to give *49* parallels the behavior of *45* [33].

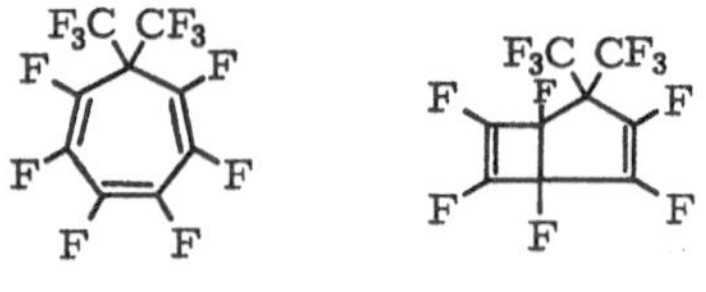

48   49

4. Irradiation of Tropylium Ions

Irradiation in 5% aqueous sulfuric acid of tropylium ion (tetrafluoroborate salt) *50* [34] was found to give rise to bicyclo[3.2.0]hepta-3,6-dien-2-ol *(52)* and the corresponding ether *53*. The authors proposed that irradiation of *50* produces the valence bond isomer *51*, a highly reactive

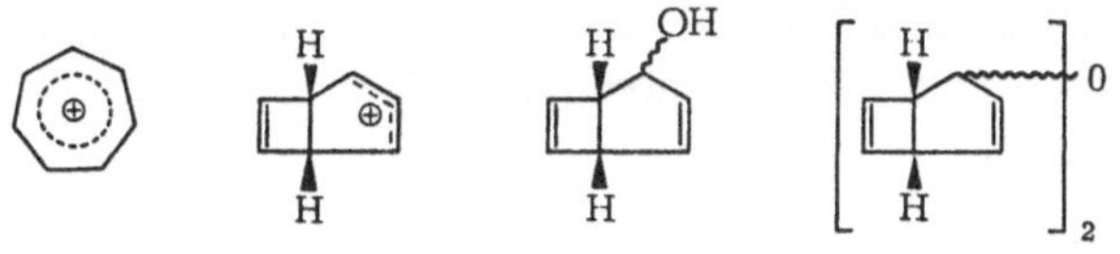

50   51   52   53

carbonium ion, which reacts with solvent to give *52*. Irradiation of tropylium perchlorate or bromide in ethanol without added acid produced tropyl ethyl as the perdominant product. Further irradiation gave a complex mixture including ditropyl and a secondary product which on the basis of catalytic reduction and spectral properties was regarded to be *54*, formed by a two-fold photochemical suprafacial 1,7-hydrogen migration in ditropyl.

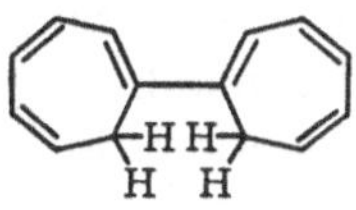

54

5. Photolysis of Other Cyclic Conjugated Trienes

Barton has suggested [35] that in photochemical reactions of cyclic conjugated olefins, ring fission will predominate in rings of n annular atoms containing (n/2)-1 double bonds and valence tautomerization will occur in other systems. Consequently, it was anticipated that 1,3,5-cyclooctatriene upon photolysis should give rise to an acyclic tetraene in the same manner that 1,3-cyclohexadiene opens [35, 36] to 1,3,5-hexatriene. When 1,3,5-cyclooctatriene *(55)* was irradiated in solution (ether or pentane) [37, 39] two isomerization products were isolated, bicyclo[4.2.0]-octa-2,7-diene *(56)* and tricyclo[5.1.0.0 [4, 8]]oct-2-ene *(57)*. The formation of *56* is not exceptional. The formation of *57* has been visualized by

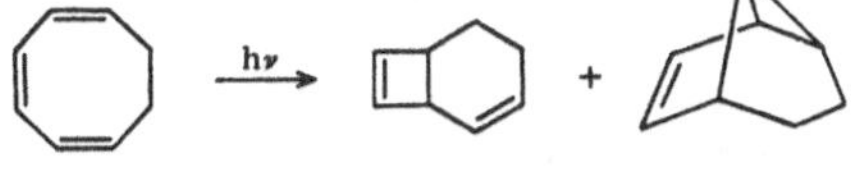

55   56   57

Chapman et al. [39] to occur in the stepwise manner indicated below. Formally, this transformation is analogous to the bridging which occurs in the photosensitized isomerization of 1,5-cyclooctadiene [40]. Ring fission products could not be detected.

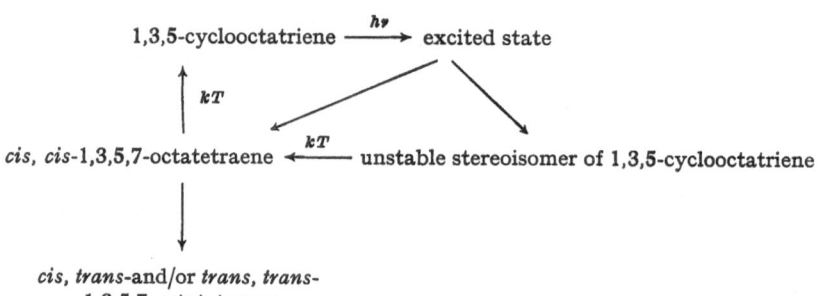

The failure of previous workers [37-39] to isolate or detect 1,3,5,7-octatetraene as a major photolysis product of *55* prompted *Goldfarb* and *Lindqvist* [41] to study the flash photolysis of *55* in cyclohexane and n-hexane. This study resulted in the detection of two transient species, both of which decay by first-order processes with lifetimes (at 25° in *n*-hexane) of 91 msec and 23 sec. A fivefold increase in the concentration of the long-lived species occurred simultaneously with the disappearance of the short-lived species. The long-lived species was identified on the basis of spectroscopic and kinetic evidence as *cis, cis*-1,3,5,7-octate-traene. The instability of this tetraene results from a low activation energy (17 kcal/mole) for recyclization to 1,3,5-cyclooctatriene *(55)*. The short-lived transient was pressumed to be a strained, cyclic stereo-isomer of 1,3,5-cyclooctatriene. The opening of this strained cyclooctate-traene to form octatetraene has an Arrhenius activation energy of 16 kcal/mole. The formation of stable 1,3,5,7-octatetraene (*cis, trans* and/or *trans, trans*) was observed to require the absorption of a second photon of light. As a result, the formation of a stable octatetraene from *55* is a biphotonic process. Presumably, the lower intensity light used in the previous studies [37-39] prevented absorption of the second photon and, as a result, the isolated products arose from other reaction of *55*.

1,3,5-cyclooctatriene $\xrightarrow{h\nu}$ excited state

kT

cis, cis-1,3,5,7-octatetraene $\xleftarrow{kT}$ unstable stereoisomer of 1,3,5-cyclooctatriene

cis, trans-and/or *trans, trans*-
1,3,5,7-octatetraene

The irradiation of cyclooctatetraene epoxide *(58)* in pentane or methanol with unfiltered light led only to rapid polymerization [42].

On the other hand, if *58* is irradiated in pentane solution with filtered light ($\lambda > 3100$Å) to prevent polymerization of products compounds *59—62* could be isolated [43]. Presumably, *59—61* arise from excited

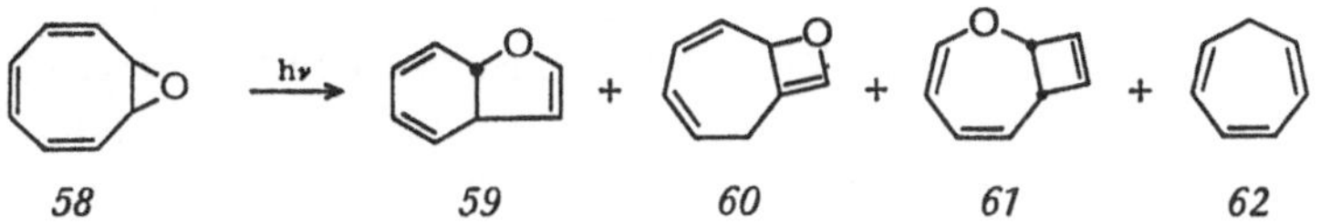

singlet state reactions of *63* which is initially produced upon photolysis of *58*. The origin of cycloheptatriene is uncertain.

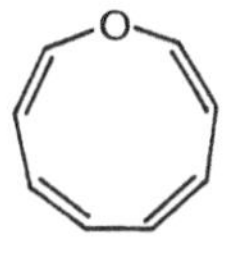

63

    Oxepin *(64)*, formally a conjugated triene, has recently been investigated [44]. Oxepin *(64)* is in thermal equilibrium with benzene oxide *(65)*. This equilibrium is markedly solvent dependent, so that it is possible to study solutions enriched either in *64* or *65*. Studies in ether solution indicate that if only *64* is excited ($\lambda > 310$mμ), the sole product is 2-oxabicyclo[3.2.0]hepta-3,6-diene *(66)*. By contrast, excitation of both *64* and *65* (λ 2537 Å) gives 11% of *66* 15% of benzene, and 74% of phenol. At this wavelength, the observed products are derived largely from *65*. Acetone as photosensitizer produced only phenol. Consequently, the authors [44] concluded that *65* intersystem crosses with a high enough efficiency to complete with the chemical reaction, while *64* crosses to the triplet manifold with extremely low efficiency. The singlet reaction of *65* is the formation of benzene, while the singlet state reaction of *64* produces *66*. Triplet *65* leads to phenol and triplet *64* most likely also leads to phenol formation.

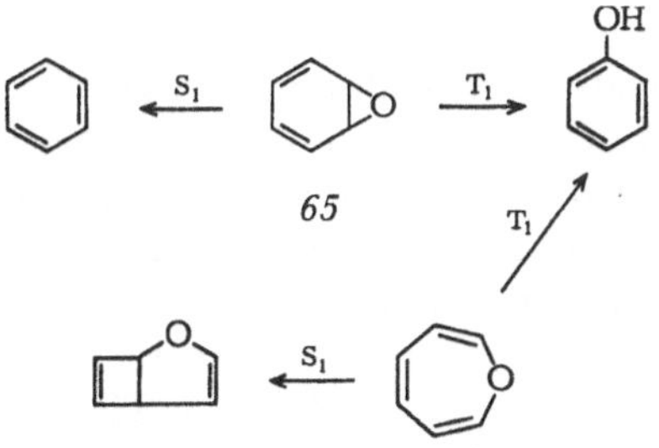

65

66   *64*

6. Electrocyclization Reactions of Substituted Benzenes

Benzene and its substituted analogues also undergo electrocyclization reactions characteristic of other conjugated triene systems. When 1,2,4-tri-tert-butylbenzene *(67)* was irradiated in ether solution using a Hanovia Type L lamp and Vycor filter, the corresponding bicyclo[2.2.0]-hexa-2,5-diene (Dewar benzene) *68* was isolated [45]. Nonbonded interactions of the *tert*-butyl substituents are partially relieved upon cycliza-

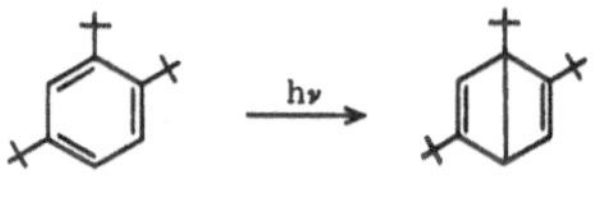

67   68

tion to *68*. Dewar benzene (unsubstituted *68*) has also been prepared and characterized [46].

Other polyalkylbenzenes have also been studied [47-54]. 1,2-Di-*tert*-butylbenzene in ether solution upon irradiation produces a mixture of 1,3- and 1,4-di-*tert*-butylbenzene [48]. o-*tert*-Butyltoluene also gives a mixture of the m- and p-isomers but it is less reactive than the di-*tert*-butylbenzene [48]. o- and *m*-xylene in ether solution gave no detectable isomerization reaction. When a nitro, acetyl or methoxy substituent was placed in the 4-position of the o-di-*tert*-butylbenzene, the isomerization reaction did not occur [48]. Using higher intensity light, however, o- and *m*-xylene were found to isomerize to the other isomeric xylenes [47].

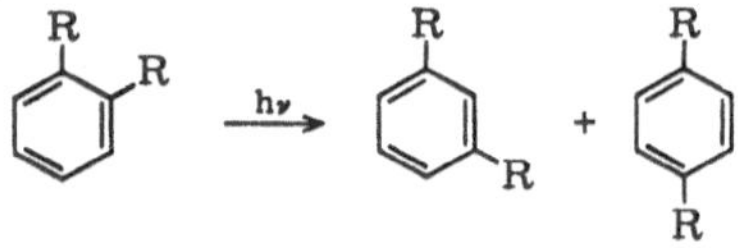

Labelling studies [51] indicate that these rearrangements are intramolecular and occur by ring carbon interchange rather than by alkyl group migrations.

In their investigations of benzene isomerization reactions *Wilzbach* and *Kaplan* [52] established that a substituted tricyclo[2.1.1.0^{5,6}]hex-2-ene (a benzvalene) *69* is an intermediate in the photoconversion of 1,2,4-tri-tert-butylbenzene *(70)* and the 1,3,5-isomer *(71)*. They also demonstrated that the irradiation of either *70* or *71* at 2537Å yields a photostationary mixture in which the principal component is prismane *(72)* produced from the Dewar benzene *73*. These authors pictured the reaction sequence as shown below.

322

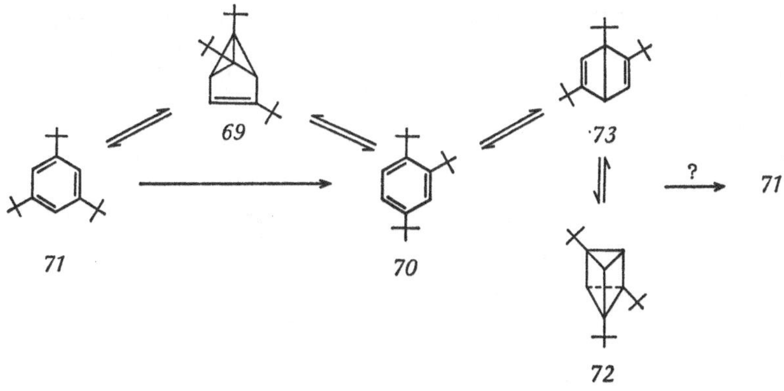

Consistent with the above scheme is the observation that 1,3,5-trideuteriobenzene *(74)* is isomerized in the vapor phase or in solution (2500 Å) to an isomeric trideuteriobenzene. Again a benzvalene intermediate was suggested. The wavelength dependence in the vapor phase suggests that the assumed benzvalene intermediate is formed, at least in part, from a vibrationally excited species [55].

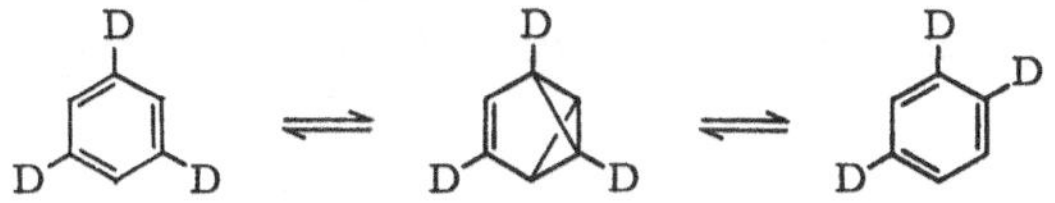

The vapor phase photolysis of *o*-xylene using light of wavelength 1600—2100 Å [53] produced small amounts of benzocyclobutene and toluene, *m*-xylene, *p*-xylene and *o*-ethyltoluene. With the exception of benzocyclobutene, all of the other products are also produced using 2537 Å light. This suggests the possibility of a common set of reactive intermediates. Fluorescence studies, as well as the effect of added inert gases on the reaction, indicated that the first excited singlet and lowest lying triplet state are not suitable choices for such common intermediates. It was suggested that highly vibrationally excited ground electronic states are the most likely condidates for such intermediates. Isotope labelling studies indicated that benzocyclobutene *(75)* was generated by the loss of two hydrogen atoms. The mechanism of formation of toluene and *o*-ethyltoluene produced in the vacuum ultraviolet photolysis of *o*-xylene was found to be radical in nature [56].

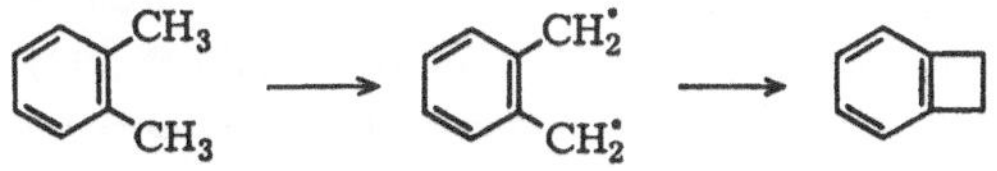

Perfluorobenzene *76* when irradiated in the gas phase (35 mm) at 25° produced the Dewar benzene *77* [57–58]. The triplet state reaction of *76* gives only higher molecular weight materials.

The photochemistry of unsubstituted benzene has recently been summarized by *Bryce-Smith* [59] and will not be repeated here. Several very recent reports should be mentioned, however. *Ward* and *Wishnok* [60] studied the liquid phase vacuum ultraviolet photolysis of benzene and were able to identify Dewar benzene *78*, benzvalene *79* and fulvene *80* in the relative amounts of 1:5:2, respectively. The quantum yields for the formation of *78—80* were estimated to be 0.006, 0.03 and 0.012, respectively. The low quantum yields suggest that radiationless transitions to

the ground electronic state may be important in this system, since neither fluorescence nor phosphorescence is significant when neat liquid benzene is photolyzed in the vacuum ultraviolet.

The irradiation of benzene vapor at 1849 Å is reported to produce *cis*-1,3-hexadien-5-yne at about one-fourth the rate of fulvene formation [61]. It was also noted that fulvene vapor produces benzene when irradiated at 2537 Å. This does not occur in solution or in the gas phase with 1849 Å light. Irradiation of the 1,3-hexadien-5-yne vapor at 2537 Å produced a 2:1 mixture of benzene and fulvene. At 1849 Å polymeric material was produced.

It thus appears that benzene derivatives, like other conjugated trienes, undergo cyclization reactions to produce bicyclic or tricyclic intermediates. In some instances these intermediates have been isolated and characterized.

7. Photolysis of Acyclic Conjugated Trienes

Acyclic conjugated trienes can undergo several modes of photoisomerization, of which the most common is reversible ring closure to a 1,3-cyclohexadiene [62]. There are at least three other reaction pathways open to triene which can be regarded as irreversible. Bond crossing of the two terminal bonds could occur to give a bicyclo[2.1.1]hexene system. A second type of bond crossing reaction can lead to a bicyclo[3.1.0]hexene. Migration of a hydrogen can lead to an allene. The latter two reactions, in addition to 1,3-cyclohexadiene formation, are those observed most often experimentally [62].

Irradiation of 1,3,5-hexatriene *(81)* in the vapor phase leads to the formation of 1,3-cyclohexadiene *(82)*, benzene, hydrogen, and 1,2,4-hexatriene *(83)*, in addition to a liquid polymer [63]. Determination of the quantum yield [64] led to the conclusion that *83* may originate from an electronically excited molecule, while benzene and hydrogen occur from a vibrationally excited ground state molecule (1,3-cyclohexadiene) formed by internal conversion from the electronically excited molecule.

Irradiation of 1,3-cyclohexadiene *(82)* in ether solution leads initially to the formation of 1,3,5-hexatriene *(81)* and extended irradiation yields a mixture (ca. 1:1) of bicyclo[3.1.0]hex-2-ene *(84)* and 3-vinylcyclobutene *(85)* [65].

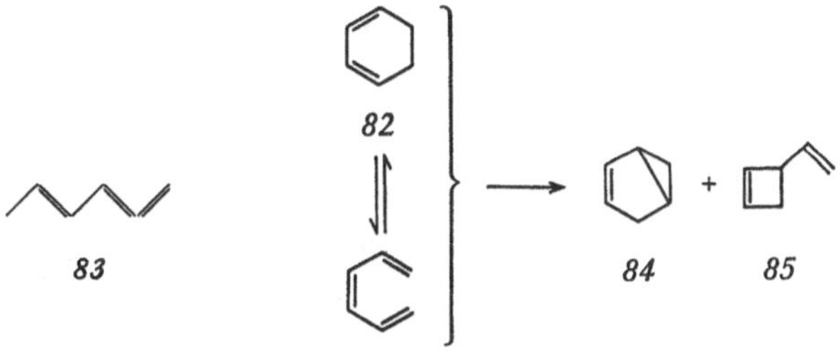

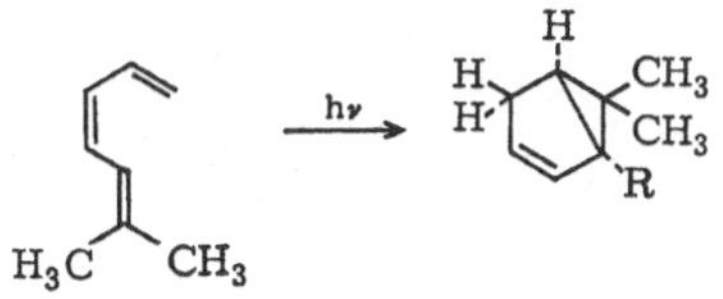

Irradiation of 6-methyl-1,3,5-heptatriene *(86)* gives the expected 6,6-dimethylbicyclo[3.1.0]hex-2-ene *(87)* as the only monomeric product [66]. Irradiation of 4-deuterio-6-methyl-1,3,5-hexatriene *(88)* afforded the labelled 6,6-dimethylbicyclo[3.1.0]hex-2-ene *(89)* on the basis of its nmr spectrum. The data suggest that in simple cases, photoisomerization of 1,3,5-hexatrienes or 1,3-cyclohexadienes to bicyclo[3.1.0]hexenes proceeds via a bond switching process from the cyclohexadiene (path a) or an electrocyclic reaction of the hexatriene, formally analogous to an intramolecular Diels-Alder reaction (path b), and that the hypothetical intermediate, 2-vinylbicyclo[1.1.0]butane (path c) is not formed. The latter path would have resulted in the formation of equivalent amounts of *89* and *90*.

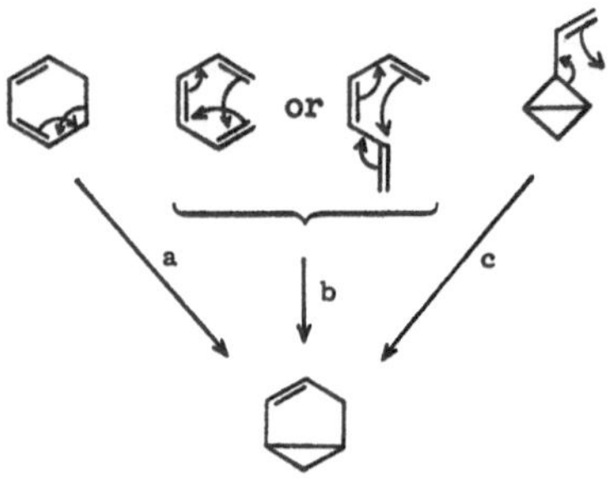

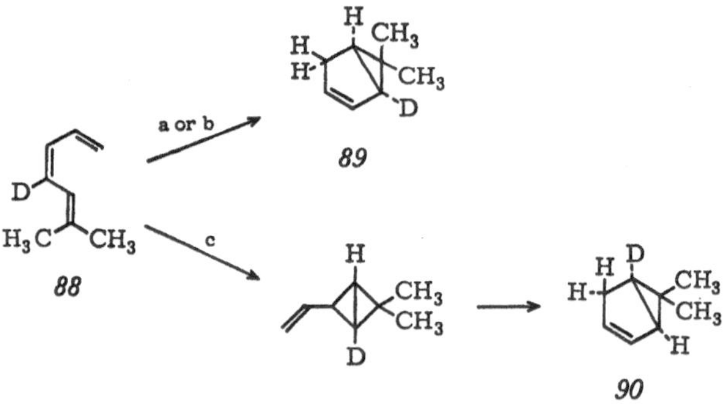

Irradiation of α-phellandrene *(91)* in ether solution gives a mixture of acyclic trienes *(92)* [36]. Further irradiation leads chiefly to two monomeric products, *93* (ca. 50%) and *94* (ca. 10%) [67].

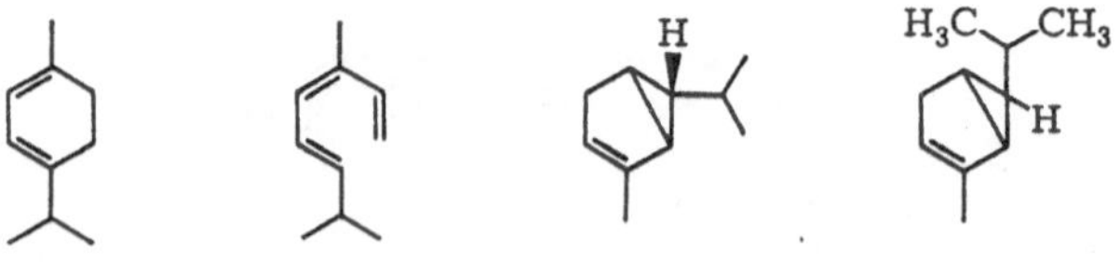

Irradiation of 1,6-dimethyl-1,3,5-hexatriene *(95)* in ether solution gives a polymer, recovered *95*, and 5,6-dimethylcyclohexa-1,3-diene *(96)* [68]. The recovered monomeric products consisted of 90—95% of *95* and 5—10% of *96*. Irradiation of *96* gave a similar mixture.

95   96

Irradiation of 1,3,4,6-tetraphenyl-1,3,5-hexatriene *(97)* through a Pyrex filter yielded a single photoproduct in high yield [69]. The structure proposed by these authors for the photoproduct was *98*. Subsequent investigation indicated that the structure of the photoproduct was in fact *99* [70]. There was insufficient evidence to permit determination of the immediate precursor of *99*.

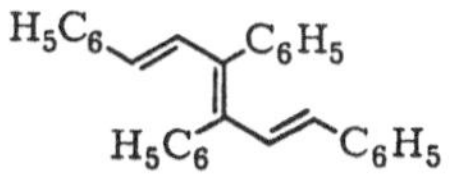

97 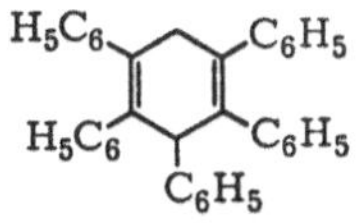   98 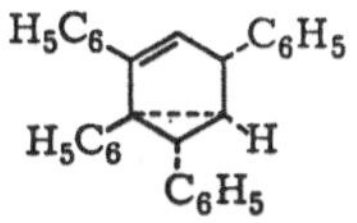   99

When a mixture of the two 4-trans-alloocimenes (10% of *100a*, 90% of *100b*) was irradiated, six main photoproducts were obtained *(100c, 100d, 101—104)*. Two of these *(100c* and *100d)* disappeared, together with *100a* and *100b*, on prolonged irradiation [71]. Triene *102* arises

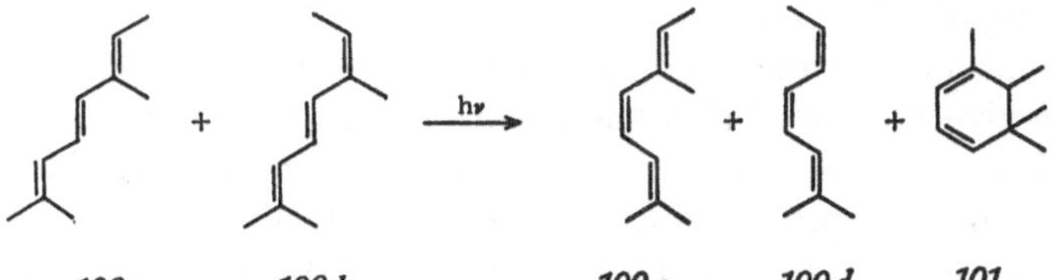

100a   100b   100c   100d   101

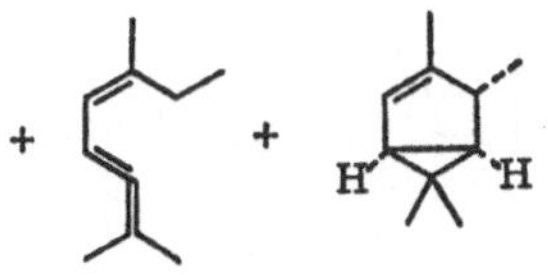

102   103

from a 1,5-hydrogen migration. Available evidence indicated that product *103* probably arises from the lowest excited singlet state of *100d* rather than from *101*. The structure of product *104* is probably similar to that of *103*, but the compound could not be isolated in sufficient quantity to allow for an unambiguous structural determination.

Transformations of other acyclic conjugated trienes (particularly in the vitamin series) have been summarized previously [1].

8. New Developments in Tropone Photochemistry

As indicated in the Introduction, the photochemistry of tropones proceeds *via* three different modes, called A, B, and C, with type C occurring in complex polycyclic troponoids. *Mukai* and *Miyashi* [72] have observed the first example of a photoreaction of a simple troponoid system to undergo the type C isomerization. Irradiation of 5-phenyltropolone methyl ether *(104)* in methanol yielded 3-methoxy-6-phenyl-$\Delta^{3,6}$-bicyclo-[3.2.0]heptadien-2-one *(105)*. The effect of the substituent is important,

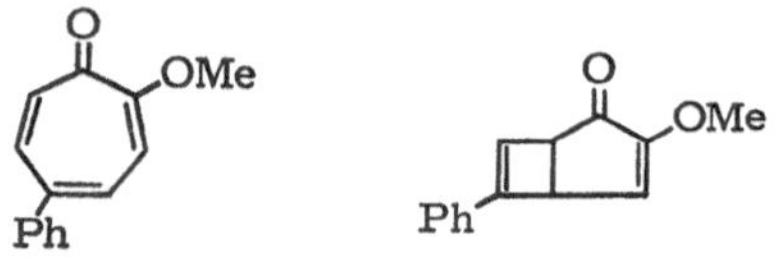

as *Dauben* [3b] has demonstrated that irradiation of 5-isopropyltropolone methyl ether resulted in type B cyclization, indicating that the phenyl and isopropyl groups appear to have different effects on the photochemistry of troponoid systems.

Irradiation of a methanol solution of 5-phenyltropolone *(106)* was found to give product *107* in addition to resinous products. Since *107* is assumed to arise from the 1,3-diketone *109* by way of formation of the type B cyclization product *108* followed by rearrangement, *106* properly belongs to the class of type B cyclizations. IR data of the resinous photo-

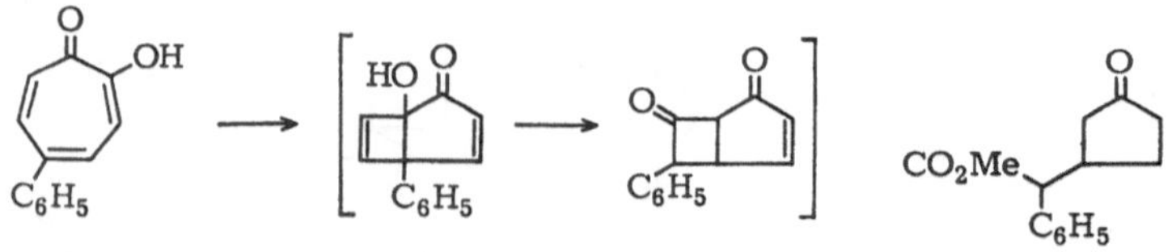

products indicated the presence of the α-diketone *110*, the type C cyclization product, but *110* could not be isolated. However, irradiation of *106*
in methanol containing a trace of acid yielded the expected product *111*
in low yield. Although irradiation of *106* gave a complex mixture of

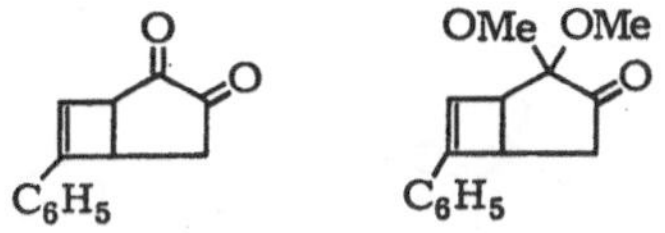

110   111

products, it can be considered that irradiation of *106* results in both
types B and C cyclizations.

For a mechanistic interpretation of the photoinduced valence isomerization of tropolone methyl ethers, *Mukai* and *Miyashi* have assumed that
two excited state species *112* and *114* need to be considered. If *112* is
stabilized by substituents in the tropolone ring, cyclization will give the
product *113*, but if *114* is more stable, product *115* will be favored. In the

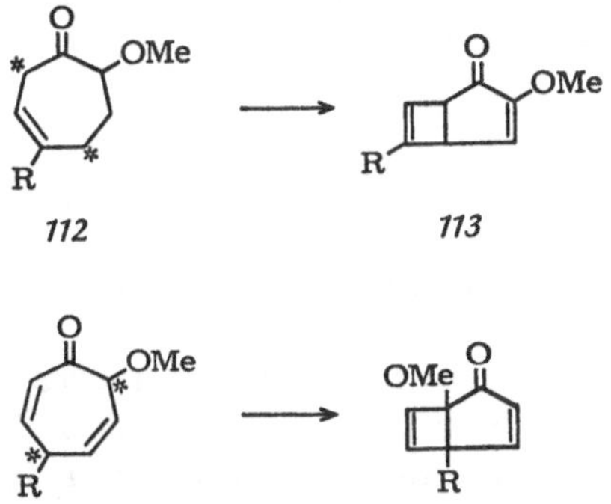

112   113

114   115

case of R=C$_6$H$_5$, species *112* will be more stabilized by the styrene
chromophore, giveing rise to type C cyclization product *113*.

The photochemistry of 5-chlorotropolone *117* and its methyl ether
116 were studied to investigate the effect of a chlorine substituent in the
photoinduced valence isomerization [73]. Irradiation of *116* gave as
products *118, 119,* and *120*.

Irradiation of *117* gave *121, 122, 123,* and *124*. Studies showed that
117 afforded the type C cyclization product *121* as the primary reaction

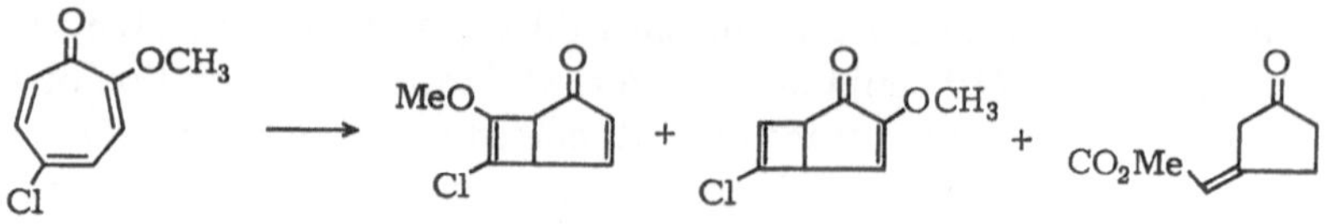

which proceeded by photochemical change to the dimer *124*, and non-photochemically to *119* and *123* by reaction with solvent. Compound *122* arises from a type B cyclization followed by rearrangement and ring opening. Thus, the introduction of a chlorine substituent at C_5 is tropolone and in the methyl ether leads to a change in cyclization from type B to a mixture of B and C.

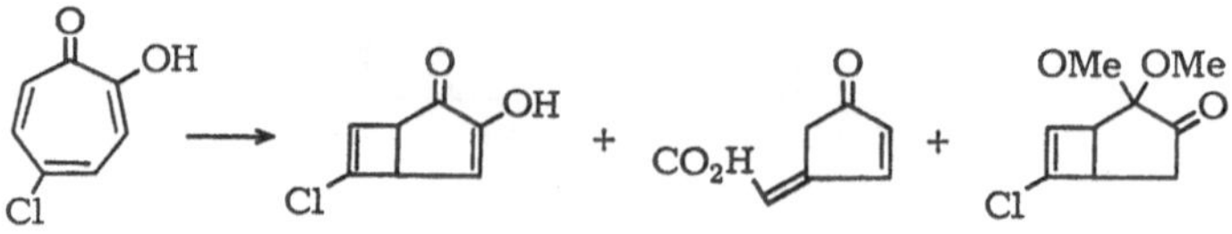

A recent review [74] summarizes many photochemical conversions of monocyclic tropolones, colchicine, isocolchicine, and benzotropolones to bicyclo[3.2.0]hepta-3,6-dien-2-one derivatives.

9. Summary

It is evident that substituents play an important role in the photochemically induced cyclization and migration reactions of conjugated trienes. Cycloheptatrienes isomerize to other trienes by migration and the direction of migration is frequently controlled by substituents present in the ring. The direction of cyclization to the bicyclo[3.2.0]-heptadiene system is also dependent upon substituents. The relative competition of these two types of reactions has been shown to depend on the type of substituent present in the ring system. Most evidence indicates that these reactions are most likely excited singlet state reactions.

Cyclooctatrienes undergo both ring opening and cyclization reactions. The cyclization reactions produce both bicyclic and tricyclic derivative. Similarly, benzene derivatives cyclize to bicyclic derivatives (Dewar benzenes) or tricyclic analogues (benzvalenes). Tropolones undergo cyclization in one of three directions depending upon the substituents present.

Ɔ. M., W. J. Middleton, and C. G. Krespan: J. Am. Chem. Soc. 87, 657 (1965).
Ɔrg. Chem. 33, 2536 (1968).
amelen, E. E., T. M. Cole, R. Greeley, and H. Schumacher: J. Am. Chem.
0, 1372 (1968).
ᴉ, D. H. R.: Helv. Chim. Acta 42, 2604 (1959).
ᴋ, R. J., N. G. Minnaard, and E. Havinga: Rec. Trav. Chim. 79, 922 (1960).
W. R., u. B. Peltzer: Angew. Chem. 76, 378 (1964).
, J., and S. Winstein: Proc. Chem. Soc. 235 (1964).
ᴉan, O. L., G. W. Borden, R. W. King, and B. Winkler: J. Am. Chem. Soc.
ᴪ0 (1964).
ᴉsan, R.: J. Am. Chem. Soc. 85, 819, 3048 (1963).
ᴩb, T. D., and L. Lindqvist: J. Am. Chem. Soc. 89, 4588 (1967).
G., and E. M. Burgess: J. Am. Chem. Soc. 84, 3104 (1962).
ᴉa, J. M., P. D. Gardner, C. B. Strow, M. L. Hill, and T. V. Van Auken:
Chem. Soc. 90, 5042 (1968).
Am. Chem. Soc. 89, 6390 (1967).
ᴉmelen, E. E., and S. P. Pappas: J. Am. Chem. Soc. 84, 3789 (1962).
Am. Chem. Soc. 85, 3297 (1963).
ᴉh, K. E., and L. Kaplan: J. Am. Chem. Soc. 86, 2307 (1964).
ᴉhler, A. W., et al.: J. Am. Chem. Soc. 86, 2940 (1964).
ᴉhler, W., et al.: J. Am. Chem. Soc. 86, 5281 (1964).
E. M., and J. M. Bollinger: Tetrahedron Letters 3803 (1964).
ᴉ, L., K. E. Wilzbach, W. G. Brown, and S. S. Yang: J. Am. Chem. Soc.
ᴉ(1965).
ᴉh, K. E., and L. Kaplan: J. Am. Chem. Soc. 87, 4004 (1965).
ᴉ. R.: J. Am. Chem. Soc. 89, 2367 (1967).
ᴉ. I. Barta: J. Am. Chem. Soc. 90, 5347 (1968).
ᴉ, K. E., A. L. Harkness, and L. Kaplan: J. Am. Chem. Soc. 90, 1116 (1968).
ᴉ. R., and C. I. Barta: J. Am. Chem. Soc. 90, 5347 (1968).
ᴉ.: J. Am. Chem. Soc. 88, 2070 (1966).
G., F. Gozzo, C. Cevidialli, and E. Kendrick: Chem. Ind. (London) 855

ᴉith, D.: Pure Appl. Chem. 16, 47 (1968).
. L., and J. S. Wishnok: J. Am. Chem. Soc. 90, 1085 (1968).
L., S. P. Walch, and K. E. Wilzbach: J. Am. Chem. Soc. 90, 5646 (1968).
W. G., and W. T. Wipke: Pure Appl. Chem. 9, 539 (1964), and references
ᴉrein.
ᴉn, R.: J. Am. Chem. Soc. 83, 2806 (1961).
. Chem. Soc. 84, 3982 (1962).
ᴉ, J., and P. H. Mazzocchi: J. Am. Chem. Soc. 88, 2850 (1966).
ᴉell, and K. L. Erickson: J. Am. Chem. Soc. 87, 3532 (1965).
ᴉ. J.: Tetrahedron Letters 549 (1962).
J., and R. E. Dessy: J. Org. Chem. 31, 4248 (1966).
W. G., and J. H. Smith: J. Org. Chem. 32, 3244 (1967).
ᴋ. J.: J. Org. Chem. 33, 3679 (1968).
, and T. Miyashi: Tetrahedron 23, 1613 (1967).
Shishido: J. Org. Chem. 32, 2744 (1967).
ᴉF.: Advan. Alicyclic Chem. 1, 257 (1966).

ᴉary 24, 1969

While the reactions of conjugated triene
pletely understood at present, it seems clear t
undoubtedly responsible for the high degree
reactions. Orbital symmetry consideration
within the excited state molecule and steri
rationalization of many of the observed proce
standing of these reactions must await furt'

10. References

1) *Chapman, O. L.*: In: Advances in Photoch
   W. A. Noyes, G. S. Hammond, and J. N. Pi
   and references cited therein.
2) a) —, and *D. J. Pasto*: J. Am. Chem. Soc. 80,
   b) — — J. Am. Chem. Soc. 82, 3642 (1960).
3) a) *Dauben, W. G., K. Koch, O. L. Chapman*, a
   83, 1768 (1961);
   b) — —, *S. L. Smith*, and *O. L. Chapman*: J
4) *Forbes, E. J.*: J. Chem. Soc. 3864 (1955).
5) *Chapman, O. L., H. G. Smith*, and *R. W. Kin*
6) *Dauben, W. G.*, and *D. A. Cox*: J. Am. Chen
7) *Chapman, O. L., H. G. Smith*, and *P. A. F*
   (1963).
8) *Dauben, W. G., K. Koch*, and *W. E. Thiessen*
9) *Forbes, E. J.*, and *R. A. Ripley*: Chem. Ind
10) — — J. Chem. Soc. 2770 (1959).
11) *Mukai, T., T. Tezuka*, and *Y. Akasaki*: J.
12) *Chapman, O. L.*, and *H. G. Smith*: J. Am. C
13) — —, *R. W. King, D. J. Pasto*, and *M. R.*
    (1963).
14) *Woodward, R. B.*, and *R. Hoffmann*: J. An
15) — — J. Am. Chem. Soc. 87, 2511 (1965).
16) *Dauben, W. G.*, and *R. L. Cargill*: Tetrah
17) *Srinivasan, R.*: J. Am. Chem. Soc. 84, 343
18) *ter Borg, A. P.*, and *H. Kloosteriziel*: Rec.
19) *Roth, W. R.*: Angew. Chem. 75, 921 (196:
20) *Doering, W. von E.*, and *P. P. Gaspar*: J.
21) *Jones, L. B.*, and *V. K. Jones*: unpublish
22) *Egger, K. W.*, and *W. R. Moser*: J. Phys
23) *Borden, G. W., O. L. Chapman, R. Swind*
    89, 2979 (1967).
24) *Weth, E.*, and *A. S. Dreiding*: Proc. Che
25) *ter Borg, A. P., E. Razenberg*, and *H. Klo*
26) *Jones, L. B.*, and *V. K. Jones*: J. Am. C
27) — — J. Am. Chem. Soc. 90, 1540 (1968).
28) *Allinger, N. L.*, and *M. A. Miller*: J. An
29) *Jones, L. B.*, and *V. K. Jones*: to be pu
30) *Chapman, O. L.*, and *S. L. Smith*: J. Or
31) *Jones, L. B.*, and *V. K. Jones*: unpublis

32) *Gale,*
33) — J.
34) *van 1*
    Soc.
35) *Barto*
36) *de Ko*
37) *Roth,*
38) *Zirne*
39) *Chap*
    86, 26
40) *Sriniu*
41) *Goldfe*
42) *Büchi*
43) *Holov*
    J. Am
44) — — J
45) *van To*
46) — — J
47) *Wilzba*
48) *Burgst*
49) *Burgst*
50) *Arnett,*
51) *Kaplar*
    87, 675
52) *Wilzba*
53) *Ward,*
54) —, and
55) *Wilzba*
56) *Ward,*
57) *Haller,*
58) *Camegg*
    (1966).
59) *Bryce-S*
60) *Ward, F*
61) *Kaplan,*
62) *Dauben,*
    cited the
63) *Srinivas*
64) — J. Am
65) *Meinwal*
67) —, *A. Ea*
68) *Fonken,*
69) *Theis, R*
70) *Dauben,*
71) *Crowley,*
72) *Mukai, 7*
73) —, and *T*
74) *Koch, K.*

Received Jar

Strahlenchemie von Alkoholen

Dr. C. v. Sonntag

Institut für Strahlenchemie, Kernforschungszentrum Karlsruhe

Inhalt

Einleitung

Es wird ein Überblick über die Radiolyse und UV-Photolyse von aliphatischen Alkoholen gegeben[a].

Einer geringen Zahl von Arbeiten über die UV-Photolyse von Alkoholen stehen viele Arbeiten mit ionisierender Strahlung (Co—60—γ—, Röntgenstrahlung, beschleunigte Elektronen u. a.) gegenüber. Daher soll hier zuerst die Radiolyse, dann die Photolyse behandelt werden.

---

[a] Eine Zusammenfassung über die Strahlenchemie der Alkohole soll als Bericht des Kernforschungszentrums Karlsruhe (KFK-Bericht) veröffentlicht werden.

Radiolyse von Alkoholen

1. Primärprozesse (Ionisation, elektronische Anregung, Ionen-Molekül-reaktionen)

Für die meisten strahlenchemischen Untersuchungen wird die bequem zu handhabende Co—60—γ-Strahlung verwendet. Die Energien der Co—60—γ-Quanten betragen 1,17 und 1,32 MeV. Bei ihrer Einwirkung auf das Medium entstehen über den Compton-Effekt in der Materie hochenergetische Elektronen, die Sekundärelektronen unterschiedlicher Energie und Reichweite erzeugen. Energiereiche Sekundärelektronen können weitere Ionisationen verursachen. Bei Alkoholen wird mit einem Energieaufwand von 25 eV ein Ionenpaar gebildet (gemessen in der Gasphase [5]), d.h. ein einziges γ-Quant vermag etwa 50 000 Ionisationen hervorzurufen. Da das Ionisationspotential der Alkohole bei 10 eV liegt, so folgt daraus, daß nur ein Teil der abgegebenen Energie in den Ionisationsakten verbraucht wird. Der Rest wird in elektronische und thermische Anregungsenergie umgewandelt.

Die für die Strahlenchemie wichtigen Primärakte sind demnach *Ionisation* (Reaktion (1)) und *elektronische Anregung* (Reaktion (2)).

$$ROH \xrightarrow{\hspace{1cm}} ROH^+ + e^- \tag{1}$$

$$ROH \xrightarrow{\hspace{1cm}} ROH^* \tag{2}$$

ROH* soll hierbei keinen spezifisch angeregten Zustand symbolisieren. Nach der optischen Näherung [98] werden hochangeregte Zustände mit größerer Wahrscheinlichkeit gebildet als niedrig angeregte Zustände. Auch überangeregte Zustände (*superexcited states*) sind zu erwarten [97,98]. So angeregte Moleküle können in molekulare oder radikalische Bruchstücke zerfallen, mit einem benachbarten Molekül eine Reaktion eingehen oder einfach desaktiviert werden. Auch die in Reaktion (1) gebildeten Radikalionen können elektronisch und/oder thermisch angeregt sein und analog den nicht ionisierten angeregten Molekülen fragmentieren. Ein Teil der bei der Radiolyse entstehenden Radikale hat einen Energieinhalt, der über dem der thermischen Energie liegt; die Radikale sind „heiß" [132]. Nur unter günstigen experimentellen Bedingungen gelingt es, Reaktionen heißer Radikale von molekularen Fragmentierungsreaktionen zu trennen.

Massenspektrometrische Untersuchungen am Methanol haben gezeigt, daß das nach Reaktion (1) entstandene Radikalkation die Ionenmolekülreaktionen (3) und (4) eingeht.

$$CH_3OH^+ + CH_3OH \rightarrow CH_3O^{\cdot} + CH_3OH_2^+ \tag{3}$$

$$CH_3OH^+ + CH_3OH \rightarrow CH_3OH_2^+ + CH_2OH \tag{4}$$

Beide Reaktionen treten unter massenspektrometrischen Bedingungen zu etwa gleichen Teilen auf [146,107]. Die Reaktionsgeschwindigkeitskonstante beträgt für die Gesamtreaktion $(k_3 + k_4) = 14{,}6 \cdot 10^{-10}$ cm³/ Molekül·s [146]. In Reaktion (3) entsteht neben dem protonierten Alkohol ein Oxy-Radikal, in Reaktion (4) ein Hydroxyalkyl-Radikal. *Basson* [18] konnte bei der Radiolyse des Äthanols in flüssiger Phase das Äthoxy-Radikal abfangen. Die Hydroxyalkyl-Radikale sind durch die Dimerisierungsprodukte, die Glykole, sowie durch ESR-Messungen nachgewiesen worden. α-Hydroxyalkyl-Radikale entstehen nicht nur über Reaktion (4) sondern auch noch über H·-Übertragungsreaktionen (s. Kap. 6).

Das in Reaktion (1) gebildete Elektron kann in polaren Lösungsmitteln solvatisiert werden. Das Verhalten der solvatisierten Elektronen in alkoholischen Medien soll im nächsten Kapitel behandelt werden.

2. Das solvatisierte Elektron

Nach den Vorstellungen von *Platzman* [96] und *Stein* [136] wird das bis auf thermische Energie abgebremste Elektron in polaren Medien solvatisiert. *Hayon* und *Weiss* [65] gelang es, das solvatisierte Elektron in seiner Reaktion mit der Chloressigsäure (Dissoziativer Elektroneneinfang, Cl⁻-Bindung) von der Reaktion des Wasserstoffradikals (H·-Übertragung) zu trennen. *Dainton* et al. [34,26] haben durch Messungen des kinetischen Salzeffekts die einfach negative Ladung dieser Spezies bewiesen. *Schulte-Frohlinde* und *Eiben* [116] konnten zeigen, daß in alkalischen Gläsern und Alkoholaten bei der Temperatur des flüssigen Stickstoffs das Elektron in der Matrix stabilisiert werden kann. Das Elektron bedingt in diesen Gläsern eine starke Absorption im Sichtbaren; die Gläser sind blau gefärbt. Im ESR-Spektrum zeigt das Elektron das erwartete Singlett. Im selben Jahr (1964) wurde das solvatisierte Elektron auch in flüssigem Wasser [64,72] und später in den Alkoholen [2,141] mit Hilfe der Pulsradiolyse nachgewiesen. Aufgrund von Radikalfängerversuchen war sein Auftreten bei der Radiolyse von Alkoholen schon früher postuliert worden [21]. Solvatisierte Elektronen in Alkoholen sind außerdem auf anderen Wegen herstellbar (z.B. durch Blitzphotolyse von Jodiden [53,43] oder durch Aufdampfen von Alkalimetallen im rotierenden Kryostaten bei tiefer Temperatur [23,24]).

Tabelle 1 zeigt die Absorptionsmaxima, die Halbwertsbreiten und die Temperaturkoeffizienten der $e_{solv.}^-$-Bande sowie die pulsradiolytisch bestimmte Ausbeute $(G(e_{solv.}^-))$ [b] ein- und mehrwertiger Alkohole. Offenbar wird mit sinkender statischer Dielektrizitätskonstante (DK)

---

[b] Der *G*-Wert ist das Maß der strahlenchemischen Ausbeute. Er ist definiert als Zahl der gebildeten Moleküle (Radikale etc.) pro 100 eV absorbierte Energie.

des Alkohols die Absorptionsbande des $e^-_{solv.}$ zu längeren Wellen verschoben. *Sauer, Arai* und *Dorfman* [113] fanden eine lineare Beziehung zwischen der DK des Alkohols und der Energie des Übergangs. Eine Ausnahme macht das n-Butanol, bei dem nach dieser Beziehung das Maximum der $e^-_{solv.}$-Bande bei 800 nm liegen sollte jedoch bei 680 nm liegt [12]. Die Maxima der $e^-_{solv.}$-Banden sind stark temperaturabhängig. Mit sinkender Temperatur werden sie zu höheren Energien verschoben (s. Tabelle 1).

Tabelle 1. *Absorptionsmaxima, Halbwertsbreiten, Temperaturkoeffizienten der $e^-_{solv.}$-Bande sowie G($e^-_{solv.}$) von bestrahlten ein- und mehrwertigen Alkoholen (nach Arai und Sauer [12] und Sauer, Arai und Dorfman [113])*

Alkohol	DK	$\lambda_{max.}$ (nm)	$E_{max.}$ (eV)	$W_{1/2}$ (eV)	$-dE_{max.}/dT$ (eV/°)	G($e^-_{solv.}$)
Glycerin	42,5	525	2,35	1,5		
Glykol	39	580	2,14	1,35		1,2
Methanol	34	630	1,96	1,29	$2,2 \cdot 10^{-3}$	1,1
Äthanol	25	700	1,77	1,55	$3,4 \cdot 10^{-3}$	1,0
n-Propanol	21	740	1,67			1,0
Isopropanol	19	820	1,51	1,22	$3,5 \cdot 10^{-3}$	1,0
n-Butanol	18	680	1,83		$4,5 \cdot 10^{-3}$	

In unpolaren Lösungsmitteln (z. B. Cyclohexan) genügt ein Zusatz von nur 4% Methanol, um das charakteristische Spektrum des solvatisierten Elektrons zu erzeugen [73]. Das Maximum ist etwas zu längeren Wellen verschoben; die Ausbeute ($G \cdot \varepsilon$) ist sehr viel höher, als dem Anteil des Methanols an der Mischung entspricht. Eine Methanol-Clusterbildung im unpolaren Cyclohexan wird diskutiert. In Alkohol-Wasser- und Alkohol-Alkohol-Mischungen zeigt sich nur eine Absorptionsbande. Bei Mischungen liegt das Maximum der Bande zwischen denen der Komponenten [12]. In Äthanol-Wasser- und Isopropanol-Wasser-Mischungen verläuft die Halbwertsbreite der Absorptionsbande des solvatisierten Elektrons nicht linear mit dem Molenbruch der Komponenten. Bereits ein geringer Wasserzusatz senkt die Halbwertsbreite und bei einem Anteil von 35 Mol% H_2O wird schon die gleiche Halbwertsbreite wie in reinem Wasser gefunden. Dieser Befund deutet darauf hin, daß in diesen Mischungen das Wasser einen wesentlichen Beitrag zur *Struktur der Solvathülle* des Elektrons leistet.

Die *Ausbeute* an solvatisierten Elektronen wurde über die Konkurrenz mit Diphenyl (es entsteht das Diphenyl-Radikalanion) und mit Triphenylmethanol (es entsteht das Triphenylmethyl-Radikal) gemessen [140].

Für das Diphenylanion und das Triphenylmethyl-Radikal sind die molaren Extinktionskoeffizienten bekannt. Der so bestimmte Wert $(G(e_{solv.}^-))$ liegt bei 1,1. Der molare Extinktionskoeffizient des solvatisierten Elektrons beträgt ca. 10^4. Die Radikalanionen der Aromaten sind in den Alkoholen wenig stabil. Sie werden rasch ($\tau/2 = 0,4 - 175$ µsec) protonisiert und gehen in entspr. Cyclohexadienyl-Radikale über [11].

Die *Reaktivität* des Elektrons gegenüber mono- und disubstituierten Benzolderivaten in Konkurrenz zu der Reaktion mit dem N_2O hat *Sherman* [119,120] in Methanol und Isopropanol untersucht. Analog zu der Reaktion in Wasser [8] findet sich ein linearer Zusammenhang zwischen den *Hammett*schen σ-para-Werten und dem Logarithmus der Reaktionsgeschwindigkeitskonstanten. Der ϱ-Wert ist positiv. Dieses Ergebnis kann als eine bimolekulare nukleophile Reaktion, S_{N2}, gedeutet werden. Der ϱ-Wert der Reaktion in Methanol unterscheidet sich mit 4,7 kaum von dem in Wasser (4,8), während der ϱ-Wert in Isopropanol nur 3,1 beträgt. Die Halogenderivate des Benzols (Chlorbenzol, Brombenzol und Jodbenzol) erfüllen die obengenannte Beziehung nicht so gut wie andere Benzolderivate (z.B. Phenol, Toluol oder Nitrobenzol). Die Reaktivität der Halogenderivate gegenüber dem solvatisierten Elektron ist größer als aufgrund dieser Beziehung zu erwarten wäre. Ein dissoziativer Elektroneneinfang unter Halogenidion-Bildung könnte zusätzliche Reaktionswege öffnen.

Nicht nur in Alkoholaten [116], sondern auch in reinen Alkoholen kann das Elektron bei *tiefer Temperatur* stabilisiert werden [32,102,25,31]. Schon bei der Temperatur des flüssigen Stickstoffs reagiert es in Äthanol langsam ($\tau/2$ ca. 7 Tage) mit seiner Umgebung unter Bildung von H-Radikalen [48]. Bei 77° K werden in der Matrix mehr Elektronen stabilisiert, als bei Zimmertemperatur mit der Methode der Pulsradiolyse solvatisierte Elektronen erhalten werden (vgl. Tabelle 1 und 2). Ebenso wie in flüssiger Phase wird mit steigender DK das Absorptionsmaximum nach kürzeren Wellen verschoben. Diese Verschiebung ist mit einer Linienverbreiterung des ESR-Signals verbunden [47] (s. Tabelle 2).

Das von *Bennett*, *Mile* und *Thomas* [23,24] durch Aufdampfen von Alkalimetallen und Alkoholdämpfen an einem rotierenden Kryostaten erhaltene, in der Alkoholmatrix stabilisierte Elektron zeigt ähnliche spektroskopische Eigenschaften wie bestrahlte Alkohol-Gläser.

Durch Licht ($\lambda > 540$ nm) läßt sich die Elektronenbande ausbleichen. Im Gegensatz zu den Elektronen, die in Medien wie Triäthylamin [36], 3-Methylpentan [123], 2-Methyl-tetrahydrofuran [102,45] und neutralem [126,152] oder alkalischem Eis [14,118] stabilisiert sind, wird in den Alkohol-Gläsern das Elektron nicht aus seiner Elektronfalle befreit und beweglich, sondern das Elektron reagiert mit seiner Umgebung unter Bildung eines H-Radikals [38].

Tabelle 2. *Absorptionsmaxima der e_t^--Banden und die Ausbeuten $(G(e_t^-))$ bzw. $(G \cdot \varepsilon)$ und Linienbreiten des ESR-Signals verschiedener Alkohol-Gläser*

Alkohol	$\lambda_{max.}$ (nm)	$G \cdot \varepsilon \cdot 10^{-4}$	$G(e_t^-)$	ΔH Gauss	Lit.
Glycerin	495				[47]
Glykol	515			15	[47]
Methanol	~516				[6]
	520				[121]
	533	3,46	$3 \pm 0,1$		[37]
	526	$3 \pm 0,3$	$2,7 \pm 0,3$		[40]
	522			14	[47]
			2,2*)		[33]
Äthanol	~516				[6]
	540		~3		[102]
	~556				[122]
	530			12	[47]
			2,3*)		[33]
n-Propanol	555	$2,15 \pm 0,2$			[40]
	560			12	[47]
Isopropanol	615	3,0			[75,76]
	645			10	[47]
			1,1*)		[33]
n-Butanol	506				[6]
	562			8	[47]
Isobutanol	705			9	[47]
n-Pentanol	570			7	[47]

*) mit ESR bestimmt

Die Elektronenfallen sind nicht einheitlich in ihrer Struktur. Die große Breite der Absorptionsbande rührt daher, daß die Elektronen in verschieden tiefen Fallen sitzen. Die in flachen Fallen festgehaltenen Elektronen können durch Bestrahlen am langwelligen Ende der Absorptionsbande ausgebleicht werden. Auch werden sie beim langsamen Erwärmen der bestrahlten Gläser als erste beweglich. In beiden Fällen verschiebt sich dabei das Maximum der Bande nach kürzeren Wellen. Besonders deutlich kann dies am n-Propanol-Glas gezeigt werden [41].

Mit (B-10(n,α)Li-7)-Teilchen bestrahlte Äthanol-Gläser zeigen gegenüber der γ-Strahlung $(G(e_t^-) \approx 3)$ eine sehr viel geringere Ausbeute an stabilisierten Elektronen $(G(e_t^-) \approx 0,14)$ [154]. Nur die Elektronen, die in δ-Bahnen erzeugt werden, scheinen stabilisiert zu werden. Eine Erweichung des Glases und eine damit verbundene Mobilisierung der

Elektronen im und in der Nähe des *Ionisationszylinders* (s.u.) wird diskutiert.

Die Reaktionen des solvatisierten Elektrons mit verschiedenen Substanzen studierten insbes. die Arbeitskreise um *Hamill* [102,59,121,151] und *Dainton* [37-41]. ESR-spektroskopische Untersuchungen wurden von *Chachaty* und *Hayon* [31,33,30] unternommen (vgl. auch Kap. 10).

Über die Reaktionen des solvatisierten Elektrons in anderen Medien, hauptsächlich Wasser und Ammoniak, sind eine große Zahl zusammenfassender Artikel erschienen [145,46,74,114,149,35,63,20,28]. Eine Zusammenstellung der Reaktionsgeschwindigkeiten des solvatisierten Elektrons in Wasser geben *Anbar* und *Neta* [10], Daten über Absorptionsspektren und sonstige Charakteristika solvatisierter Elektronen in verschiedenen Medien haben *Habersbergerová*, *Janovský* und *Teplý* [54] zusammengetragen.

3. Der Mechanismus der Wasserstoffbildung

Der bei der Radiolyse von Alkoholen entstehende Wasserstoff kann verschiedene Vorläufer haben: solvatisierte Elektronen, „thermische" H-Radikale und „heiße" H-Radikale. Zudem kann Wasserstoff in einer molekularen Fragmentierungsreaktion entstehen.

Die Elektronen reagieren nach Gl. (5) (protonierter Alkohol entsteht aus der Ionenmolekülreaktion des Alkoholradikalkations mit einem Alkoholmolekül, s. Reaktionen (3) und (4)) oder (6) zu H-Radikalen, die dann nach Gl. (7) zu molekularem Wasserstoff führen.

$$ROH_2^+ + e_{solv.}^- \quad \rightarrow ROH + H\cdot \tag{5}$$

$$ROH \; + e_{solv.}^- \quad \rightarrow RO^- + H\cdot \tag{6}$$

$$H\cdot \; + RCH_2OH \rightarrow H_2 \; + R\dot{C}HOH \tag{7}$$

Mit der Pulsradiolyse wurde für solvatisierte Elektronen ein G-Wert von 1,0—1,2 erhalten. Dieser G-Wert entspricht etwa dem Anteil der Elektronen, der dem Coulomb-Feld des Gegenions entkommt. *Freeman* [67,105,104] hat in Untersuchungen über Alkohol/N_2O-Systeme zeigen können, daß ein sehr viel größerer Anteil, nämlich $G(e_{solv.}^-) = 4,0$ bis $4,6$ gebildet wird. Aber nicht alle Elektronen führen zu H-Radikalen und damit zu H_2, sondern ein Teil von ihnen ($G \approx 1$) reagiert in der Spur mit schon gebildetem Aldehyd oder Keton bzw. mit Alkoholradikalen, ohne daß Wasserstoff entsteht. Ein Protonen-Zusatz kann mit diesen Reaktionen in Konkurrenz treten, und gemäß Reaktion (5), gefolgt von Reaktion (7), bildet sich zusätzlicher Wasserstoff.

Der nicht über Elektronen als Vorläufer entstehende Anteil beträgt nur etwa 20—25% des Gesamtwasserstoffs.

Falls die in Wasser gemessenen Reaktionsgeschwindigkeitskonstanten auf die Alkohole übertragen werden dürfen, dann reagieren thermische H-Radikale (H·) mit den Alkoholen mit Reaktionsgeschwindigkeitskonstanten von $1{,}6\cdot10^6$ 1/mol·s (mit Methanol) bis $5\cdot10^7$ 1/mol·s (mit Isopropanol) [10]. D. h. ein sehr guter Fänger ($k = 10^{10}$ 1/mol·s) kann erst in Konzentrationen ab 10^{-2} mol/l ,thermische' H-Radikale abfangen. ,Heiße' H-Radikale können nicht mehr abgefangen werden und sind in den Angaben über die „molekulare" Ausbeute in der Regel mit enthalten.

Für Methanol und Äthanol werden nicht abfangbare Ausbeuten [1,3,17] von $G(H_2)_{na} = 1{,}7$ gefunden. Nach *Freeman* [103,67] beträgt der Anteil des Wasserstoffs, der nicht über Elektronen als Vorläufer gebildet wird, für Methanol $G = 1{,}8$ und für Äthanol $G = 2{,}2$. Diese Werte liegen nahe den Werten für nicht-abfangbaren Wasserstoff, so daß entweder die Radikalfänger (Benzochinon, FeCl$_3$ [1,3]), Chloressigsäure [17]) nicht effektiv genug sind, um die thermischen H-Radikale abzufangen, oder thermische H-Radikale nur in untergeordnetem Maße als primäre Wasserstoffvorläufer auftreten.

4. Untersuchungen mit Radikalfängern; homogenkinetische Betrachtung

Die homogenkinetische Betrachtung von Radikalabfangreaktionen geht von der Annahme aus, daß die Radikalkonzentrationen über die gesamte Lösung statistisch ist. Diese Annahme ist für die Radiolyse nur sehr bedingt gültig [86]. In den Bereichen größerer Energieabgabe (*spurs, blobs, tracks* [86,93]), in denen nicht nur ein einziges Radikalpaar gebildet wird, herrscht eine gegenüber der übrigen Lösung sehr viel größere Radikaldichte mit entsprechend hohen Kombinationsraten. In diesen Bereichen tritt dann nach kurzer Zeit (ca. 10^{-8} s) [86] eine lokale Verarmung an Radikalfängern ein und Radikal-Kombinationsreaktionen bekommen eine größere Bedeutung, als bei statistischer Verteilung von Radikalpaaren zu erwarten wäre. Trotzdem hat die homogenkinetische Betrachtung in Ermangelung guter, physikalisch begründeter und mathematisch einfach zu behandelnder Modelle vielfach Anwendung gefunden und teilweise auch gute Ergebnisse erzielt.

Das einfachste Modell geht von einem Radikal aus, welches z. B. mit dem Lösungsmittel zu einem bestimmten Produkt führt. Bei Zugabe eines Radikalfängers kann dieser mit dem Radikal reagieren und das Produkt wird nicht gebildet. Hierfür lassen sich die Gleichungen (8) und (9) schreiben, wobei X das Radikal, RH das Lösungsmittel, S der Radikalfänger und P das Produkt bedeuten soll.

$$X + RH \xrightarrow{k_1} P \tag{8}$$

$$X + S \xrightarrow{k_2} \text{nicht } P \tag{9}$$

Dann ist

$$G(P) = G(P)_0 \cdot \frac{k_1\,[\text{RH}]}{k_1\,[\text{RH}] + k_2\,[\text{S}]} \qquad (10)$$

wobei $G(P)$ der bei der Fängerkonzentration $[S]$ gemessene, $G(P)_0$ der ohne Zusatz von S gemessene G-Wert von P ist.
Aus (10) und (11)

$$\Delta\, G(P) = G(P)_0 - G(P) \qquad (11)$$

folgt (12)

$$\Delta\, G(P) = G(P)_0 - G(P)_0 \frac{k_1\,[\text{RH}]}{k_1\,[\text{RH}] + k_2\,[\text{S}]} \qquad (12)$$

deren Inverses (13) liefert

$$\frac{1}{\Delta\, G(P)} = \frac{1}{G(P)_0} \left(1 + \frac{k_1\,[\text{RH}]}{k_2\,[\text{S}]}\right) \qquad (13)$$

Trägt man $1/\Delta G(P)$ gegen $[\text{RH}]/[\text{S}]$ auf, so erhält man eine Gerade. Entsteht P noch über einen weiteren Prozeß, der durch S nicht beeinflußt wird, dann verläuft die Gerade nicht durch den Nullpunkt und der Abschnitt auf der y-Achse ergibt den reziproken Wert der nicht abfangbaren Ausbeute an P.

Diese Ausbeute wird in der Literatur häufig als „molekulare" Ausbeute („*molecular yield*") bezeichnet. Sie sollte wohl besser mit „nicht abfangbare" Ausbeute bezeichnet werden, da die „molekularen" Ausbeuten nicht nur aus echt molekularen Fragmentierungsreaktionen, sondern auch aus Reaktionen „heißer" Radikale stammen [132].

Betrachtungen für den komplizierten Fall, daß ein Radikalfänger mit zwei verschiedenen Vorläufern des gleichen Produktes reagiert, geben *Teplý* und *Habersbergerová* [144,142,143].

Bei der Radiolyse von Alkoholen mit verschiedenen Fängern zeigt sich, daß bei der oben beschriebenen Auftragungsart für die Wasserstoffausbeute zwei Geraden erhalten werden: eine für den Konzentrationsbereich von 10^{-5} bis etwa $5 \cdot 10^{-3}$ mol/1 Fänger und eine zweite für höhere Fängerkonzentrationen. Von mehreren Arbeitskreisen [4,17,143] wird angenommen, daß dieser Effekt durch die Reaktion des Fängers mit zwei verschiedenen reaktiven Wasserstoffvorläufern, dem solvatisierten Elektron und dem H-Atom entsteht. Eine Überschlagsrechnung mit bekannten (in Wasser gemessenen) Reaktionsgeschwindigkeitskonstanten [10] zeigt, daß diese Deutung wohl nicht richtig ist. Auch die Fängerversuche von *Freeman* [67,105,104] mit N_2O sprechen gegen diese Deutung.

Freeman nimmt dagegen nur einen abfangbaren Wasserstoffvorläufer an, das solvatisierte Elektron, und erklärt die experimentellen Ergebnisse mit einer nicht-homogenen Kinetik (s. folgendes Kapitel). Doch ist nach *Hamill* [58] auch mit Reaktionen nicht-solvatisierter Elektronen zu rechnen. Da über ihre Eigenschaften noch zu wenig bekannt ist, soll hier nur auf diese Möglichkeit hingewiesen werden.

5. Nicht-homogenkinetische Deutung von Elektronenabfangreaktionen

Zur Deutung der Abhängigkeit der G-Werte der bei der γ-Radiolyse entstehenden freien Ionen von der Dielektrizitätskonstanten des Mediums hat *Freeman* [52] ein Modell entwickelt, welches berücksichtigt, daß bei der γ-Radiolyse Ionenpaare mit verschiedenen Abständen gebildet werden. Die Verteilungsfunktion der Entfernung der Elektronen von ihrem Mutterion wurde abgeschätzt [52]. Dem elektrischen Feld des Mutterions entkommen die Elektronen, deren Abstand $> r$ ist, wobei r der kritische Radius ist, bei dem die Coulomb'sche Anziehung gleich der thermischen Energie (kT) wird. Die Wahrscheinlichkeit $(\phi(y))$, mit der ein Ionenpaar des Abstandes y seiner Rekombination entkommt und „freie Ionen" gebildet werden, ist dann

$$\phi(y) = e^{-r/y}.$$

Der G-Wert der „freien Ionen" ergibt sich sodann zu

$$G(\text{freie Ionen}) = (\int N(y)\phi(y)dy / \int N(y)dy) \cdot G(\text{totale Ionisation}).$$

Dabei bedeutet $N(y)$ die relative Zahl von Elektronen, die einen Abstand y von ihrem Mutterion erreichen.

Eine ausführlichere Darstellung der *Freeman*'schen Berechnungen findet sich anderenorts [52,67,103,106].

Da die N_2O-Versuche von *Freeman* und ihre nicht-homogenkinetische Deutung eine wesentliche Alternative zu der homogenkinetischen Betrachtung bringen und zeigen, daß der größte Teil der Wasserstoffvorläufer Elektronen und nicht, wie aus dem homogenkinetischen Ansatz abgeleitet wurde, H-Atome sind, soll zusammen mit einigem experimentellen Material auf den einfachsten Fall, die Wasserstoffbildung beim Methanol, eingegangen werden. Vorangeschickt werden soll:

1. Protonenzusatz läßt bei Alkoholen den $G(H_2)$-Wert nach den Reaktionen (5) und (14) steigen.

$$e^-_{\text{solv.}} + ROH_2^+ \rightarrow H\cdot + ROH \qquad (5)$$

$$H\cdot \;\; + ROH \rightarrow H_2 + \text{Prod.} \qquad (14)$$

2. N_2O reagiert rasch mit solvatisierten Elektronen (k(in H_2O) = $5,6 \cdot 10^9$ 1/mol$\cdot s$[10])) (Reaktion (15)), langsam jedoch mit H-Radikalen (k(in H_2O) = 10^5 1/mol$\cdot s$[10]).

$$e_{solv.}^- + N_2O \rightarrow N_2 + O^- \tag{15}$$

Bei den Alkoholen ist eine Abnahme an H_2 mit einer Zunahme an N_2 verbunden.

Beim Methanol [67)] steigt bei Zugabe von Säure der $G(H_2)$-Wert von 5,4 auf 6,4. Die gleiche Konzentrationsabhängigkeit zeigt die Summe von H_2 und N_2 bei Zusatz von N_2O. Auch hier wird ein $G(H_2 + N_2)$ von 6,4 erreicht, der Anteil des N_2 beträgt 4,6. $G(H_2)$ = 1,8 kann durch den Elektronenfänger N_2O auch in molarer Konzentration nicht unterdrückt werden. In der Auftragung $G(N_2)$ gegen $\log[N_2O]$ zeigt die Kurve bei $[N_2O] \approx 10^{-2}$ mol/1 eine Art Plateau.

Jha und *Freeman* [67)] geben den Befunden folgende Deutung:

1. $G(N_2)_{max}$ = 4,6 stellt den G-Wert der „totalen Ionisation"[c)] dar. Damit liegt der G-Wert der „totalen Ionisation", der in früheren Arbeiten mit $G \approx 3$ diskutiert wurde [103,52)], näher an dem Wert, der aus dem W-Wert der Alkohole [5)] zu erwarten wäre (G(totale Ionisation) = 4).

2. Sämtliche Elektronen werden solvatisiert, aber nur ein gewisser Teil (G = 2,0) entkommt der Rekombination in der Spur. Diese freien solvatisierten Elektronen reagieren, weil sie der Spur entkommen sind, homogenkinetisch mit zugesetzten Fängern.

3. Der $G(H_2)$-Wert bei 10^{-3}–10^{-2} mol/1 Säure und $G(H_2 + N_2)$ bei 10^{-3}–10^{-2} mol/1 N_2O beträgt 5,5. Dieser Wert entspricht dem $G(H_2)$-Wert von „reinem" Methanol, da den Reaktionen von geringen Verunreinigungen (z.B. Aceton) mit den solvatisierten Elektronen durch die Reaktion des H^+ bzw. N_2O der Rang abgelaufen wird. In die Spur-Reaktionen greifen das Proton und N_2O bei diesen Konzentrationen jedoch nicht ein.

4. Bei höheren Protonenkonzentrationen steigt $G(H_2)$ an. Das Ansteigen kann u.a. auf die innerhalb der Spur ablaufenden Konkurrenzreaktionen (17) und (16) zurückgeführt werden.

$$e_{solv.}^- + CH_3O \cdot \rightarrow CH_3O^- \tag{16}$$

---

c) Die „totale Ionisation" ist zwar in der Gasphase, nicht jedoch in der flüssigen Phase eindeutig definiert. In der flüssigen Phase wird von *Freeman* der G-Wert der durch Elektronenfänger maximal abfangbaren Elektronen mit G(totale Ionisation) gleichgesetzt.

C. v. Sonntag

Das Methoxyradikal entsteht nach Gl. (3).

$$CH_3OH^+ + CH_3OH \rightarrow CH_3O\cdot + CH_3OH_2^+ \tag{3}$$

Den G-Wert für Reaktion (16) oder andere elektronenabfangende, nicht zu H_2 führende Spurreaktionen geben die Autoren mit 0,9 an.

5. Die restlichen solvatisierten Elektronen, die ebenfalls der Spur nicht entkommen, reagieren mit dem nach Gl. (3) gebildeten protonierten Alkohol zu H-Radikalen (Gl. (17)). Die Reaktion (18) ist nicht auszuschließen.

$$CH_3OH_2^+ + e_{solv.}^- \rightarrow CH_3OH + H\cdot \tag{17}$$

$$CH_3OH + e_{solv.}^- \rightarrow CH_3O^- + H\cdot \tag{18}$$

$G(H_2)$ aus dieser Quelle beträgt 1,7.

6. Ein $G(H_2)$-Wert von 1,8 ist durch Elektronenfänger nicht zu unterdrücken. Dieser nicht abfangbare $G(H_2)$-Anteil könnte über die Reaktionen (19) bis (21) entstehen.

$$CH_3OH \longrightarrow H\cdot + (CH_3O\cdot \text{ oder } \cdot CH_2OH) \tag{19}$$

$$CH_3OH \longrightarrow H_2 + CH_2O \tag{20}$$

$$CH_3OH \longrightarrow CH_2OH^+ + H\cdot + e^- \tag{21}$$

6. H·-Übertragungsreaktionen und die Bildung der Glykole und Carbonylverbindungen

Die während der γ-Radiolyse gebildeten H-Radikale (H·) entziehen Alkoholen bevorzugt den zur Hydroxylgruppe α-ständigen Wasserstoff und bilden mit ihm H_2. Der Angriff an der Hydroxyl-Gruppe ist — zumindest bei O-deuterierten Alkoholen — vernachlässigbar klein. Die D_2-Ausbeute saurer O-deuterierter n-Alkohole liegt bei 1 bis 2% D_2 [117]. Doch ist bei den deuterierten Alkoholen für den H·-Entzug durch thermische Wasserstoffatome ein hoher Isotopieeffekt einzurechnen. *Anbar* und *Meyerstein* [9] fanden beim Methanol für den H-Entzug an der Methylgruppe $k_H/k_D = 20$, für den reaktiveren tertiären Wasserstoff des Isopropanols noch $k_H/k_D = 7,5 \pm 1$. Dieser Wert ist in guter Übereinstimmung mit dem von *Vacek* und *v. Sonntag* [148] bestimmten Isotopieeffekt von $k_H/k_D = 7,1 \pm 0,2$. Für den Isotopieeffekt der Wasserstoff-Übertragung auf thermische Methylradikale wurde beim Isopropanol ein Wert von 8 gefunden [132]. Es konnte geschätzt werden, daß im undeuterierten Isopropanol zu etwa 92,5% das tertiäre H, zu 5%

der Wasserstoff der Hydroxylgruppe und nur zu 2,5% ein Wasserstoff der Methylgruppen auf das Methylradikal übertragen wird [132]. Ähnliche Verhältnisse sind für die n-Alkohole zu erwarten. Hier muß aber wegen der geringeren Reaktivität der α-ständigen Wasserstoffe eine geringere Selektivität angenommen werden. Die Reaktionsgeschwindigkeitskonstante von H-Radikalen mit den Alkoholen wurde in Wasser gemessen. Sie nehmen vom Isopropanol ($5 \cdot 10^7$ l/mol·s) über das Äthanol ($1,1 \cdot 10^7$ l/mol·s) und Methanol ($1,6 \cdot 10^6$ l/mol·s) zum tert.-Butanol (10^5 l/mol·s) hin ab [10]. Tert.-Butanol hat keinen zur Hydroxylgruppe α-ständigen Wasserstoff, sein Verhalten entspricht etwa dem der Methylgruppe beim Isopropanol. Über H·-Übertragungsreaktionen werden demnach bei primären und sekundären Alkoholen durch sämtliche reaktiven Radikale (z.B. ·OH, H·, RO· und ·CH$_3$) bevorzugt α-Hydroxyalkyl-Radikale gebildet (vgl. Reaktionen (7) und (3)). Diese Radikale können nach Gl. (22) und (23) entweder zu Alkohol und Carbonylverbindungen disproportionieren (22), oder zu Glykolen dimerisieren (23).

$$2\ R_1R_2\dot{C}OH \xrightarrow{\ k_1\ } R_1R_2CHOH + R_1R_2CO \tag{22}$$

$$2\ R_1R_2\dot{C}OH \xrightarrow{\ k_2\ } (R_1R_2COH)_2 \tag{23}$$

Exakte Disproportionierungs-Dimerisierungsquotienten (k_1/k_2) sind nicht bekannt, jedoch wird angenommen, daß k_1/k_2 in der Reihenfolge Methanol $<$ Äthanol $<$ Isopropanol zunimmt. Diese Annahme stimmt mit den Befunden überein, daß bei der γ-Radiolyse der Alkohole in der gleichen Reihenfolge das Verhältnis G(Carbonylverbindungen) zu G(Glykole) von 0,57 über 1,9 auf 6,6 zunimmt. Diese Werte dürfen aber nicht als relative Disproportionierungs-Dimerisierungsquotienten angesehen werden. Aufgrund der LET-Effekte (s. folgendes Kapitel) ist vielmehr anzunehmen, daß das durch den Primärakt (1) gefolgt von (24) gebildete Alkoxy-Radikal einen wesentlichen Beitrag zur Carbonylbildung leistet.

$$ROH \xrightarrow{\ \ \sim\!\!\sim\!\!\sim\ \ } ROH^+ + e^- \tag{1}$$

$$ROH^+ + ROH \longrightarrow RO\cdot + ROH_2^+ \tag{24}$$

Dieser Beitrag kann von Alkohol zu Alkohol schon allein wegen der unterschiedlichen Reaktivität der Alkohole und Alkoxy-Radikale verschieden groß sein. Da die Oxy-Radikale bevorzugt disproportionieren[42], entstehen nach (25) und auch nach (26) nur Carbonylverbindungen.

$$2\ R_1R_2CHO\cdot \longrightarrow R_1R_2CHOH + R_1R_2CO \tag{25}$$

$$R_1R_2CHO\cdot + R_1R_2\dot{C}OH \rightarrow R_1R_2CHOH + R_1R_2CO \tag{26}$$

Die entsprechende Dimerisierungsreaktion (27)

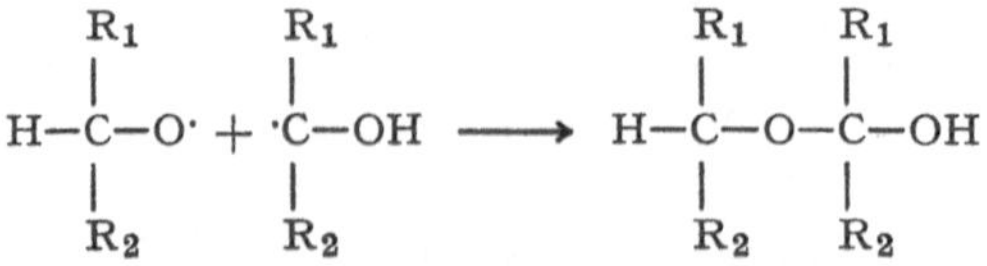

$$
\begin{array}{ccc}
\overset{\displaystyle R_1}{\underset{\displaystyle R_2}{|}} & \overset{\displaystyle R_1}{\underset{\displaystyle R_2}{|}} & \\
H-C-O\cdot + \cdot C-OH & \longrightarrow & \overset{\displaystyle R_1}{\underset{\displaystyle R_2}{|}} \; \overset{\displaystyle R_1}{\underset{\displaystyle R_2}{|}} \\
& & H-C-O-C-OH
\end{array} \tag{27}
$$

liefert das Halbacetal der Carbonylverbindung, welches leicht hydrolytisch in die Carbonylverbindung und den Alkohol übergeht. Die Disproportionierung der α-Hydroxyalkyl-Radikale verläuft nicht nach Reaktion (28) durch Übertragung des Hydroxylwasserstoffs, sondern

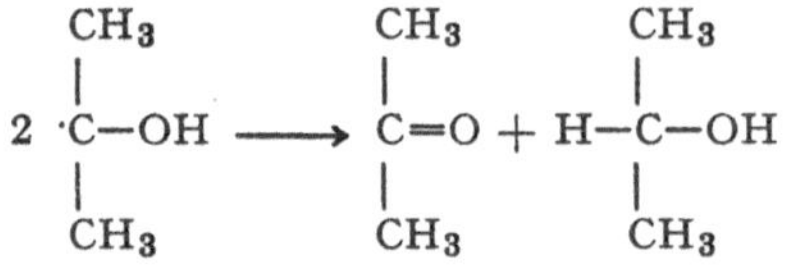

$$
2 \; \overset{\displaystyle CH_3}{\underset{\displaystyle CH_3}{|}} \cdot C-OH \longrightarrow \overset{\displaystyle CH_3}{\underset{\displaystyle CH_3}{|}} C=O + \overset{\displaystyle CH_3}{\underset{\displaystyle CH_3}{|}} H-C-OH \tag{28}
$$

hauptsächlich durch Übertragung des Wasserstoffs der benachbarten Alkylgruppe und unter Ausbildung der Enolform der entsprechenden Carbonylverbindung (Reaktion (29)). Der Mechanismus wurde am Isopropanol untersucht [84].

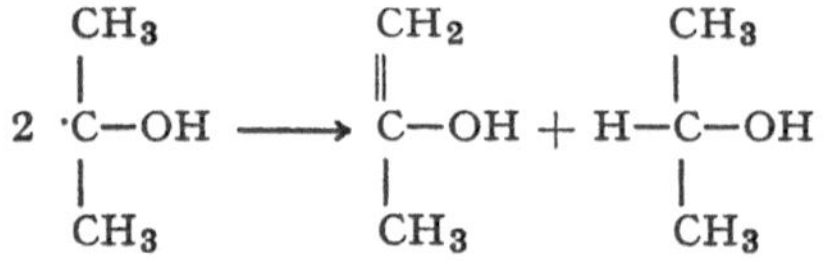

$$
2 \; \overset{\displaystyle CH_3}{\underset{\displaystyle CH_3}{|}} \cdot C-OH \longrightarrow \overset{\displaystyle CH_2}{\underset{\displaystyle CH_3}{\|}} C-OH + \overset{\displaystyle CH_3}{\underset{\displaystyle CH_3}{|}} H-C-OH \tag{29}
$$

7. LET-Effekte

Unter LET (*Linearer Energie Transfer*) wird die von dem ionisierenden Teilchen pro durchlaufener Wegstrecke abgegebene Energie verstanden. Er wird in eV/Å angegeben. Co—60—γ-Strahlen und hochenergetische Elektronen haben einen *LET* von ca. 0,03 eV/Å. Der Abstand der einzelnen Spuren, in denen ein Energiebetrag von durchschnittlich 60 eV abgegeben wird, ist so groß, daß sie bei genügend niedriger Dosisleistung der Quelle nicht überlappen. Bei Teilchen mit hohem *LET*, z. B. B-10(n, α)Li-7-Rückstoßstrahlung (*LET* = 30 eV/Å), bildet sich ein zusammenhängender *Ionisationszylinder*. Die Radikalkonzentration innerhalb dieses Ionisationszylinders ist größer als in den bei niedrigem *LET* gebildeten kleinen Spuren. Nach dem Durchgang des Teilchens beginnen die gebildeten Radikale zu diffundieren. Auch während der Diffusion bleibt wegen der zylindrischen Anordnung der Spur bei hohem *LET* eine hohe Radikalkonzentration länger erhalten als bei der Diffusion aus den kleinen voneinander getrennten Einheiten, die bei der Bestrahlung mit niedrigem *LET* entstehen.

Daraus folgt, daß bei hohem *LET* Radikal-Radikal-Reaktionen den bei niedrigem *LET* ablaufenden Radikal-Lösungsmittel-Reaktionen den Rang ablaufen können.

Bei niedrigem *LET* können die Alkoxy-Radikale mit dem Lösungsmittel unter Bildung von α-Hydroxyalkyl-Radikalen reagieren (Reaktion (30)).

$$R_1R_2CHO\cdot + R_1R_2CHOH \rightarrow R_1R_2CHOH + R_1R_2\dot{C}OH \qquad (30)$$

Die große Reaktionsfreudigkeit der Alkoxy-Radikale hat *Wijnen* [155,156] am Methoxy-Radikal gezeigt.

Bei hohem *LET*, d.h. hoher Radikalkonzentration, reagieren die Alkoxy-Radikale miteinander oder mit α-Hydroxyalkyl-Radikalen unter Disproportionierung zu Carbonylverbindungen (s. o.). Bei niedrigem *LET* entstehen aus der Dimerisierung der α-Hydroxyalkyl-Radikale Glykole. Dies erklärt, warum bei hohem *LET* das Verhältnis *G*(Carbonylverbindungen) zu *G*(Glykol) zugunsten der Carbonylverbindungen verschoben ist.

Das Auftreten sehr viel größerer Mengen CO bei der Radiolyse von Methanol mit hohem *LET* wird durch Radikal-Radikal-Reaktionen höherer Ordnung erklärt [66].

8. C—C-Brüche und ihre Abhängigkeit von der Struktur des Alkohols

Der Einfluß der Struktur auf die Zersetzung organischer Substanzen ist am besten bei den Aliphaten untersucht [60,115]. *Schuler* und *Kuntz* [115] stellten eine empirische Formel auf (Inkrementenmethode), durch die Methylradikalbildung und Strukturelemente korreliert werden können. Eine quantenmechanisch begründete Arbeit liegt noch nicht vor. Bei den Aliphaten ist eine Korrelation der Struktur und der C—C-Brüche einfacher als bei den Alkoholen, und zwar deswegen, weil die Elektronen (nach den bisherigen Kenntnissen) in den Aliphaten nicht solvatisiert werden und zu dem Mutterion zurückkehren. Dabei bildet sich ein hochangeregter Zustand, welcher unter Produktbildung zerfällt [44]. Jedoch wird nicht in allen Fällen der Mechanismus so einfach sein, insbesondere dann nicht, wenn vor der Rückkehr des Elektrons Ionen-Molekülreaktionen eingeschoben sind.

Bei den Alkoholen ist wegen der Solvatisierung des Elektrons und der raschen Protonenübertragungsreaktion des Mutterions eine Fragmentierung aus angeregten Zuständen, die durch die Rückkehr des Elektrons zu seinem Mutterion entstehen könnten, nicht wahrscheinlich. Andererseits zeigt sich, daß auch bei den Alkoholen wie bei den Aliphaten mit zunehmendem Verzweigungsgrad die C—C-Brüche zunehmen und die Wasserstoffbildung abnimmt. Die stark erniedrigte Wasserstoffausbeute bei der γ-Radiolyse des tert.-Butanols ($G(H_2) = 1,1$) [70,134] kann nur schwer mit dem von *Freeman* [105,104,67] vorgeschlagenen Mechanismus der Alkoholradiolyse in Einklang gebracht werden. Jedoch ist zu prüfen, ob dieser niedrige $G(H_2)$-Wert ein initialer G-Wert ist.

Die hohe Fragmentierung verzweigter Alkohole (tert.-Butanol: G(Methan) + 2 G(Äthan) + G(Neopentan) + G(tert.-Amylalkohol) = 3,35) [134] spricht dafür, daß evtl. auch das im Primärakt gebildete (angeregte?) Mutterion fragmentiert. α-Hydroxyalkyl-Radikale sind durch den Einfluß der Hydroxylgruppe stabilisiert. Es ist daher verständlich, daß besonders leicht Methylgruppen in α-Stellung zur Hydroxylgruppe eliminiert werden. Ein zusätzlicher Faktor kann die bevorzugte Ionisation der nicht bindenden Elektronenpaare am Sauerstoff sein [157]. Oxy-Radikale zeigen eine hohe Fragmentierungsbereitschaft [150].

In Tabelle 3 ist die Methanbildung, die in erster Näherung ein Maß für die Spaltung der C—CH$_3$-Bindung ist, für mehrere Alkohole zusammengestellt. Die Methanausbeuten zeigen, daß die zur Hydroxylgruppe

Tabelle 3. *G-Werte der Methanbildung verschiedener Alkohole bei der γ-Radiolyse in flüssiger Phase. 27° C. Sauerstoff-frei (nach v. Sonntag* [131]*). (Nicht sämtliche Alkohole konnten gaschromatographisch rein erhalten werden)*

Alkohol	G(CH$_4$)
Methanol	0,36
Äthanol	0,61
n-Propanol	0,052
Isopropanol	1,55
n-Butanol	0,03
Isobutanol	0,13
sek.-Butanol	0,69
tert.-Butanol	2,75
2-Methyl-butanol	0,052
2,2-Dimethyl-propanol	0,61
tert.-Amylalkohol	1,29

α-ständigen Methylgruppen tatsächlich bevorzugt abgespalten werden. Sek.-Butanol (2-Methylgruppen, eine davon α-ständig) hat eine höhere Methanausbeute als Isobutanol (ebenfalls 2-Methylgruppen), tert.-Butanol eine höhere als tert.-Amylalkohol.

9. Radiolyse in der Gasphase

Die G-Werte der Produkte sind in der Gasphase fast doppelt so hoch wie in der flüssigen Phase. Da angeregte Moleküle nicht so rasch wie in der flüssigen Phase desaktiviert werden können, ist die Fragmentierung größer. Darüberhinaus sind Fragmente, die durch den Bruch von mehr als einer Bindung entstehen, z.B. Kohlenmonoxyd, in der Gasphase stark erhöht. In der Gasphase wurde der G-Wert der Ionisation zu 4 bestimmt [5]. Ein entsprechender $G(H_2)$-Wert kann im Methanol durch die Elektronenfänger CH_3Br und CCl_4 abgefangen werden [91], wenn man annimmt, daß die einzige Neutralisierungsreaktion die Reaktion (5) ist.

$$ROH_2^+ + e^- \rightarrow ROH + H \tag{5}$$

Massenspektrometrische Untersuchungen haben gezeigt, daß diese Annahme gerechtfertigt ist.

Es wurde versucht, massenspektrometrische Daten mit den Ergebnissen der Gasphasenradiolyse zu korrelieren [22,88,158,101]. Theoretisch am besten begründet ist die Arbeit von *Prášil* [101].

Die noch nicht ausreichende Kenntnis der Verteilungsfunktion angeregter Zustände und die Unsicherheit in der Übertragung der bei den niedrigen Drucken des Massenspektrometers beobachteten Reaktionen auf die vergleichsweise hohen Drucke der Gasphasen-Radiolyse behaften die Berechnungen mit großen Unsicherheiten und die erzielten guten Übereinstimmungen mit dem Experiment können nicht als Beweise für die vorgeschlagenen Mechanismen angesehen werden [101].

10. Radiolyse im kristallinen und glasigen Zustand bei 77° K

Die Ausbeute an Produkten und die Fähigkeit, Vorläufer dieser Produkte, z.B. Elektronen zu stabilisieren, hängt stark von dem Aggregatzustand ab. So kann das Elektron in Alkohol-Gläsern bei der Temperatur des flüssigen Stickstoffs in der Matrix stabilisiert werden, nicht jedoch in der kristallinen Phase [16]. Wegen der Schwierigkeit, Alkoholkristalle zu ziehen, und den großen Vorteilen, die die Gläser aufgrund ihrer Transparenz für die Spektroskopie bieten, sind bisher fast ausschließlich Alkohol-Gläser untersucht worden. Die niedrigen Alkohole lassen sich meist leicht durch rasches Abkühlen in die glasige Form bringen. Beim Methanol bedarf es allerdings eines geringen Zusatzes an Wasser [37] oder an n-Propanol [121]. Auf diese Weise konnten die Reaktionen der Elektronen, die mit hohen Ausbeuten ($G(e_t^-) \approx 3$) in den Alkohol-Gläsern stabilisiert werden, mit den Methoden der Absorptions- und ESR-Spektroskopie verfolgt werden. Die Elektronen haben in den Alkohol-Gläsern Absorptionsmaxima im sichtbaren Bereich (s. Kap. 2), während die in der Matrix ebenfalls stabilisierten Alkoholradikale im ultravioletten Teil des Spektrums absorbieren [37]. Bestrahlte Alkohol-Gläser zeigen im ESR-Spektrum neben der Bande des Elektrons Signale, die den α-Hydroxyalkyl-Radikalen oder deren Ionen zuzuordnen sind [90,124,160,6,7,33]. Durch Ausbleichen der Elektronenbande mit sichtbarem Licht verschwindet das ESR-Signal des Elektrons, die Signale der α-Hydroxyalkyl-Radikale werden erhalten. In manchen Fällen sind diese Signale allerdings noch von einem Singlett überlagert, das einem Alkoxy-Radikal zugeordnet wurde [33]. Die α-Hydroxyalkyl-Radikale wurden auch über andere Wege dargestellt [89], und ihre ESR-Spektren lassen sich mit den durch die γ-Radiolyse erhaltenen vergleichen. Die α-Hydroxyalkyl-Radikale werden durch UV-Licht zersetzt [29,137,70,69,92]. Der Mechanismus der Zersetzung ist noch nicht völlig geklärt. Aus dem Hydroxymethyl-Radikal entsteht ein Formyl-Radikal (Reaktion (31)).

$$\cdot CH_2OH \xrightarrow{h\nu} \cdot CHO + H_2 \qquad (31)$$

Indessen treten bei der Photolyse auch intermediär Methylradikale und beim Äthanol Äthylradikale auf. Eine durch das α-Hydroxyalkyl- oder Formyl-Radikal sensibilisierte Zersetzung des Alkohols wird diskutiert (Reaktion (32) und (33)).

$$\cdot CH_2OH + CH_3OH \xrightarrow{h\nu} \cdot CH_3 + H_2O + CH_2O \qquad (32)$$

$$\cdot CHO \quad + CH_3OH \xrightarrow{h\nu} \cdot CH_3 + H_2O + CO \qquad (33)$$

Geringe Verunreinigungen im Alkohol können bei der UV-Bestrahlung mit dem Alkohol zu α-Hydroxyalkyl-Radikalen reagieren, welche dann die soeben besprochenen Reaktionen zeigen. Das erklärt nach *Kevan* [74] die Bildung von Methyl- und Äthylradikalen, die von *Sullivan* und *Koski* [138,139] bei der UV-Photolyse ($\lambda = 254$ nm) von Methanol und Äthanol bei 77° K erhalten wurden.

Beim Erwärmen der bei 77° K bestrahlten Alkoholgläser verschwinden zuerst die Elektronen. Erst bei höheren Temperaturen diffundieren dann auch die α-Hydroxyalkyl-Radikale. Beim thermischen Ausbleichen scheinen die Elektronen nicht nur mit den durch Bestrahlung erzeugten Protonen, sondern auch mit Radikalen und Carbonylverbindungen eine Reaktion einzugehen. Bei —196° C reagieren die Elektronen im Äthanol-Glas langsam ($\tau/2$ ca. 7 Tage) mit dem Alkohol zu einem H-Radikal [48], das unter diesen Bedingungen nicht in der Matrix stabilisiert wird. Selbst bei 4,2° K sind in Alkohol-Gläsern H-Atome nicht stabilisierbar [125]. Beim Ausbleichen mit sichtbarem Licht reagiert das Elektron mit seiner Umgebung unter Bildung eines Wasserstoff-Radikals. Die Quantenausbeute für diese Reaktion beträgt in Methanol 0,07 [37]. Die Reaktionen der H-Radikale in den verschiedenen Alkoholgläsern scheint unterschiedlich zu sein. *Johnson, Hagopian* und *Yun* [71] zeigten, daß thermische H-Atome bei 77° K nicht mit Äthanol, wohl aber mit den 1-Hydroxyäthyl-Radikalen zu Acetaldehyd und Wasserstoff reagieren. Im bestrahlten Äthanol-Glas (vgl. Tabelle 4) wird demgemäß auch eine hohe Ausbeute an Acetaldehyd gefunden. Da im Methanol-Glas kein Formaldehyd entsteht (vgl. Tabelle 4) müssen die H-Atome in diesem Glas entweder dimerisieren, mit dem Alkohol zu H_2 und Hydroxymethyl-Radikalen reagieren oder an andere Radikale (z.B. Hydroxymethyl-Radikale) addieren.

Tabelle 4. *G-Werte der wichtigsten Radiolyseprodukte von Methanol, Äthanol und Isopropanol bei Zimmertemperatur und 77° K. Es wurden diejenigen Arbeiten zugrunde gelegt, die aufgrund ihrer guten Materialbilanz und niedrigen Dosis initiale G-Werte erwarten lassen*

	Methanol		Äthanol		Isopropanol	
	25° C	77° K	25° C	77° K	25° C	77° K
Wasserstoff	5,4	3,2	5,0	4,4	4,5	3,5
Methan	0,4	0,35	0,6	0,3	1,55	0,8
Formaldehyd	1,95	0,1				
Acetaldehyd			3,2	3,0	0,9	
Aceton					4,0	3,1
Glykole	3,5	2,9	1,7	0,85	0,6	

An einem heißen Wolframdraht erzeugte und durch Stoß thermalisierte H-Atome reagieren selbst bei 87° K nur äußerst selten mit Isopropanol und Äthanol. Eine Reaktion mit Methanol war nicht nachzuweisen [57]. In Schwefelsäure-Gläsern hingegen reagieren H-Atome bei dieser Temperatur quantitativ mit Alkoholen unter Bildung der α-Hydroxyalkyl-Radikale [83,147,148].

UV-Photolyse von Alkoholen

1. Absorptionsspektren und Lichtquellen

Die Alkohole beginnen bei $\lambda \approx 250$ nm zu absorbieren; das erste Absorptionsmaximum liegt bei $\lambda \approx 182$ nm [61,62,56]. Die niedrigste Absorptionsbande wird einem n-σ^*-Übergang zugeschrieben [27]. Seine Oszillatorstärke (gemessen in der Gasphase) nimmt in der Reihenfolge Methanol, n-Propanol, Äthanol, Isopropanol von 0,005 auf 0,015 zu [61]. Die erste Absorptionsbande ist strukturlos [56,61], die nächst höheren zeigen Feinstruktur [56]. Das Fehlen einer Feinstruktur bei dem niedrigsten Übergang kann evtl. auf die ungenügende Auflösung des Spektrometers zurückgeführt werden [56].

Daß die direkte UV-Photolyse der Alkohole nur wenig untersucht ist, mag damit zusammenhängen, daß die Zersetzung bei der bequem zugänglichen Hg-Linie der Wellenlänge $\lambda = 254$ nm noch nicht sehr groß ist ($\phi < 10^{-3}$ mol/einstein) und für die ebenfalls von dem Hg-Niederdruckbrenner emittierte Linie bei $\lambda = 185$ nm bis vor einigen Jahren nicht genügend reines Quarz für das Küvettenmaterial zur Verfügung stand. Bei der Verwendung der Mischstrahlung $\lambda = 254/185$ nm muß der Umsatz äußerst gering gehalten werden, da die Hg-Linie $\lambda = 254$ nm etwa 8 mal intensiver ist als die Linie $\lambda = 185$ nm und die Produkte der UV-Photolyse der Alkohole, insbesonders die äußerst photoreaktiven Carbonylverbindungen, dieses Licht mit vergleichsweise hohen molaren Extinktionskoeffizienten absorbieren. Bei den Arbeiten in flüssiger Phase muß darauf geachtet werden, daß die Lösung ausreichend gut gerührt wird, da die Eindringtiefe des Lichtes der Wellenlänge $\lambda = 185$ nm nur einige Hundertstel mm beträgt. Aus den erwähnten Gründen sind häufig nicht die Primärprodukte, sondern ihre Photolyseprodukte isoliert worden. Auf der anderen Seite konnten leicht photolysierbare Primärprodukte nicht mehr nachgewiesen werden (s. Tabelle 5). Durch ein Filter aus γ-bestrahltem LiF kann die Linie $\lambda = 254$ nm weitgehend eliminiert werden [153]. Allerdings wird auch die Ausbeute an $\lambda = 185$ nm auf etwa die Hälfte herabgesetzt.

Die Photolyse mit energiereicherem Licht als $\lambda = 185$ nm wurde bisher nur in der Gasphase untersucht [56]. Den Bau der Lampen für diesen Bereich haben *Ausloos* und *Lias* [13] beschrieben.

2. UV-Photolyse in flüssiger Phase

Über die UV-Photolyse in flüssiger Phase können wegen der geringen Zahl vorhandener Arbeiten [87,108-112,128-130,133,135,159] bisher erst wenige generelle Aussagen gemacht werden.

Das Hauptprodukt der Photolyse von Alkoholen ist *Wasserstoff* H_2. Bei stark verzweigten Alkoholen wie z. B. tert.-Butanol kann die Methanbildung an die erste Stelle treten (s. Tabelle 5).

Der Wasserstoff kann über eine intramolekulare Abspaltung von molekularem Wasserstoff und/oder über H-Radikale als Vorläufer entstehen. Auch scheint noch ein dritter Weg, eine intermolekulare Ab-

Tabelle 5. *UV-Photolyse (λ 254/185 nm) von Alkoholen; Sauerstoff-frei; flüssige Phase*

Literatur	Methanol		Äthanol		Isopropanol		tert.-Butanol	
	159)*)	131)**)	159)*)	131)**)	159)*)	130)**)	159)*)	108–9)**)
Wasserstoff	2,97	0,83	2,54	0,845	1,72	0,75	kein	0,11
Kohlenmonoxid	0,098	kein	0,75	kein	1,27	<0,0015	0,42	kein
Methan	0,012	0,05	0,81	0,01	0,68	0,046	5,10	0,265
Äthan				0,011	0,05	0,0023		0,013
Propan						0,026		
Propen						0,003		
iso-Butan								0,015
iso-Buten								0,013
Neopentan								0,005
Formaldehyd	0,32	0,058						
Acetaldehyd			0,10	0,54	kein	0,04		
Propenoxid						0,016		
Aceton					0,23	0,72	0,35	0,205
Isopropanol								0,057
tert.-Amylalkohol								0,050
2,2,3-Trimethyl-butanol-(3)								0,003
2,2,4-Trimethyl-pentanol-(4)								0,012
3-Hydroxy-3-methyl-2-butanon							0,33	kein
1-Butoxi-2-methyl-propanol-(2)								0,03
Isobutenoxid								0,08
Äthylenglykol	1,92	0,78						
Butandiol-(2,3)			1,63	0,305				
Pinakon					1,52	0,036	1,25	0,015
2,4-Dimethyl-pentan-diol-(2,4)								0,032
2,5-Dimethyl-hexan-diol-(2,5)							1,74	0,044
Glycerin	0,26							

*) Ausbeute in mmol/h, Bestrahlung bei Rückflußtemperatur, 2—3% Umsatz.
**) Initiale Quantenausbeuten bezogen auf die Quantenausbeute des Äthanolaktinometers ϕ (H_2) = 0,4. Bestrahlung bei 26° C. Umsatz <0,01%.

spaltung von molekularem Wasserstoff, beschritten zu werden [108,111].
Diese intermolekulare H_2-Abspaltung führt beim tert.-Butanol zu dem
Äther 1-Butoxy-2-methyl-propanol-(2) (I) (Reaktion (34)) [108,111]. Bei
den primären und sekundären Alkoholen würde eine vergleichbare Reaktion zu den entsprechenden Halbacetalen der Carbonylverbindungen (II)
führen (Reaktion (35)), die rasch hydrolytisch in Carbonylverbindung
und Alkohol spalten.

$$
\begin{array}{ccccc}
& CH_3 & & CH_3 & \\
& | & & | & \\
CH_3-\overset{|}{\underset{|}{C}}-O- & \boxed{H \quad H} & -CH_2-\overset{|}{\underset{|}{C}}-O-H & \xrightarrow{h\nu} & \\
& CH_3 & & CH_3 &
\end{array}
$$

$$
\begin{array}{ccc}
& CH_3 & CH_3 \\
& | & | \\
H_2 + CH_3-\overset{|}{\underset{|}{C}}-O-CH_2-\overset{|}{\underset{|}{C}}-O-H & & (I) \\
& CH_3 & CH_3 \qquad (34)
\end{array}
$$

$$
\begin{array}{ccccc}
R_1 & & R_1 & & \\
| & & | & & \\
H-\overset{|}{\underset{|}{C}}-O- & \boxed{H \quad H} & -\overset{|}{\underset{|}{C}}-O-H & \xrightarrow{h\nu} & \\
R_2 & & R_2 & &
\end{array}
$$

$$
\begin{array}{cc}
R_1 & R_1 \\
| & | \\
H_2 + H-\overset{|}{\underset{|}{C}}-O-\overset{|}{\underset{|}{C}}-O-H & (II) \qquad (35) \\
R_2 & R_2
\end{array}
$$

Diese intermolekulare Wasserstoffabspaltung ist daher in diesen
Fällen von einer intramolekularen Abspaltung nicht zu trennen.

Am Sauerstoff deuterierte n-Alkohole liefern in flüssiger Phase über
80% HD unabhängig von der Kettenlänge [135]. Dieses deutet darauf
hin, daß entweder eine molekulare Abspaltung unter der Beteiligung der
OD-Gruppe (der rein radikalische Mechanismus nach *Yang* et al. [159] ist
in Frage zu stellen [130]) und/oder eine primäre radikalische Spaltung der
O—D-Bindung der Grund für den hohen HD-Anteil des Wasserstoffs ist.
Ähnliche Verhältnisse wurden bei deuterierten Isopropanolen gefunden
[128]. Beim tert.-Butanol, bei dem der Wasserstoff einer am Sauerstoff
deuterierten Probe zu 95% aus HD besteht, wird der Wasserstoff nicht
über Radikale, sondern in molekularen Eliminierungsprozessen gebildet,
die sowohl intramolekular unter Epoxydbildung als auch intermolekular
unter Bildung des Äthers 1-Butoxy-2-methyl-propanol-(2) (I) (s. o.) verläuft [108,109,111].

Bei sekundären und tertiären Alkoholen treten bei der Photolyse
Epoxyde auf, nicht jedoch bei den primären Alkoholen [133]. Die bevorzugte Konformation der primären Alkohole zeigt keine für die Epoxydbildung günstige Stellung der OH-Gruppe zu einem β-Wasserstoffatom.
Stattdessen ist eine molekulare Aldehydbildung zu erwarten.

Ein deutlicher Unterschied zwischen γ-Radiolyse und UV-Photolyse
($\lambda = 185$ nm) wird beim Isopropanol [132,129] sichtbar. Bei der γ-Radio-

lyse wird Methan mit einem hohen G-Wert ($G(CH_4) = 1.6$, vgl. $G(H_2) = 4.5$) gebildet, während bei der UV-Photolyse Methan nur ein Nebenprodukt ist ($\phi(CH_4) = 0,046$, vgl. $\phi(H_2) = 0,75$). Auch die Bildungsmechanismen sind verschieden. Bei der γ-Radiolyse entsteht das Methan zu etwa 95% über Methylradikale als Vorläufer, deren Anteil bei der UV-Photolyse nur etwa 8% ausmacht.

Die Zersetzung der Alkohole ist stark von dem umgebenden Medium abhängig. Bei der Photolyse von tert.-Butanol z.B. wird Methan in reiner Phase zu etwa 35% über Methylradikale als Vorläufer gebildet, während in wäßrigem Medium Methan fast ausschließlich molekular abgespalten wird [108,112].

In tert.-Butanol-Isopropanol-Mischungen wird das tert.-Butanol in weit geringerem Maße zersetzt als seinen in reiner Phase gemessenen Quantenausbeuten und dem von ihm absorbierten Lichtanteil entsprechen würde. Die Zersetzung des Isopropanols ist dagegen um den gleichen Betrag (etwa 40%) erhöht [108,110]. Dies kann deutlich an den Photolyseprodukten des tert.-Butanols, Methan ($\phi = 0,265$) und Isobutenoxyd ($\phi = 0,08$) und dem Hauptprodukt der Isopropanol-Photolyse, dem Wasserstoff ($\phi = 0,75$) gezeigt werden. Die geringe Methanbildung bei der UV-Photolyse des Isopropanols ($\phi = 0,046$) und die vergleichsweise geringe Wasserstoffbildung bei der UV-Photolyse des tert.-Butanols ($\phi = 0,11$) ermöglichen die Messung des wohl mit einer Art Energieübertragung zu deutenden Effektes.

Bei der UV-Photolyse des tert.-Butanols wurde eine weitere von der Umgebung des tert.-Butanols abhängige Reaktion gefunden. Reines tert.-Butanol bildet beim Bestrahlen mit Licht der Wellenlänge $\lambda = 185$ nm 1-Butoxy-2-methyl-propanol-(2) (I) und Wasserstoff (s. Reaktion (34)) in einer Reaktion, die nicht über freie Radikale verläuft [108, 111]. Für die Entstehung des Produktes sind zwei direkt benachbarte tert.-Butanol-Moleküle nötig. Mit zunehmender Verdünnung durch Wasser, Isopropanol oder n-Hexan nimmt die Ausbeute an (I) nach einer Funktion ab, die dieser Nächsten-Nachbar-Bedingung Rechnung trägt [108,111].

In reinem Isopropanol wird eine Wasserstoff-Quantenausbeute von 0,75 [130] bzw. 0,63[d] [127] gefunden; in wäßriger Umgebung beträgt sie nur 0,29 [15] bzw. 0,25[d] [127].

Diese Beispiele mögen zeigen, daß bei der UV-Photolyse von Alkoholen die Umgebung eine außerordentliche Rolle spielt und z.B. im wäßrigen Medium beobachtete Verhältnisse nicht auf die reine Phase übertragen werden dürfen.

---

[d] Um die Messungen vergleichbar zu machen, wurden die Werte auf eine Wasserstoff-Quantenausbeute des Äthanol-Aktinometers von $\phi(H_2) = 0,4$ umgerechnet.

3. UV-Photolyse in der Gasphase

In der Gasphase wurde bisher eingehender nur die UV-Photolyse von Methanol, diese aber sehr gründlich untersucht [55,56,94,99]. Insbesondere die neueren Arbeiten von *Porter* und *Noyes* [99] sowie *Hagège* et al. [55,56] haben den Photolysemechanismus zu großen Teilen aufklären können. Danach ist das Hauptprodukt der Photolyse Wasserstoff.

Die Spaltung in $\cdot CH_3 + \cdot OH$ (Reaktion (39)) hat nur einen Anteil von ca. 1%. Der Wasserstoff wird nach diesen Autoren zu etwa 20% molekular (Reaktion (36)) und zu etwa 80% über H-Radikale als Vorläufer (Reaktionen (37) und (38)) gebildet, wobei die Reaktion (38) kaum eine Rolle spielt [99].

$$CH_3OH + h\nu \longrightarrow CH_2O \quad + H_2 \quad 20\% \tag{36}$$

$$\longrightarrow CH_3O\cdot \; + H\cdot \; \left.\right\} \; 79\% \tag{37}$$

$$\longrightarrow \cdot CH_2OH + H\cdot \tag{38}$$

$$\longrightarrow \cdot CH_3 \quad + \cdot OH \quad 1\% \tag{39}$$

Die Bestrahlungsprodukte, Glykol, Methan und Wasser, laufen linear mit der Dosis. Formaldehyd wird schon nach kurzer Bestrahlungsdauer wegen seines hohen Absorptionskoeffizienten seinerseits photolysiert und es entsteht in dieser Sekundärreaktion neben weiterem Wasserstoff Kohlenmonoxyd. Glykol ist nach Wasserstoff das wichtigste Produkt. Es wird mit 70% der Ausbeute des Wasserstoffs gebildet [55]. Die Ausbeuten selbst sind stark druckabhängig und zwar ergibt sich eine lineare Beziehung zwischen der reziproken Quantenausbeute und dem reziproken Druck [56]. Sie kann durch die Gl. (40) dargestellt werden:

$$\phi^{-1} = \phi_\infty^{-1} + (kp)^{-1} \tag{40}$$

Bei den drei untersuchten Wellenlängen ($\lambda = 124$, 147 und 185 nm) wird die Beziehung (40) eingehalten, wobei ϕ_∞ und k für jede Wellenlänge eigene Werte haben.

Die Steigerung der Quantenausbeuten kann auch durch zugesetzte Fremdgase erreicht werden, deren Wirksamkeit von ihrer Polarisierbarkeit abhängt [56]. Die Autoren folgern, daß die Zersetzung über eine stoßinduzierte Prädissoziation erfolgt.

4. Quecksilber-sensibilisierte Photolyse ($\lambda = 254$ nm) in der Gasphase

Die Hg-sensibilisierte Photolyse von Alkoholen in der Gasphase führt in den Alkoholen zur Wasserstoffbildung [77-82,95,100].

Die Quantenausbeute der Wasserstoffbildung steigt von Methanol über Äthanol zum Isopropanol hin an, wird jedoch bei intermittierender Bestrahlung für diese Alkohole nahezu 1 [78-80)] (s. Tabelle 6).

Tabelle 6. *Quantenausbeute der Wasserstoffbildung bei der Hg-sensibilisierten Photolyse von Alkoholen in der Gasphase bei Dauerbestrahlung und intermittierender Bestrahlung; 25° C (nach Knight und Gunning [78-80,82)])*

Alkohol	H_2-Quantenausbeute	
	Dauerbestrahlung	Intermittierende Bestrahlung*)
Methanol	0,46	0,89
Äthanol	0,53	0,96
Isopropanol	0,72	1,0
tert.-Butanol	0,045	0,1

*) Bestrahlungsperiode 0,2 ms, Dunkelperiode 160 ms.

Eine Ausnahme macht jedoch das tert.-Butanol, bei dem die H_2-Quantenausbeute zwar bei intermittierender Bestrahlung ebenfalls ansteigt, den Wert 1 aber bei weitem nicht erreicht [82)].

Als Primärprozeß wird von *Knight* und *Gunning* [77-82)] die Spaltung der O—H-Bindung angenommen. Das Auftreten von Nitriten bei Anwesenheit von NO als Fänger stützt diese Annahme. Nach *Phibbs* und *Darwent* [95)] entstehen jedoch bei der Hg-sensibilisierten Photolyse von O-deuteriertem Methanol (Deuterierungsgrad 81%) nur 6,6% HD. Dieser Befund stellt die Deutung von *Knight* und *Gunning* (Primärprozeß: Ausschließlich Bruch der O—H-Bindung) in Frage.

Zur Deutung ihrer Ergebnisse nehmen *Knight* und *Gunning* [81,82)] die Bildung energiereicher Isopropoxy- und tert.-Butoxy-Radikale an. Bei der Bildung energiereicher Isopropoxy- und tert.-Butoxy-Radikale ist jedoch zu erwarten, daß auch das H-Radikal bei dem Bruch der O—H-Bindung einen (wahrscheinlich sogar höheren) Energieanteil übernimmt. Diese „heißen" H-Radikale sollten bevorzugt unter H_2-Bildung mit dem Alkohol reagieren und nicht, wie von *Knight* und *Gunning* angenommen wird, zu hohen Anteilen an andere Radikale addieren. Noch nicht diskutiert wird eine durch Komplexbildung (z. B. durch die intermediäre Bildung von HgH) veränderte Reaktivität der H-Radikale, wie sie von *Kuntz* und *Mains* [85)] bei der Hg-sensibilisierten Photolyse von flüssigen Alkanen erwogen wird.

C. v. Sonntag

Literatur

1) *Adams, G. E.,* and *J. H. Baxendale:* Radical and Molecular Yields in the γ-Irradiation of Liquid Methanol. J. Am. Chem. Soc. *80,* 4215 (1958).

2) — —, and *J. Boag:* Electron Attachment in Irradiated Solutions. Proc. Roy. Soc. (London) *A 277,* 549 (1964).

3) — —, and *R. D. Sedgwick:* Some Radical and Molecular Yields in the γ-Irradiation of Some Organic Liquids. J. Phys. Chem. *63,* 854 (1959).

4) —, and *R. D. Sedgwick:* Mechanism of Hydrogen Formation in the Radiolysis of Liquid Ethanol. Trans. Faraday Soc. *60,* 865 (1964).

5) *Adler, P.,* u. *K. H. Bothe:* Ionisationsaufwand W organischer Moleküle und seine Abhängigkeit von der Molekelstruktur. Z. Naturforsch. *20 a,* 1700 (1965).

6) *Alger, R. S., T. H. Anderson,* and *L. A. Webb:* Irradiation Effects in Simple Organic Solids. J. Chem. Phys. *30,* 695 (1959).

7) — — — Trapped Radicals in Irradiated n-Propanol at 77° K. J. Chem. Phys. *35,* 49 (1961).

8) *Anbar, M.,* and *E. J. Hart:* The Reactivity of Aromatic Compounds toward Hydrated Electrons. J. Am. Chem. Soc. *86,* 5633 (1964).

9) —, and *D. Meyerstein:* Isotope Effects in the Hydrogen Abstraction from Aliphatic Compounds. J. Phys. Chem. *68,* 3186 (1964).

10) —, and *P. Neta:* A Compilation of Specific Bimolecular Rate Constants for the Reactions of Hydrated Electrons, Hydrogen Atoms and Hydroxyl-Radicals. Intern. J. Appl. Radiation Isotopes *18,* 493 (1967).

11) *Arai, S.,* and *L. M. Dorfman:* Pulse Radiolysis Studies. VI. The Lifetimes of Aromatic Anions in Aliphatic Alcohols. J. Chem. Phys. *41,* 2190 (1964).

12) —, and *M. C. Sauer, Jr.:* Absorption Spectra of the Solvated Electron in Polar Liquids: Dependence on Temperature and Composition of Mixtures. J. Chem. Phys. *44,* 2297 (1966).

13) *Ausloos, P.,* and *S. G. Lias:* Gasphase Photolysis of Hydrocarbons in the Photoionization Region. Radiation Res. Rev. *1,* 75 (1968).

14) *Ayscough, P. B., R. G. Collins,* and *F. S. Dainton:* Some Elementary Processes in Radiation- and Photochemistry Revealed by Electron Spin Resonance. Nature *205,* 965 (1965).

15) *Barrett, J., M. F. Fox,* and *A. L. Mansell:* Reactions of the Hydrated Electron in the Photochemistry of Aqueous Sulfate and Bisulfate Ions. J. Chem. Soc. (A) 483 (1967).

16) *Barzynski, H.,* and *D. Schulte-Frohlinde:* On the Nature of the Electron Traps in Alkaline Ice. Z. Naturforsch. *22 a,* 2131 (1967).

17) *Basson, R. A.:* Effect of Temperature on the Yields of the Precursors of Hydrogen in the Radiolysis of Ethanol. J. Chem. Soc. (A) 1179 (1967).

18) — Mechanism of the Formation of Acetaldehyd and Butanediol in the Radiolysis of Ethanol. J. Chem. Soc. (A) 1989 (1968).

19) *Bates, R.,* and *H. S. Taylor:* Studies in Photosensitation-I. J. Am. Chem. Soc. *49,* 2436 (1927).

20) *Baxendale, J. H.:* Electrons in Solution. In: Current Topics in Radiation-III, S. 1; *M. Ebert* und *A. Howard* (Hrsg.) (1967).

21) —, and *F. W. Mellows:* The γ-Radiolysis of Methanol and Methanol Solutions. J. Am. Chem. Soc. *83,* 4720 (1961).

22) —, and *R. D. Sedgwick:* Radiolysis of Methanol Vapour. Trans. Faraday Soc. *57,* 2137 (1961).

358

23) *Bennett, J. E., B. Mile*, and *A. Thomas:* Trapped Electrons Produced by the Deposition of Alkali-Metal Atoms on Ice and Solid Alcohols at 77° K. Part. I. Optical Spectra and Electron Spin Resonance Spectra. J. Chem. Soc. (A) 1393 (1967).

24) — — — Trapped Electrons Produced by the Deposition of Alkali-Metal Atoms on Ice and Solid Alcohols at 77° K. Part. II. Chemical Reactions during Thermal and Photolytic Bleaching. J. Chem. Soc. (A) 1399 (1967).

25) *Blandamer, M. J., L. Shields*, and *M. C. R. Symons:* Unstable Intermediates Part XXVIII. Solvated Electrons: Rigid Organic Solvents. J. Chem. Soc. 1127 (1965).

26) *Buxton, G. V., F. S. Dainton*, and *M. Hammerli:* Kinetic Salt Effects on Reactions of the Solvated Electron in Methanol. Trans. Faraday Soc. *63*, 1191 (1967).

27) *Calvert, J. G.*, and *J. N. Pitts, Jr.:* Photochemistry, S. 441. New York: Wiley 1966.

28) *Catterall, R.*, and *M. C. R. Symons:* Unstable Intermediates. Part XXXIV. Solvated Electrons: A Model for the Spin-Paired Species in Liquid Ammonia. J. Chem. Soc. (A) 13 (1966).

29) *Chachaty, C.:* Photolyse des radicaux libres produits par irradiation de quelques alcools en C_3 et C_4 a 77° K. Compt. Rend. *259*, 2423 (1964).

30) — Etude par résonance paramagnétique électronique des radicaux et des ions produits par irradiation a 77° K de verres organiques en présence de capteurs d'électrons. J. Chim. Phys. *64*, 614 (1967).

31) —, and *E. Hayon:* Electron Spin Resonance Evidence of Radiation-Induced Electrons Trapped in Organic Glasses at 77° K. Nature *200*, 59 (1963).

32) — — Spectres de résonance paramagnetique de quelques alcools irradiés a 77° K par les rayons-γ. In: Electron Magnetic Resonance and Solid Dielectrics, S. 570; *R. Servant* und *A. Charru* (Hrsg.). Amsterdam: North-Holland Publishing Comp. 1964.

33) — — Etude par résonance paramagnétique éléctronique des éléctrons et des radicaux piégés dans les alcools irradiés par les rayons γ à 77° K. J. Chim. Phys. *61*, 1115 (1964).

34) *Collinson, E., F. S. Dainton, D. R. Smith*, and *S. Tazuké:* Evidence for the Unit Negative Charge on the "Hydrogen Atom" formed by the Action of Ionizing Radiation on Aqeous Systems. Proc. Chem. Soc. (London) 140 (1962).

35) *Dainton, F. S.:* The Simplest Free Radicals. In: Free Radicals in Solution, S. 15. London: Butterworth 1967.

36) —, *G. A. Salmon*, and *C. von Sonntag:* Radiolysis of Triethylamine-3-Methylpentane Glasses. Proc. Roy. Soc. (London) *A 313*, 31 (1969).

37) — —, and *J. Teplý:* The Radiation Chemistry of Low Temperature Methanol Glasses. Proc. Roy. Soc. (London) *A 286*, 27 (1965).

38) — —, and *P. Wardman:* Processes in γ-Irradiated Methanol Glass. Chem. Commun. 1174 (1968).

39) — — — The Radiation Chemistry of Liquid and Glassy Methanol. Proc. Roy. Soc. (London) *A 313*, 1 (1969).

40) — — —, and *U. Zucker:* Radiolysis of Glassy Methanol and n-Propanol at 77° K. Proc. 2nd. Tihany Symp. on Radiat. Chem. Akademiai Kiado, Budapest 247 (1967).

41) — —, and *U. F. Zucker:* Distinguishable Electrons Traps in γ-Irradiated n-Propanol Glass. Chem. Commun. 1172 (1968).

42) *Denver, D. F.*, and *J. G. Calvert:* Rate Studies of the Oxidation of Methyl Radicals in Oxygen-Rich Media at 25° C. J. Am. Chem. Soc. *84*, 1362 (1962).

C. v. Sonntag

43) *Dobson, G., and L. I. Grossweiner:* Primary Processes in the Photooxidation of Iodide Ion in Ethanol. Radiation Res. *23*, 290 (1964).
44) *Dyne, P. J.:* Charge Transfer and Electron Capture in the Radiolysis of Aliphatic Hydrocarbons. Can. J. Chem. *43*, 1080 (1965).
45) —, and *O. A. Miller:* Photochemistry of Electrons Trapped in Glasses of Methyltetrahydrofuran at 77° K. Can. J. Chem. *43*, 2696 (1965).
46) *Eiben, K.:* Solvatisierte und stabilisierte Elektronen in der Strahlenchemie. Angew. Chem., im Druck.
47) *Ekstrom, E., and J. E. Willard:* Effects of Matrix Polarity on the Optical and ESR Spectra of Trapped Electrons in Organic Glasses. J. Phys. Chem. *72*, 4599 (1968).
48) *Fletcher, J. W., and G. R. Freeman:* The Radiolysis of Ethanol VI. Hydrogen Precursors at — 196° C. Can. J. Chem. *45*, 635 (1967).
49) *Freeman, G. R.:* Yield of "Free Ions" in Gamma Irradiation Liquid Saturated Hydrocarbons. J. Chem. Phys. *39*, 988 (1963).
50) — Charge Scavenging during the Radiolysis of Liquid Cyclohexane. J. Chem. Phys. *43*, 93 (1965).
51) — Kinetics of Positive-Charge and Electron Scavenging and the Kinetics of Charge Neutralization in the Radiolysis of Dielectric Liquids. J. Chem. Phys. *46*, 2822 (1967).
52) —, and *J. M. Fayadh:* Influence of the Dielectric Constant on the Yield of Free Ions Produced during Radiolysis of a Liquid. J. Chem. Phys. *43*, 86 (1965).
53) *Grossweiner, L. I., E. F. Zwicker, and G. W. Swenson:* Photochemical Production of the Solvated Electron in Ethanol. Science *141*, 1180 (1963).
54) *Habersbergerová, A., I. Janovský, and J. Teplý:* Absorption Spectra of Intermediates Formed during Radiolysis and Photolysis. Radiation Res. Rev. *1*, 109 (1968).
55) *Hagège, J., P. C. Roberge, and C. Vermeil:* Methanol Photochemistry: Collision-Induced Predissociation. Ber. Bunsenges. Physik. Chem. *72*, 138 (1968).
56) —, *S. Leach,* et *C. Vermeil:* Photochimie du méthanol en phase vapeur à 1236 et à 1849 Å. J. Chim. Phys. *62*, 736 (1965).
57) *Hagopian, A. K. E., and R. H. Johnsen:* The Reaction of Thermal Hydrogen Atoms with Frozen Organic Substrates. J. Phys. Chem. *72*, 1949 (1968).
58) *Hamill, W. H.:* Electrons in Aqeous and Organic Media and a Model for Radiolysis. J. Chem. Phys. *49*, 2446 (1968).
59) —, *J. P. Guarino, M. R. Ronayne,* and *J. A. Ward:* Electronic Spectra and Related Ionic Effects in Gamma-Irradiated Organic Glassy Solids. Discussions Faraday Soc. *36*, 169 (1963).
60) *Hardwick, T. J.:* The Radiolysis of Saturated Hydrocarbons. J. Phys. Chem. *66*, 1611 (1962).
61) *Harrison, A. J., B. J. Cederholm,* and *M. A. Terwilliger:* Absorption of Acyclic Oxygen Compounds in the Vacuum Ultraviolet. I. Alcohols. J. Chem. Phys. *30*, 355 (1959).
62) —, and *J. S. Lake:* Photolysis of Low Molecular Weight Oxygen Compounds in the Far Ultraviolet Region. J. Phys. Chem. *63*, 1489 (1959).
63) *Hart, E. J.:* The Hydrated Electron. In: Actions Chimiques et Biologiques des Radiations—10, S. 1; Haissinsky, Ed. Paris: Masson 1966.
64) —, and *J. W. Boag:* Absorption Spectrum of the Hydrated Electron in Water and in Aqueous Solutions. J. Am. Chem. Soc. *84*, 4090 (1962).
65) *Hayon, E., and J. Weiss:* Primary Products in the Irradiation of Aqueous Solutions with X or Gamma Rays. Proc. 2nd Intern. Conf. Peaceful Uses At. Energy, Geneva *29*, 80 (1958).

66) *Imamura, M.,* and *N. N. Lichtin:* The Radiolysis of Methanol and Methanolic Solutions II. Comparison of Radiolysis by Co^{60} γ-Rays and by B^{10} (n,a)Li^7 Recoils. J. Am. Chem. Soc. *85,* 3565 (1963).

67) *Jha, K. N.,* and *G. R. Freeman:* I Kinetics of Reactions of Electrons during Radiolysis of Liquid Methanol II Reactions of Electrons with Liquid Alcohols and with Water. J. Chem. Phys. *48,* 5480 (1968).

68) *Johnsen, R. H.:* Photolysis of γ-Ray Produced Free Radicals in Ethanol at Low Temperatures. J. Phys. Chem. *63,* 2088 (1959).

69) — The Photolysis of Trapped Free Radicals Produced by Ionizing Radiation. J. Phys. Chem. *65,* 2144 (1961).

70) —, and *D. A. Becker:* The Radiation Chemistry of Some Higher Aliphatic Alcohols. — Further Studies on Radicals Trapped at Low Temperatures. J. Phys. Chem. *67,* 831 (1963).

71) —, *A. K. E. Hagopian,* and *H. B. Yun:* Reactions of Thermal Hydrogen-Atoms with Ethanol and Ethanol Free Radicals at 77° K. J. Phys. Chem. *70,* 2420 (1966).

72) *Keene, J. P.:* The Absorption Spectrum and Some Reactions Constants of the Hydrated Electron. Radiation Res. *22,* 1 (1964).

73) *Kemp, T. J., G. A. Salmon,* and *P. Wardman:* Some Aspect of the Quenching of Excited States Formed in Pulse Radiolysis. In: Pulse Radiolysis, S. 255; *M. Ebert, J. P. Keene, A. J. Swallow,* and *J. H. Baxendale* (Eds.). New York: Academic Press 1965.

74) *Kevan, L.:* Radiation Chemistry of Frozen Polar Systems. In: Actions chimiques et Biologiques des Radiations; *M. Haissinsky* (Ed.). Paris: Masson (erscheint in Band 13) 1969.

75) *Kiss, F.,* u. *Kh. S. Bagdasaryan:* Radiolyse von Isopropanol und Benzophenon- und Naphtalin-Lösungen in Isopropanol im flüssigen Zustand bei 30° C und im glasartigen Zustand bei — 196° C. Zh. Fiz. Khim. *40,* 1339 (1966).

76) — — Radiolysis of Isopropanol in Glassy State. Proc. 2nd. Tihany Symposium on Radiation Chemistry S. 257, Akademiai Kiado, Budapest (1967).

77) *Knight, A. R.,* and *H. E. Gunning:* Primary Methoxy Radical Formation in the Reaction of Methanol Vapor with Hg 6($^3P_1$) Atoms. Can. J. Chem. *39,* 1231 (1961).

78) — — Primary Quantum Yield Determination by Intermittent Illumination in the Reaction of Methanol Vapor with Hg 6 ($^3P_1$) Atoms. Can. J. Chem. *39,* 2251 (1961).

79) — — The Reaction of Ethanol Vapor with Hg 6 ($^3P_1$) Atoms. Can. J. Chem. *39,* 2466 (1961).

80) — — The Reaction of Isopropanol Vapor with Hg 6 ($^3P_1$) Atoms. Part. I. Pure Substrate. Can. J. Chem. *40,* 1134 (1962).

81) — — The Reaction of Isopropanol Vapor with Hg 6 ($^3P_1$) Atoms. Part. II. Reaction in the Presence of Nitric Oxide. Can. J. Chem. *41,* 763 (1963).

82) — — The Reaction of t-Butyl-Alcohol Vapor with Hg 6 ($^3P_1$) Atoms. Can. J. Chem. *41,* 2849 (1963).

83) *Köhnlein, W.,* and *D. Schulte-Frohlinde:* Thermal Reactions of Hydrogen Atoms with Organic Compounds in Sulfuric Low Temperature Glasses. Radiation Res. *38,* 173 (1969).

84) *Koltzenburg, G., K. Gorzny,* and *G. O. Schenck:* Semipinacol Radicals as Intermediates in Liquid Phase Photochemical Reactions. International Conf. on Photochemistry, Tokyo (1965), Preprints p. 183.

85) *Kuntz, R. R.,* and *G. J. Mains:* The Mercury ($^3P_1$)-sensitized Photolysis of Some Liquid Alcanes at 25°. J. Am. Chem. Soc. *85,* 2219 (1963).

C. v. Sonntag

86) *Kuppermann, A.:* Diffusion Kinetics. Nucleonics *19*, Nr 10, 38 (1961).
87) *Leuschner, G.,* u. *K. Pfordte:* Intermolekulare Dehydrierungen durch UV-Licht. Photoreaktionen der Alkohole. Liebigs Ann. Chem. *619*, 1 (1958).
88) *Lindholm, E.,* and *P. Wilmenius:* Ion-Molecule Reactions in the Radiolysis of Methanol. Arkiv Kemi *20*, 255 (1963).
89) *Livingston, R.,* and *H. Zeldes:* Paramagnetic Resonance Study of Liquids during Photolysis: Hydrogen Peroxide and Alcohols. J. Chem. Phys. *44*, 1245 (1966).
90) *Luck, C. F.,* and *W. Gordy:* Effects of X-Irradiation upon Some Organic Substances in the Solid State: Simple Alcohols, Amines, Amides and Mercaptans. J. Am. Chem. Soc. *78*, 3240 (1956).
91) *Meaburn, M.,* and *F. W. Mellows:* Effect of Hydrogen Atom and Electron Scavengers on the Radiolysis of Methanol Vapor. Trans. Faraday Soc. *61*, 1701 (1965).
92) *Milliken, S. B.,* and *R. H. Johnsen:* Ultraviolet Photolysis of X-Irradiated Methanol at 77° K. J. Phys. Chem. *71*, 2116 (1967).
93) *Mozumder, A.,* and *J. L. Magee:* Theory of Radiation Chemistry. VII Structure and Reactions in Low LET Tracks. J. Chem. Phys. *45*, 3332 (1966).
94) *Patat, F.,* u. *H. Hoch:* Der Photochemische Zerfall von Methyl- und Äthylalkohol. Z. Elektrochem. *41*, 494 (1935).
95) *Phibbs, M. K.,* and *B. deB. Darwent:* The Mercury Photosensitized Reactions of Methyl Alcohol. J. Chem. Phys. *18*, 495 (1950).
96) *Platzman, R. L.:* Physical and Chemical Aspects of Basic Mechanisms in Radiobiology. U. S. Natl. Res. Council Publ. Nr. 305, p. 34 (1953).
97) — Superexcited States of Molecules. Radiation Res. *17*, 419 (1962).
98) — Superexited States of Molecules, and the Primary Action of Ionizing Radiation. Vortex *23*, 372 (1962).
99) *Porter, R. P.,* and *W. A. Noyes, Jr.:* Photochemical Studies. LIV. Methanol Vapor. J. Am. Chem. Soc. *81*, 2307 (1959).
100) *Pottie, R. F., A. G. Harrison,* and *F. P. Lossing:* Free Radicals by Mass Spectrometry XXII. Primary Decomposition Steps in the Mercury-Photosensitized Decomposition of Methanol and Dimethyl Ether. Can. J. Chem. *39*, 102 (1961).
101) *Prášil, Z.:* To the Theory of Radiation Chemistry II. Calculation of Relative Radiation Chemical Yields of the Gas Phase Radiolysis of Methanol. Collection Czech. Chem. Commun. *31*, 3263 (1966).
102) *Ronayne, M. R., J. P. Guarino,* and *W. H. Hamill:* Electron Attachment and Solvation in Gamma Irradiated Organic Glasses at −196°. J. Am. Chem. Soc. *84*, 4230 (1962).
103) *Russell, J. C.,* and *G. R. Freeman:* The Radiolysis of Ethanol. V. Reactions of the Primary Reducing Species in the Liquid Phase. J. Phys. Chem. *71*, 755 (1967).
104) — — Reactions of the Primary Reducing Species in the Radiolysis of Liquid 2-Propanol. J. Phys. Chem. *72*, 808 (1968).
105) — — The Yields of the Primary Reducing Species in the Radiolysis of Liquid Ethanol. J. Phys. Chem. *72*, 816 (1968).
106) — — Yields of Solvated Electrons in the γ-Radiolysis of Water + 10% Ethanol: Nonhomogenious Kinetics of Electron Scavenging in Water. J. Chem. Phys. *48*, 90 (1968).
107) *Ryan, K. R., L. W. Sieck,* and *J. H. Futrell:* Ion-Molecule Reactions in Methanol and Ethanol. J. Chem. Phys. *41*, 111 (1964).

108) *Sänger, D.:* Die Photolyse von tert.-Butanol und seinen binären Mischungen. Dissertation Karlsruhe (1969).

109) —, and *C. v. Sonntag:* Radiation Chemistry of Alcohols — IX. UV-Photolysis of tert.-Butanol. In Vorbereitung.

110) — — Radiation Chemistry of Alcohols — X. UV-Photolysis of the System tert.-Butanol-Isopropanol. tert.-Butanol-Sensitized Decomposition of Isopropanol. In Vorbereitung.

111) — — Radiation Chemistry of Alcohols — XI. The Formation of 1-Butoxi-2-Methyl-Propanol-(2). A Nearest-Neighbour-Reaction in the UV-Photolysis and γ-Radiolysis of tert.-Butanol. In Vorbereitung.

112) — — Radiation Chemistry of Alcohols — XII. The Methane-Formation in the UV-Photolysis of tert.-Butanol-Water Mixtures. Change from a Radical- to a Molecular-Fragmentation Process. In Vorbereitung.

113) *Sauer, M. C., Jr., S. Arai,* and *L. M. Dorfman:* Pulse Radiolysis Studies. VII. The Absorptionspectra and Radiation Chemical Yields of the Solvated Electron in the Aliphatic Alcohols. J. Chem. Phys. *42*, 708 (1965).

114) *Schindewolf, U.:* Bildungsreaktionen und Eigenschaften solvatisierter Elektronen. Angew. Chem. *80*, 165 (1968).

115) *Schuler, R. H.,* and *R. R. Kuntz:* Methyl Radical Production in the Radiolysis of Hydrocarbons. J. Phys. Chem. *67*, 1004 (1963).

116) *Schulte-Frohlinde, D.,* u. *K. Eiben:* Solvatisierte Elektronen in eingefrorenen Lösungen. Z. Naturforsch. *17a*, 445 (1962).

117) —, *G. Lang,* u. *C. v. Sonntag:* Strahlenchemie von Alkoholen. IV. Einfluß der Kettenlänge auf die Wasserstoffbildung bei der γ-Radiolyse von gesättigten n-Alkoholen. Ber. Bunsenges. Physik. Chem. *72*, 63 (1968).

118) *Seddon, W. A.,* and *D. R. Smith:* Electron Spin Resonance Studies of γ-Irradiated Aqueous Alkaline Solutions of Acrylamide at 77° K. Can. J. Chem. *45*, 3083 (1967).

119) *Sherman, W. V.:* The Reactivity of Electrons Produced in the γ-Radiolysis of Aliphatic Alcohols. J. Am. Chem. Soc. *88*, 1567 (1966).

120) — The γ-Radiolysis of Liquid 2-Propanol. II. The Reaction of Solvated Electrons with Mono- and Disubstituted Benzenes. J. Phys. Chem. *70*, 2872 (1966).

121) *Shida, T.,* and *W. H. Hamill:* Molecular Ions in Radiation Chemistry V. Intermediates in γ-Irradiated Glassy Solutions of Methanol. J. Am. Chem. Soc. *88*, 3689 (1966).

122) *Shields, L.:* Electron Trapping in Rigid Ethanol-Methyl-2-Tetrahydrofuran Mixtures. J. Phys. Chem. *69*, 3186 (1965).

123) *Skelly, D. W.,* and *W. H. Hamill:* Trapped Electrons in γ-Irradiated 3-Methyl-Pentane at −196°. J. Chem. Phys. *44*, 2891 (1966).

124) *Smaller, B.,* and *M. S. Matheson:* Paramagnetic Species Produced by γ-Irradiation of Organic Compounds. J. Chem. Phys. *28*, 1169 (1958).

125) *Smith, D. R.,* and *J. J. Pieroni:* Detection of Trapped Electrons in Organic Glasses after Gamma Irradiation at 4.2° K by Electron Spin Resonance Spectroscopy. Can. J. Chem. *45*, 2723 (1967).

126) —, *W. A. Seddon,* and *P. E. Bindner:* Electron spin resonance detection of radicals and electrons condensed from water vapor after irradiation with 1 MeV helium ions. Can. J. Chem. *46*, 1747 (1968).

127) *Sokolov, U.,* and *G. Stein:* Photolysis of Liquid Water at 1849 Å. J. Chem. Phys. *44*, 3329 (1966).

128) *Sonntag, C. v.:* Strahlenchemie von Alkoholen I. Die Wasserstoffbildung bei der UV-Photolyse deuterierter Isopropanole. Tetrahedron *24*, 117 (1968).

129) — Strahlenchemie von Alkoholen — VII. Die Methanbildung bei der UV-Photolyse deuterierter Isopropanole. Intern. J. Radiation Phys. Chem. *1*, 33 (1969).
130) — Die UV-Photolyse ($\lambda = 185$ nm) von Isopropanol. In Vorbereitung.
131) — unveröffentlicht.
132) —, u. W. *Brüning*: Strahlenchemie von Alkoholen — V. Die Methanbildung bei der γ-Radiolyse von Hepta-Deutero-Isopropanol. Intern. J. Radiation Phys. Chem. *1*, 25 (1969).
133) —, u. D. *Sänger*: Strahlenchemie von Alkoholen — VI. Epoxidbildung bei der UV-Photolyse von sekundären und tertiären Alkoholen. Tetrahedron Letters 4515 (1968).
134) — — unveröffentlicht.
135) —, u. D. *Schulte-Frohlinde*: Strahlenchemie von Alkoholen — II. Die Wasserstoffbildung bei der UV-Photolyse O-deuterierter n-Alkohole. Z. Physik. Chem. N. F. *55*, 329 (1967).
136) *Stein, G.*: Some Aspects of the Radiation Chemistry of Organic Solutes. Discussions Faraday Soc. *12*, 227 (1952).
137) *Sullivan, P. J.*, and W. S. *Koski*: Electron Spin Resonance Study of Irradiated Methanol: J. Am. Chem. Soc. *84*, 1 (1962).
138) — — An Electron Spin Resonance Study of the Relative Stabilities of Free Radicals Trapped in Irradiated Methanol at 77° K. J. Am. Chem. Soc. *85*, 384 (1963).
139) An Electron Spin Resonance Study of Free Radicals in Ultraviolet Irradiated Ethanol at 77° K. J. Am. Chem. Soc. *86*, 159 (1964).
140) *Taub, I. A., D. A. Harter, M. C. Sauer, Jr.*, and *L. M. Dorfman*: Pulse Radiolysis Studies. IV. The Solvated Electron in the Aliphatic Alcohols. J. Chem. Phys. *41*, 979 (1964).
141) —, *M. C. Sauer, Jr.*, and *L. M. Dorfman*: Pulse Radiolysis Studies of the Reactivity of the Solvated Electron in Ethanol and Methanol. Discussions Faraday Soc. *36*, 206 (1963).
142) *Teplý, J.*, and *A. Habersbergerová*: The Hydrogen Formation in the Radiolysis of Liquid Methanol. UJV 1553 Prag (1966).
143) — — Hydrogen Formation in the Radiolysis of Liquid Methanol and Ethanol. Collection Czech. Chem. Commun. *32*, 1350 (1967).
144) — — Hydrogen Formation in the Radiolysis of Liquid Methanol. Proc. 2nd Tihany Symposium on Radiation Chemistry, Akademiai Kiado, Budapest, S. 239 (1967).
145) *Thomas, J. K.*: Methods of Production of Solvated Electrons and their Chemical and Physical Properties. Radiation Res. Rev. *1*, 183 (1968).
146) *Thynne, J. G. J., F. K. Amenn-Kpodo*, and *A. G. Harrison*: Ion-Molecule Reactions in Methyl Alcohol and Methyl-d_3-Alcohol. Can. J. Chem. *44*, 1655 (1966).
147) *Vacek, K.*, and D. *Schulte-Frohlinde*: The Kinetics of the Reaction of Trapped Hydrogen Atoms in Sulfuric Acid Glasses. J. Phys. Chem. *72*, 2686 (1968).
148) —, u. C. v. *Sonntag*: H-D Isotope Effect in the Reaction of Hydrogen Radicals with Isopropyl Alcohol in 6M—H_2SO_4 in the Liquid and in the Glassy State. Chem. Commun. *1969*, 1256.
149) *Walker, D. C.*: The Hydrated Electron. Quart. Rev. (London) *21*, 79 (1967).
150) *Walling, C.*: Some Aspects of the Chemistry of Alkoxy Radicals. In: Free Radicals in Solution, S. 69. London: Butterworth 1967.
151) *Ward, J. A.*, and W. H. *Hamill*: The Radiation Chemistry of Benzylacetate at $-196°$ and 20° in Polar Solvents. J. Am. Chem. Soc. *87*, 1853 (1965).

152) *Wardman, P.,* and *W. A. Seddon:* Electron spin resonance studies of radicals condensed from irradiated water vapor: reactions of the radicals. Can. J. Chem. *47*, 2149 (1969).

153) *Weeks, J. L., S. Gordon,* and *G. M. A. C. Meaburn:* Irradiated Lithium Fluoride as an Optical Filter in the Far Ultraviolet. Nature *191*, 1186 (1961).

154) *Wendenburg, J.,* u. *A. Henglein:* Die Ausbeute solvatisierter Elektronen bei der Einwirkung von γ-Strahlung und von He/Li-Teilchen auf organische Gläser bei −196° C. Z. Naturforsch. *19b*, 995 (1964).

155) *Wijnen, M. H. J.:* Reactions of Alkoxy-Radicals. I. Photolysis of Methyl Acetate. J. Chem. Phys. *27*, 710 (1957).

156) — Reactions of Alkoxy Radicals. III. Photolysis of Methyl-d₃ Acetate below 100° C. J. Chem. Phys. *28*, 271 (1958).

157) *Williams, T. Ff.:* Specific Elementary Processes in the Radiation Chemistry of Organic Oxygen Compounds. Nature *194*, 348 (1962).

158) *Wilmenius, P.,* and *E. Lindholm:* Dissociation of Methanol Molecule Ions Formed in Charge Exchange Collisions with Positive Ions. Ion-Molecnle Reactions of Methanol with very Slow Positiv Ions. Arkiv Phys. *21*, 97 (1962).

159) *Yang, N. C., D. P. C. Tang, D. M. Thap,* and *J. S. Sallo:* Liquid Phase Photolysis of Simple Alcohols. J. Am. Chem. Soc. *88*, 2851 (1966).

160) *Zeldes, H.,* and *R. Livingston:* Paramagnetic Resonance Study of Irradiated Glasses of Methanol and Ethanol. J. Chem. Phys. *30*, 40 (1959).

Eingegangen am 25. März 1969

Photochemistry of Metal Carbonyls, Metallocenes, and Olefin Complexes

Dr. E. Koerner von Gustorf and F.-W. Grevels

Max-Planck-Institut für Kohlenforschung, Abteilung Strahlenchemie, Mülheim/Ruhr

Contents

Introduction

Nowadays the borderlines between inorganic chemistry, organic chemistry and physical chemistry are vanishing, due to the enormous mutual interaction of these fields. Organometallic and coordination chemistry, as well as photochemistry, have been strong contributory factors in this "overlap process". Therefore, the photochemistry of organometallic and coordination compounds provides very specific examples of this general phenomenon.

While scattered reports on the photosensitivity of transition metal compounds are found in the early photochemical literature [375], more systematic investigations since 1950 coincide with the almost explosive development of organometallic chemistry [106,107,141,368,384], with the rapid progress of organic photochemistry [89,257,343,401,512] and with the impressive renaissance of inorganic chemistry [5,118]. This is mainly due to the development of molecular orbital theory [27] and ligand-field theory [363,398], as well as to the impact the discovery of ferrocene [259,328] had on the general understanding of chemical bonding.

There are several recent reviews of the photochemistry of transition metal complexes [3,4,5,29,30,45,466,514,518,580]), the most comprehensive being [4]). However, these papers deal mainly with ionic coordination compounds containing inorganic ligands; only one [466]) is devoted to photochemical substitution reactions of metal carbonyls and their derivatives, while in another [4]) this subject is discussed in a short paragraph.

We have attempted to review the photochemistry of metal carbonyls, metallocenes, and olefin complexes under two aspects:

to inform organometallic chemists about the preparative possibilities and mechanistic features of this field,

and, hopefully, to increase the interest of organic photochemists in organometallic chemistry.

In addition some work on the radiation chemistry of these compounds has been compiled.

As far as the coverage of the literature is concerned, we have searched Chemical Abstracts from 1960—1968 and a number of leading journals page by page for the year 1968. Much of the information about preparative work is hidden and often hard to find. Therefore, we did not attempt a complete coverage of all work done in the field. Work prior to 1960 has been cited only sporadically.

We have anticipated that the reader is familiar with the basic principles of photochemistry and with the photochemical "vocabulary" [374]). A short summary of some experimental photochemical procedures is intended to be helpful to the organometallic chemist who plans to initiate work in the field [a]).

A. Experimental Procedures

Most transition metal complexes are sensitive towards oxygen and/or water, especially under irradiation. This fact sometimes impedes photochemical work in this field. Detailed information about the properties and "thermal" chemistry of transition metal carbonyl compounds is found in [519,2]), and there are excellent reviews on arene complexes [177, 387,525,532]) and olefin complexes [178,221]).

The handling of organometallics (including measuring of absorption spectra) with exclusion of air and moisture, as well as standard proce-

---

[a]) *Note added in proof:* The first volume of the "Annual Survey of Photochemistry" covering the 1967 literature just appeared; it contains a comprehensive chapter by *D. Valentine* on progress in the study of inorganic and organometallic spectroscopy and photochemistry [581]). The reader's attention should be drawn to the forthcoming volumes of this very informative series.

dures of preparation are described in [272,561] b). Methods for the characterization of organometallic compounds have been recently compiled [563].

There are several important steps in the conduction of a photochemical experiment:

a) Measurement of the electronic spectrum of the starting material and assignment of the observed electronic transitions [249,254,499]. For the proper assignment of absorption bands it is often necessary to measure the spectra in solvents of different polarity and differing hydrogen bonding ability.

b) After it has been decided which transition is to be excited, an important point is the choice of a proper light source with strong emission in the region of interest. Sometimes filters or a monochromator have to be applied to select a certain emission band of a lamp for monochromatic irradiation. Extensive information on this matter is given in [89,303,320, 397,401]. The *Srinivasan-Griffin* Rayonet photochemical reactors [500] which are equipped with exchangable lamps for the 254, 300, and 350 nm region have been found extremely useful especially for preparative work in the authors' laboratory.

c) The type of light source and the amount of material used will determine the choice of the reaction vessel [89,397]. Apparatus for the photolysis of metal carbonyls have been described by *Strohmeier* [466,490]. In preparative photochemical work with organometallic compounds, precipitates are often formed which can cover the walls of the reaction vessels, thus inhibiting transmission of light into the system. To avoid this problem, a simple apparatus (Fig. 1) has been constructed in the authors' laboratory [202,285], which can easily be fitted into a Rayonet photochemical reactor.

Apparatus which allows the carrying out and working up of photochemical reactions below −100 °C has been developed by *E. Koch* [280, 281].

d) The light intensity has to be determined in order to allow evaluation of the quantum yield of a photochemical reaction. For this purpose standardized chemical actinometers e.g. potassium ferrioxalate (applicable at 250−450 nm) are most convenient [89,320]. In calculating the quantum yield from the measured light intensity and the absorption characteristics of the irradiated system, one has to consider the changes in optical density occurring during irradiation. These changes are due not only to the disappearance of the starting material; photochemically

---

b) Caution: Irradiation of metal carbonyl compounds with LASER beams for measuring Raman-spectra may be hazardous, an explosion was reported with the dimeric cyclopentadienylchromium dinitrosyl [180].

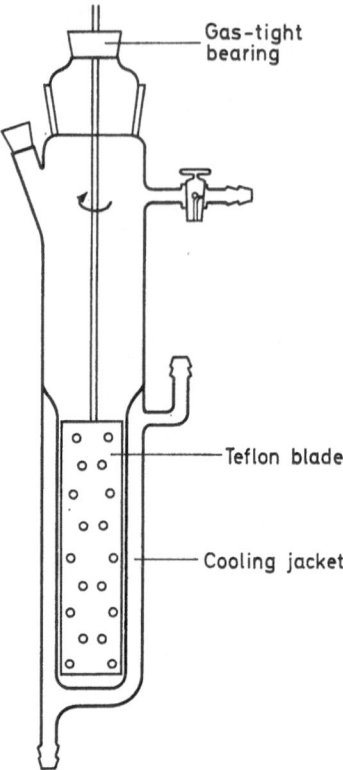

Gas-tight bearing

Teflon blade

Cooling jacket

Fig. 1. Irradiation apparatus for photochemical reactions in which a precipitate is formed [after [202,285]]

formed products can absorb light, too, and undergo further photochemical reactions. This is especially the case in photochemical substitution reactions of metal carbonyls. Pertinent examples are discussed in section E2.

e) Important methods to establish the mechanism of a photochemical reaction are:

1. Emission spectroscopy (fluorescence, phosphorescence) of the excited species allows their lifetime and multiplicity to be evaluated and a term scheme to be set up [535]. The influence of varying concentrations of reactants and additives on the quantum yields of the luminescent processes as well as on the product distribution gives information about the reaction mechanism [436]. Radiationless processes can be directly observed by studying the optical-acoustic relaxation of periodically irradiated solutions [229].

2. The absorption spectroscopy of transients (flash photolysis) is the most valuable tool for studying the decay of electronically excited species and reactive intermediates [360,379]. The limiting factor in the observability of a transient is the duration of the flash: $\sim 10^{-6}$ sec. for conventional discharge lamps and $\sim 10^{-9}$ sec. for a Laser set-up [536].

3. A complementary method for the study of transients is photolysis in rigid media. This technique often allows the application of additional methods, such as IR and ESR to identify transients. Examples from the photochemistry of metal carbonyls are discussed in section C.

B. UV Spectra and Their Interpretation

The first electronic absorption spectra of metal carbonyls were measured by *Eyber* [148] and by *Thompson* and *Garratt* [186,503] who reported continuous absorption in the gas phase for $Fe(CO)_5$ and $Ni(CO)_4$, which was correlated with a dissociation process:

$$M(CO)_n \xrightarrow{h\nu} M(CO)_{n-1} + CO$$

More recently several absorption maxima at > 40000 cm^{-1} for $Ni(CO)_4$ [412] and a shoulder at ~ 40000 cm^{-1} for $Fe(CO)_5$ [314] have been observed in solution. The spectra of $Cr(CO)_6$, $Mo(CO)_6$, $W(CO)_6$, $V(CO)_6$, and $[Mn(CO)_5]_2$ have been carefully studied in solution [8,58,130, 314,326,484,526] as well as in the gas phase [58,194,208,326]. All the metal hexacarbonyls show strong absorption maxima at >28000 cm^{-1} with $\log \varepsilon \gtrsim 3$.

Molecular orbital energy level diagrams which allow the assignment of these transitions have been developed by *Gray* et al. for $M(CO)_6$ [58,194]. Fig. 2 shows the diagram for $Cr(CO)_6$ (after [58]). The correlation with the observed transitions (in CH_3CN) is given in Table 1.

$M(CO)_6$ shows two different types of transitions:

1. d-d Transitions (Ligand-Field Transitions): These are electronic transitions between molecular orbitals which are mainly composed of metal (M) d-orbitals. According to a crude electrostatic picture (crystal-field theory), these are transitions from d-orbitals directed between the ligands to d-orbitals directed towards the ligands $[t_{2g} (d_{xy}, d_{xz}, d_{yz}) \rightarrow e_g (d_{x^2-y^2}, d_{z^2})$ in O_h symmetry]. The corresponding electronic absorption bands are expected to be weak, since these transitions $[^1A_{1g} \rightarrow {}^1T_{1g}; {}^1A_{1g} \rightarrow {}^1T_{2g}]$ are spin-allowed but forbidden for electric dipole radiation. The relative high intensity of these bands in $Cr(CO)_6$ may be explained by their proximity to strongly allowed transitions.

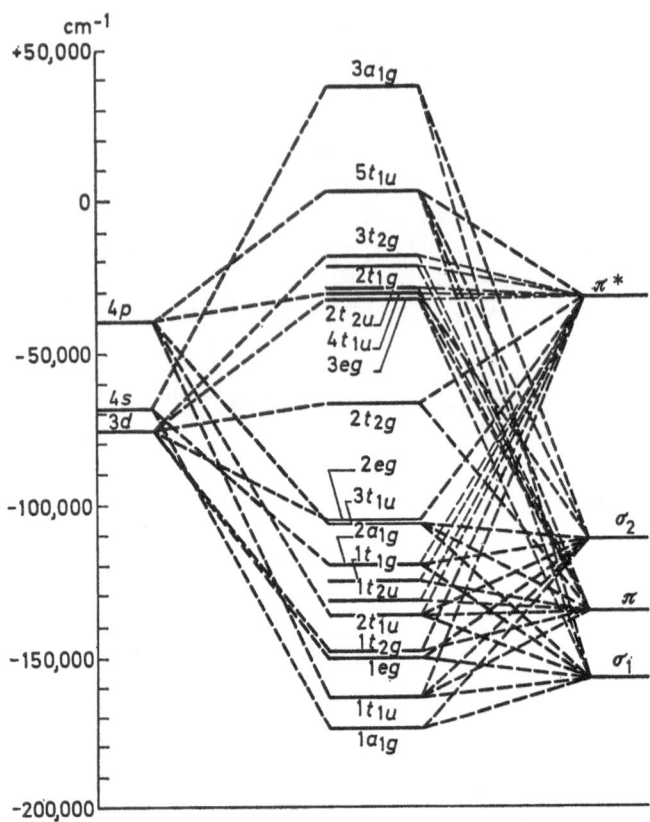

Fig. 2. Molecular orbital energy level diagram for Cr(CO)$_6$ [after [58]]

Table 1. *Electronic transitions in Cr(CO)$_6$ (after [58])*

Transition energy [cm^{-1}]	ε	Assignment		
29 500	700	[1]		
31 550	2670	$^1A_{1g} \rightarrow {}^1T_{1g}$,	$2\,t_{2g} \rightarrow 3\,e_g$,[3]	d-d
35 700	13 100	$^1A_{1g} \rightarrow c\,^1T_{1u}$,	$2\,t_{2g} \rightarrow 4\,t_{1u}$,	CTML
38 850	3500	$^1A_{1g} \rightarrow {}^1T_{2g}$,	$2\,t_{2g} \rightarrow 3\,e_g$,[3]	d-d
43 600	85 100	$^1A_{1g} \rightarrow d\,^1T_{1u}$,	$2\,t_{2g} \rightarrow 2\,t_{2u}$,	CTML
50 900		[2]		

[1] Possibly part of the vibrational structure of $^1A_{1g} \rightarrow {}^1T_{1g}$.
[2] Data from vapor phase. Possibly orbitally or spin-forbidden CT-transition.
[3] Splitting of this transition by interelectronic repulsion is not shown in Fig. 2.

The net electron density at the metal is not significantly changed but shifted into the metal-ligand area. Increased repulsion between M- and -CO results and dissociation may occur. Nucleophilic attack at the metal should be facilitated.

2. *Charge-Transfer Transitions:* These are electronic transitions between molecular orbitals mainly composed of metal (M) orbitals and those mainly composed of ligand (L) orbitals. Three different types of charge-transfer transitions are conceivable:

a) Charge-transfer from the metal to the ligand: CTML

$$M-L \xrightarrow{h\nu} M^+ + L^-$$

The $2\,t_{2g} \to 4\,t_{1u}$ and the $2\,t_{2g} \to 2\,t_{2u}$ transitions in $Cr(CO)_6$ are of this nature. Both transitions are spin- and orbitally-allowed and are observed as very intense bands. Some metal character is admixed into the $4\,t_{1u}$ MO but not into the $2\,t_{2u}$ MO. Therefore, a larger change in the electric dipole moment is associated with the $2\,t_{2g} \to 2\,t_{2u}$ transition, the $2\,t_{2u}$ being composed of ligand π^* only.

b) Charge-transfer from the ligand to the metal: CTLM

$$M-L \xrightarrow{h\nu} M^- + L^+$$

According to an estimate [194] this transition should appear at > 60000 cm^{-1} in $M(CO)_6$ and has not as yet been observed. Other workers consider the 50900 cm^{-1} band as resulting from such a transition [92].

C. Charge-transfer from the complex as a whole to the solvent: CTS

$$M-L + Solvent \xrightarrow{h\nu} [M-L]^+ + Solvent^-$$

This process is often observed in halogen containing solvents and will be discussed later in this section.

The result of CT is a decrease in electron density at the metal or the ligand, respectively. Nucleophilic attack at M^+ is facilitated and dissociation of $M-L$ can occur.

For the sake of completeness a third type of transition should be mentioned:

3. *Intra-Ligand Transitions:* These are electronic transitions between two ligand orbitals, which are not involved in $M-L$ bonding. The $M-L$ bond is usually not affected by such a transition, but internal rearrange-

ments of the ligand can occur. These transitions are observed e.g. in metal carbonyl complexes containing additional π-systems attached to a ligand. The electron density at the central metal remains unchanged in intra-ligand transitions.

The assignments given by *Gray* [58] appear to be reliable, since the orbital energies resulting from his calculations are remarkably close to the experimental vertical ionization potentials determined by photoelectron spectroscopy. However, it should be mentioned that a semiempirical MO-model for $Cr(CO)_6$, $Fe(CO)_5$, and $Ni(CO)_4$ of *Schreiner* and *Brown* [412] results in a significantly different energy level diagram for $Cr(CO)_6$. These authors in particular question the assignment of the $t_{2g} \rightarrow e_g$ transition. An energy level scheme different from *Gray*'s has also been proposed by an Italian group [504] on the basis of SCF$-$MO$-$LCAO calculations.

Gray's assignment of the CTML-band in $Cr(CO)_6$ is supported by SCCC (*Mulliken-Wolfsberg-Helmholz*) calculations [92]. Another approach did not give reliable data for the electronic spectra [93].

One specific feature of transition metal complex spectra remains to be discussed: a band in $Mo(CO)_6$ [28850 c^{-1}, ε: 350] and $W(CO)_6$ [28300 cm^{-1}, ε: 1000], which is not observed in $Cr(CO)_6$ has been assigned to a spin-forbidden d-d transition [$^1A_{1g} \rightarrow {}^3T_{1g}$] [58]. The intensity of this band increases with the atomic number of the central metal due to increased spin-orbit coupling [512]. The increased probability of intersystem crossing in transition metal complexes should be kept in mind in a discussion of their photoreactivity.

Some substituted metal carbonyls, as well as ionic carbonyls whose electronic absorption spectra have been recorded (but which are not discussed in the text), are compiled in Table 2.

Most of the listed complexes show strong absorption maxima, probably due to CT-transitions. The transitions in duroquinone-, cyclopentadienone-, and thiophene-dioxide-iron tricarbonyl have been assigned on the basis of qualitative MO considerations [408]. MO schemes have been established for $Mn(CO)_5X$ [193,195] and arene chromium tricarbonyls [92]. The band at 26670 cm^{-1} in $C_6H_6Cr(CO)_3$ has been attributed to a Cr $\rightarrow$ ring CT. We will return to this point in section E6.

Most the spectral data given in Table 2 still await a theoretical treatment.

The situation is quite different with ferrocene, which has received more theoretical consideration than any other organometallic compound. Available MO energy level diagrams are mainly based on SCF calculations of *Dahl* and *Ballhausen* [125], *Shustorovich* and *Dyatkina* [423,424,425], and of *Yamazaki* [527]. A comparison of these data with those from a *Wolfsberg-Helmholz* calculation is given in [179]. For detailed discussion the

Table 2. *Metal carbonyl compounds, whose UV data have been reported*

$M(CO)_m$	L	References
$V(CO)_6^-$	—	58)
$Cr(CO)_5$	piperidine pyridine chinoline	484)
$Cr(CO)_4$	ethylenediamine propylenediamine trimethylenediamine dipyridyl phenanthroline	393,394)
$Cr(CO)_3$	benzene	92,314,526)
	mesitylene aniline dimethyl-aniline	314)
$Mo(CO)_5$	piperidine	130,484)
	pyridine PCl_3, $P(OC_2H_5)_3$, $P(OCH_3)_3$, $P(OC_6H_5)_3$, $P(C_6H_5)_3$, $As(C_6H_5)_3$	130)
$Mo(CO)_4$	ethylenediamine propylenediamine trimethylenediamine dipyridyl phenanthroline	393,394)
$Mo(CO)_3$	mesitylene	314)
$W(CO)_5$	piperidine	484)
$W(CO)_4$	(piperidine)$_2$	484)
	ethylenediamine propylenediamine trimethylenediamine dipyridyl phenanthroline	393,394)
$Mn(CO)_6^+$	—	58)
$Mn(CO)_5$	CH_3, CF_3, CH_3CO, CF_3CO	314)
	Cl, Br, I, $-O-NO_2$	193)
$Mn(CO)_3$	C_5H_5, $CH_3-C_5H_4$, $CH_3CO-C_5H_4$ $C_6H_5-CO-C_5H_4$ and other substituted cyclopentadienyl groups	314)
$Re(CO)_6^+$	—	58)

Table 2 (continued)

$M(CO)_m$	L		References
$Fe(CO)_4$	maleic anhydride dimethyl maleate dimethyl fumarate	}	289)
	cis-1.2-dibromoethylene trans-1.2-dibromoethylene	}	202)
	1.5-cyclooctadiene		287)
$Fe(CO)_3$	butadiene		314)
	1.5-cyclooctadiene 1.3-cyclooctadiene	}	287)
	duroquinone cyclopentadienone 2.5-diphenyl-cyclopentadienone 3.4-diphenyl-cyclopentadienone tetraphenyl-cyclopentadienone 2.5-dimethyl-thiophene-dioxide tetraphenyl-thiophene-dioxide	}	408)
	tetraphenyl-cyclobutadiene		184)
$[Fe(CO)_4]^{2-}$ $[Fe(CO)_3NO]^-$ $[Fe_2(CO)_8]^{2-}$ $[Fe_3(CO)_{11}]^{2-}$ $[Fe_3(CO)_{11}H]^-$ $[Fe_4(CO)_{13}]^{2-}$ $[Fe_4(CO)_{13}H]^-$		}	231,233)

reader is referred to [387]. Some controversy still exists with respect to the order of the highest filled and lowest unfilled levels in ferrocene. An energy level diagram based on recent results from photoionization is found in [183].

The electronic absorption spectrum of ferrocene has been carefully studied by several groups [17,18,96,419,422,427,420]; Table 3 gives possible assignments of some of the observed bands (after [17]). More information is needed about the nature of the excited states. Phosphorescence resulting from excitation of the 30860 cm^{-1} band, which is not brought about by excitation in the 23000 cm^{-1} region, has been reported [422] and does not appear to be due to impurities [427], as suggested by others [18]. This finding could mean that a phosphorescent state is reached from the $S_0 \to S_2$ transition, but not from $S_0 \to S_1$. Emission from a higher triplet state would definitely be a phenomenon demanding further investigation.

Table 3. *Electronic transitions in ferrocene (after [17])*

Transition energy [cm^{-1}]	Orbital excitation	Type [R = C$_5$H$_5$]
18934	not assigned	[1]
22730	$a_{1g} \rightarrow e_{1g}$	M$\rightarrow$MR$_\pi$
30860		
37000	$e_{2g} \rightarrow e_{1g}$	MR$_\pi \rightarrow$MR$_\pi$
41200		
42700	$e_{1g} \rightarrow e_{1g}$	MR$_\pi \rightarrow$MR$_\pi$
48200		
51200	$e_{1u} \rightarrow e_{1g}$	R$_\pi \rightarrow$MR$_\pi$
53000	not assigned	

[1] Singlet-triplet transition according to [422], see also [96,419].

Solutions of ferrocene in halogen containing solvents show charge-transfer absorption at ~ 32500 cm^{-1} [73,369] of the type

$$M - L + \text{Solvent} \xrightarrow{h\nu} [M - L]^+ + \text{Solvent}^-$$

as discussed earlier in this section. For CCl$_4$ this process results in formation of the ferricenium ion, whose electronic configuration and absorption spectrum have been studied [310,421]:

$$(C_5H_5)_2Fe + CCl_4 \xrightarrow{h\nu} (C_5H_5)_2Fe^+ + Cl^- + \cdot CCl_3$$

Reports on the electronic spectra of substituted ferrocenes are quite frequent [39,140,190,191,224,314,354,358,399,400,529,578], and even the effect of pressure on the spectra of ferrocene and ferricenium ion has been reported [531]. With all this information available the ferrocenes should attract the interest of more photochemists in the future.

Other aromatic transition metal complexes for which electronic spectra and assignments have been reported are: (C$_5$H$_5$)$_2$TiX$_2$ [X = Cl, Br, I] [96], (C$_6$H$_6$)$_2$Cr [59], [(C$_6$H$_6$)$_2$Cr]$^+$I$^-$ [310,421,526], (C$_5$H$_5$)$_2$Ni [17,422], and tetramethyl- and tetraphenyl-cyclopentadienone cobalt cyclopentadienyl [406,408].

Electronic spectra and band assignments for simple olefin transition metal complexes are relatively scarce: absorption spectra (with band assignments) of *Zeise's* salt $(K[Pt(C_2H_4)Cl_3] \cdot H_2O$ have been reported [335] as well as the spectra of other platinum olefin complexes [132]. Details about the spectra of Ni(duroquinone)$_2$ and olefin-Ni-duroquinone are available [409], and the absorption data for the tetraphenylcyclobutadiene complexes of PdCl$_2$ and NiCl$_2$ have been given by *Fritz* [184].

C. Primary Processes in Metal Carbonyl Photochemistry

The intense absorption bands in metal carbonyls result from charge-transfer from an orbital mainly localized at the metal into an orbital mainly localized at the ligand (CTML) as discussed in section B. In a valence bond picture this can be rationalized as:

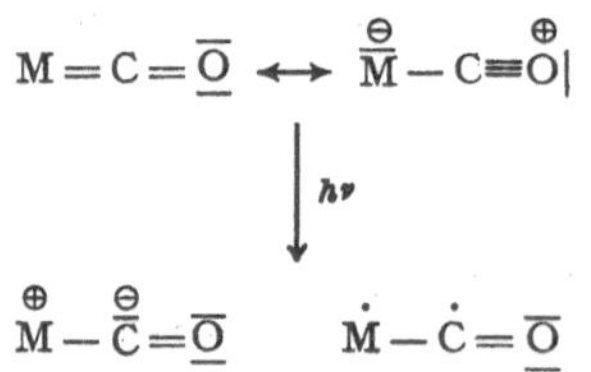

$$M = C = \overline{\underline{O}} \longleftrightarrow \overset{\ominus}{M} - C \equiv \overset{\oplus}{O}|$$

$$\downarrow h\nu$$

$$\overset{\oplus}{M} - \overset{\ominus}{C} = \overline{\underline{O}} \qquad \overset{\cdot}{M} - \overset{\cdot}{C} = \overline{\underline{O}}$$

As a result dissociation can be expected:

$$M(CO)_n \xrightarrow{h\nu} [M(CO)_n]^* \longrightarrow M(CO)_{n-1} + CO$$

This indeed is the case.

Strohmeier and *Gerlach* [490] were the first to show that photolysis of Cr(CO)$_6$ in inert solvents gave yellow Cr(CO)$_5$, which decomposed to Cr(CO)$_6$ and Cr if CO was removed from the system by streaming N$_2$.

Cr(CO)$_6$ in methyl methacrylate polymers showed photochromic behaviour: a yellow colour appeared on irradiation which faded slowly at room temperature but fast at 100 °C. The corresponding species has been regarded as Cr(CO)$_5$ [322].

R. K. Sheline et al. [137,441,442] presented IR spectroscopic evidence for Cr(CO)$_5$, Mo(CO)$_5$ and W(CO)$_5$ by low temperature irradiation of the corresponding carbonyls in isopentane-methylcyclohexane glasses. The IR data obtained are consistent with C_{4v} symmetry for all three (square pyramidal) pentacarbonyls. On warming up carefully the glass with Mo(CO)$_5$ a transformation to a trigonal bipyramidal species with D_{3h} symmetry occurs.

Transients causing a yellow color in solutions of $Cr(CO)_6$, $Mo(CO)_6$, and $W(CO)_6$ in ether-isopentane appear not to be pentacarbonyls [142] but ether complexes of these species [136].

Low temperature photolysis of $Fe(CO)_5$ in a glass produced a new band at 1834 cm^{-1} [possibly corresponding to $Fe_2(CO)_9$] if the spectra were taken from a melted sample [442]. Measurements at low temperature showed bands at 1990 (s), 1980 (w), and 1946 (s) cm^{-1} due to an intermediate iron carbonyl species which has not been identified [441].

Photolysis of $Ni(CO)_4$ in a rare-gas matrix at 15 °K [570] yields $Ni(CO)_3$ with a non-planar structure (C_{3v}, according to IR data).

Keeley and *Johnson* [260] observed an exceedingly small incorporation of ^{14}CO in $Fe_2(CO)_9$ formed in the vapor photolysis of $Fe(CO)_5$. They argued that electronically excited $Fe(CO)_5^*$ [Eq. (1)—(3)] should be the transient leading to diiron enneacarbonyl and not $Fe(CO)_4$ which they considered to be very reactive towards exchange (Eq. (4)—(6)].

$$Fe(CO)_5 \xrightarrow{h\nu} Fe(CO)_5^* \qquad (1)$$

$$Fe(CO)_5^* + Fe(CO)_5 \longrightarrow Fe_2(CO)_9 + CO \qquad (2)$$

$$Fe(CO)_5^* + {}^{14}CO \longrightarrow Fe(CO)_4{}^{14}CO + CO \qquad (3)$$

$$Fe(CO)_5 \xrightarrow{h\nu} Fe(CO)_4 + CO \qquad (4)$$

$$Fe(CO)_4 + Fe(CO)_5 \longrightarrow Fe_2(CO)_9 \qquad (5)$$

$$Fe(CO)_4 + {}^{14}CO \longrightarrow Fe(CO)_4{}^{14}CO \qquad (6^a)$$

$$Fe(CO)_4 + {}^{14}CO \longrightarrow Fe(CO)_3{}^{14}CO + CO \qquad (6^b)$$

This conclusion appears somewhat questionable, since the constant of the transient's reaction with $Fe(CO)_5$ exceeds its deactivation with CO to $Fe(CO)_5$ by at least one order of magnitude [148]. Therefore, Eq. (4)—(6^a) would fit the data as well as Eq. (1)—(3), and there is no obvious reason why (6^b) should be faster than (6^a).

Noack [568] very recently reported the light induced exchange of $Fe(CO)_5$ with gaseous $C^{18}O$ in iso-octane as well as the photochemical scrambling of $C^{18}O$ in mixtures of $Fe(CO)_5$ and $Fe(C^{18}O)_5$, and proposed $Fe(CO)_4$ as the intermediate responsible for these observations.

Flash photolysis of $Fe(CO)_5$ ($5 \cdot 10^{-3}$ mole/l) in benzene produced a transient ($\bar{\nu}_{max}$ 23800 cm^{-1}) with a half life of 0.3 sec., whose decay rate was increased by CO and very effectively by $Ni(CO)_4$ [which did not absorb light] [283]. The very long lifetime of this transient is not compatible with an electronically excited state; even a triplet state should be rather short-lived in this instance due to the internal heavy atom effect. Therefore, this species can be looked upon as $Fe(CO)_4$. The effect of $Ni(CO)_4$ is described by Eq. (7)—(9).

$$Fe(CO)_5^* \longrightarrow Fe(CO)_4 + CO \qquad (7)$$

$$Fe(CO)_4 + Ni(CO)_4 \longrightarrow Fe(CO)_5 + Ni(CO)_3 \qquad (8)$$

$$Ni(CO)_3 + CO \longrightarrow Ni(CO)_4 \qquad (9)$$

No net photochemical reaction of $Fe(CO)_5$ is observed in the presence of $Ni(CO)_4$ [289].

The following mechanism has been established for the flash photolysis of $Ni(CO)_4$ in the gas phase:

$$Ni(CO)_4 \xrightarrow{h\nu} Ni(CO)_3 + CO$$

$$Ni(CO)_3 + CO \longrightarrow Ni(CO)_4$$

$$Ni(CO)_3 \xrightarrow{h\nu} Ni(CO)_2 + CO$$

$$Ni(CO)_2 + CO \longrightarrow Ni(CO)_3$$

$$Ni(CO)_2 \longrightarrow \text{solid products}$$

The mean lifetime of $Ni(CO)_4^*$ was determined as $7 \cdot 10^{-9}$ sec. [88]. In the presence of O_2 gas-phase flash photolysis of $Ni(CO)_4$ or $Fe(CO)_5$ follows the stoichiometric Eq. (10) and (11) [87], see also [324]).

$$Ni(CO)_4 + O_2 \xrightarrow{h\nu} Ni(CO)O + CO_2 + 2\,CO \qquad (10)$$

$$2\,Fe(CO)_5 + 3\,O_2 \xrightarrow{h\nu} Fe_2O_2(CO) + 4\,CO_2 + 5\,CO \qquad (11)$$

Gas-phase flash photolysis of $Fe(CO)_5$ can yield excited iron atoms and ions [86].

The formation of addition compounds from photodissociated metal carbonyls and acetylenes has been postulated from flash spectroscopic studies of acetylene/O_2 explosions in the presence of metal carbonyls [144].

Cyclobutadiene has been observed directly by kinetic mass spectrometry in the flash photolysis of cyclobutadiene iron tricarbonyl [513].

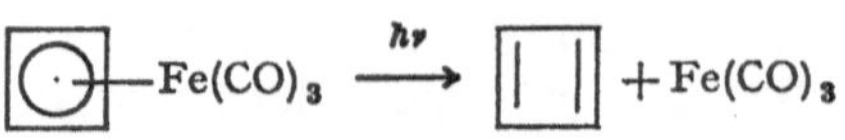

D. Reactions of $M(CO)_{n-1}$

From the discussion in section C. it became evident that the primary process in the photochemistry of simple carbonyls is the formation of a coordinatively unsaturated species $M(CO)_{n-1}$.

$$M(CO)_n \xrightarrow{h\nu} [M(CO)_n]^* \to M(CO)_{n-1} + CO$$

Stabilization of this intermediate can be brought about

a) by recombination with CO:

$$M(CO)_{n-1} + CO \to M(CO)_n \qquad (12)$$

This has been demonstrated with ^{14}CO [483].

b) by reaction with a metal carbonyl:

$$M(CO)_{n-1} + m\ M'_x(CO)_y \to M(CO)_{n-1}[M'_x(CO)_y]_m \qquad (13)$$

c) by addition of a ligand L having n- or π-donor properties:

$$M(CO)_{n-1} + L \to LM(CO)_{n-1} \qquad (14)$$

d) by reaction with a molecule $X-Y$ according to Eq. (15):

$$M(CO)_{n-1} + X-Y \to (CO)_{n-1}M \underset{Y}{\overset{X}{\diagdown}} \qquad (15)$$

If any of the products formed absorbs light of the wavelength used to photolyze $M(CO)_n$, it is obvious that further photochemical reactions can occur, e.g. Eq (16) and (17):

$$LM(CO)_{n-1} \xrightarrow{h\nu} LM(CO)_{n-2} + CO \qquad (16)$$

$$LM(CO)_{n-2} + L' \longrightarrow LL'M(CO)_{n-2} \qquad (17)$$

Examples of reactions Eq. (12)—(17) are discussed in section E.

E. Photochemical Substitution of Metal Carbonyl Compounds

1. Photochemical Reactions of Simple Metal Carbonyls with each Other

The first observed photochemical reaction with a metal carbonyl was the formation of diiron enneacarbonyl [$Fe_2(CO)_9$] from $Fe(CO)_5$ in sunlight [134,135].

$$2\ Fe(CO)_5 \xrightarrow{\ h\nu\ } Fe_2(CO)_9 + CO \qquad (18)$$

For the mechanism of Eq. (18) see section C [283,289].

$Fe_2(CO)_9$ can only be produced photochemically [74,429] and radiation-chemically (see section L.1). Thermal activation of this process would require temperatures above the thermal stability of $Fe_2(CO)_9$, which yields triiron dodecacarbonyl (Eq. (19)) above 40 °C; the latter in turn is sensitive, too.

$$3\ Fe_2(CO)_9 \rightarrow 3\ Fe(CO)_5 + [Fe(CO)_4]_3 \qquad (19)$$

Illumination of $Ru(CO)_5$ and $Os(CO)_5$ produces [$Ru(CO)_4]_3$ and [$Os(CO)_4]_3$ [85], and not the corresponding enneacarbonyls [316]. The latter are expected to be unstable [112]; low temperature studies should reveal their possible intermediacy.

Recently, a variety of polynuclear metal carbonyls (or anions) containing different metals have been synthesized by irradiation of $Fe(CO)_5$ in the presence of other metal carbonyls (or anions):

$[(CO)_5Mn]_2Fe(CO)_4$ [6,415];

$[(CO)_5Re]_2Fe(CO)_4$ [147];

$[ReFe_2(CO)_{12}]^-$ [147];

$[(CO)_5Mn]\,[Re(CO)_5]\,Fe(CO)_4$ [146];

$[TcFe_2(CO)_{12}]^-$ [313];

$[FeMn(CO)_9]^-$ [391];

$[FeCo(CO)_8]^-$ [391].

The light sensitivity of $V_2(CO)_{12}$ has been reported [380].

2. Substitution with n-Donors

As discussed in section C. illumination of simple metal carbonyls usually results in dissociation of the excited molecules. At this point one should bear in mind, that with more complex metal carbonyl compounds this is not necessarily so, due to possible transitions having no major effect on the carbonyl groups but facilitating nucleophilic attack at the metal. This point does not appear to be sufficiently emphasized (however, see [514]).

For simple metal carbonyls, a S_N1 type of substitution mechanism can be anticipated [Eq. (20) and (21)]:

$$M(CO)_n \xrightarrow{h\nu} [M(CO)_n]^* \rightarrow M(CO)_{n-1} + CO \qquad (20)$$

$$M(CO)_{n-1} + L \longrightarrow M(CO)_{n-1}L \qquad (21)$$

With more complex carbonyl compounds, especially those containing extended π-systems the alternative S_N2 type of mechanism [Eq. (22) and (23)] has to be considered in addition:

$$LM(CO)_{n-1} \xrightarrow{h\nu} [LM(CO)_{n-1}]^* \qquad (22)$$

$$[LM(CO)_{n-1}]^* + L' \longrightarrow LL'M(CO)_{n-2} + CO \qquad (23)$$

Detailed studies of the concentration dependence of reaction rates are necessary to distinguish between the two mechanisms.

Very frequently the argument has been put forward that the major difference between photochemical and thermal metal carbonyl substitution reactions would be the S_N1 character of the former and S_N2 of the latter. However, it should be noted that S_N1 mechanisms are very common in thermal carbonyl chemistry [11,46,426] and the preference for S_N1 or S_N2 often depends on the nature of the ligand alone [11]. The major difference actually appears to be the very small activation energy observed for the reaction of electronically excited molecules, which allows the preparation of sensitive compounds which could never survive the high temperatures necessary to provide the larger activation energy of the corresponding thermal process. Despite this fact the selective photochemical excitation of one specific state appears to be another advantage. Differences [101] between photochemical and thermal substitution reactions can be due as much to different mechanisms as to different thermal stabilities of the products formed [7].

The photochemical reactions of $Cr(CO)_6$, $Mo(CO)_6$, and $W(CO)_6$ with amines according to Eq. (20) and (21) have been studied extensively by *Strohmeier* et al. [466,490]. The quantum yield of the dissociative process [Eq. (20)] appears to be unity [489]. The determination is somewhat hampered by the internal light filter action (for spectra see [484]) of the products formed [Eq. (21) and (23)] [279,468,487]. The ratio of the rate constants for Eq. (21) and the back reaction Eq. (24) determines the initial rate of CO evolution, depending on the donor properties of L and the electronic situation at the central metal [482].

$$M(CO)_{n-1} + CO \rightarrow M(CO)_n \tag{24}$$

The quantum yield of Eq. (20) is wavelength independent for C_5H_5 $Mn(CO)_3$ in acetone at 366 and 436 nm [478], an argument against the involvement of a vibrationally excited ground-state molecule [4].

The wavelength-dependent quantum yield of reaction Eq. (23) has been explained by a wavelength-dependent probability of Eq. (25) versus Eq. (26) [478], possibly due to a hot ground-state molecule [4].

$$M(CO)_{n-1}L \xrightarrow{h\nu} M(CO)_{n-1} + L \xrightarrow{L'} M(CO)_{n-1}L' + L \tag{25}$$

$$M(CO)_{n-1}L \xrightarrow{h\nu} M(CO)_{n-2}L + CO \xrightarrow{L'} M(CO)_{n-2}LL' + CO \tag{26}$$

The photochemical substitution of metal carbonyls has found wide application in the preparation of a fascinating variety of transition metal n-donor complexes. In addition to the complexes listed in Table 4 a few examples are shown in more detail to illustrate the preparative possibilities.

$$M(CO)_n + P[M'(CH_3)_3]_3 \xrightarrow[-CO]{h\nu} (CO)_{n-1}MP[M'(CH_3)_3]_3 \tag{27}$$

M=Cr, Fe; M'=Si, Ge, Sn [417, 418] n=6; n=5

$$M(CO)_6 + X[M'(CH_3)_3]_2 \xrightarrow[-CO]{h\nu} (CO)_5MX[M'(CH_3)_3]_2 \tag{28}$$

M=Cr, Mo, W; M'=Si, Ge, Sn, Pb; X=S, Se [416]

Eq. (27) and (28) provide an interesting basis to study the electronic effects of different metals on the donor properties of P, S and Se in carbonyl complexes. Anyhow, one of the specific features of work done in this field is, that a major interest has been directed towards an understanding of the different factors determining the stability of a metal ligand bond (e.g. [444]).

Table 4. *Photochemical substitution of metal carbonyl compounds with n-donors*

Ammonia:

$Mo(CO)_6$ [450]; $Cr(CO)_6$ [450]; $W(CO)_6$ [450]; $CpMn(CO)_3$ [460]

Aliphatic Amines, monodentate:

$Cr(CO)_6$ [450]; $ArCr(CO)_3$ [451,476]; $Mo(CO)_6$ [450,491]; $W(CO)_6$ [13,245,450,457,491]; $(R)CpMn(CO)_3$ [164,460,464,474]; $Fe(CO)_5$ [414]

Aliphatic Amines, bidentate:

$Cr(CO)_6$ [14]; $(R)CpMn(CO)_3$ [477]

Aromatic Amines, monodentate:

$Cr(CO)_6$ [457,493,497]; $ArCr(CO)_3$ [457,458,471,476]; $Mo(CO)_6$ [492]; $W(CO)_6$ [13,496]; $Mn_2(CO)_{10}$ [534]; $(R)CpMn(CO)_3$ [457,460,464,494]; $Fe(CO)_5$ [414]

Aromatic Amines, bidentate:

$Cr(CO)_6$ [14]; $Mo(CO)_6$ [14,479]; $W(CO)_6$ [189]

Amides:

$Mo(CO)_6$ [440]; $W(CO)_6$ [440]; $CpMn(CO)_3$ [164], product: amine complex

Nitriles:

$Cr(CO)_6$ [495]; $ArCr(CO)_3$ [471]; $Mo(CO)_6$ [488]; $W(CO)_6$ [321,440,488]; $Mn_2(CO)_{10}$ [534]; $CpMn(CO)_3$ [465]; $Fe(CO)_5$ [414]

Isonitriles:

$CpMn(CO)_3$ [164]; $Fe(CO)_5$ [444]

Dicyan:

$Cr(CO)_6$ [206]; $Mo(CO)_6$ [206]; $W(CO)_6$ [206]

Dialkylcyanamides:

$Cr(CO)_6$ [66]; $Mo(CO)_6$ [66]

CN^-:

$CpV(CO)_4$ [156]; $CpMn(CO)_3$ [157]

PH_3:

$CpV(CO)_4$ [153]; $Cr(CO)_6$ [155,574]; $Cr(CO)_5C(CH_3)OCH_3$ [155];
$Mo(CO)_6$, $W(CO)_6$, $Cr[P(C_6H_5)_3](CO)_5$ [155]

Phosphines, monodentate:

$CpV(CO)_4$ [209,443,511]; $CpNb(CO)_4$ [349], mono- and disubst.; $Cr(CO)_6$ [155,574];

$ArCr(CO)_3$ [458,476]; $CpMo(CO)_3R$ [37,38]; $\boxed{\bigcirc\!\!-\!Mo(CO)_4}$ [9]; $[CpMo(CO)_3]_2$ [38,210];
$[(CO)_4MoPMe_2]_2$ [315,502]; $CpMo(CO)_3X$ [211]; $CpW(CO)_3R$ [37,38]; $Mn_2(CO)_{10}$ [364,534]
$(R)CpMn(CO)_3$ [36,312,361,413,443,481]; $Re_2(CO)_{10}$ [252,362]; $Fe(CO)_5$ [25,311,312,426,560];
$CpFe(CO)_2R$ [37,346,348,509]; Dien-$Fe(CO)_3$ [95,192]; $CpFe(CO)CF_3CNH(NCCF_3)$ [262];
$CpFe(CO)_2J$ [346,378]; $[CpFe(CO)_2]_2$ [378]; $[Fe(CO)_3PMe_2]_2$ [502];
$Me_3Sn(Si)Fe(CO)_2Cp$ [264]; $Ru(CO)_5$ [143]; $Os(CO)_5$ [143]; $Co_2(CO)_6(CF_3C_2H)$ [213];
$CF_3COOFe(CO)_2Cp$ [555]; π-pyrrolyl-$Mn(CO)_3$ [557]; $C_5H_5Cr(CO)_2NO$ [576];
$C_5H_5Mo(CO)_2NO$ [576]

Table 4 (continued)

Phosphines, bidentate:

(R)CpMn(CO)$_3$ [218,312,361]; Mo(CO)$_4$C$_6$H$_4$(PEt$_2$)$_2$ [94]; [Mo(CO)$_4$PMe$_2$]$_2$ [315,502];
CpMo(CO)$_3$X [211]; [Fe(CO)$_3$PMe$_2$]$_2$ [502]; [Bu$_2^n$SnFe(CO)$_4$]$_2$ [242,243];
Me$_3$Sn(Si)Fe(CO)$_2$Cp [264]; CpMo(CO)$_3$Cl [554]; CpFe(CO)$_2$X (X = Cl,Br) [554];
CF$_3$COOFe(CO)$_2$Cp [555]; CF$_3$COOMn(CO)$_5$ [555]; CF$_3$COOFe(CO)$_3$(C$_3$H$_5$) [555];
Fe(CO)$_5$ [311,312]

Phosphites:

CpV(CO)$_4$ [443]; Cr(CO)$_6$ [515,516]; W(CO)$_6$ [435,515,516]; [CpMo(CO)$_3$]$_2$ [210,263];
Me$_3$SnMoCp(CO)$_3$ [263]; (R)CpMn(CO)$_3$ [413,443]; Fe(CO)$_5$ [435,560];
CpFe(CO)$_2$R [345,348]; CpFe(CO)$_2$J [345,350]; CF$_3$C(NH)Fe(CO) (NCCF$_3$)Cp [262];
; CF$_3$COOFe(CO)$_2$Cp [555]; CF$_3$COOMn(CO)$_5$ [555]; CH$_3$COMn(CO)$_3$[P(OCH$_3$)$_3$]$_2$ [575]

PF$_3$ (and other PX$_3$) (Review: [305]*):*

ArCr(CO)$_3$ [104]; Mo(CO)$_6$ [103]; Mn$_2$(CO)$_{10}$ [101]; HMn(CO)$_5$ [327]; (R)CpMn(CO)$_3$ [104];
HC$_2$F$_4$Mn(CO)$_5$ [327]; Fe(CO)$_5$ [105]; Co(NO) (CO)$_3$ [102]

Arsines, Stibines, Bismutines; monodentate:

CpNb(CO)$_4$ [349]; CpMo(CO)$_3$X [211]; (R)CpMn(CO)$_3$ [36,152,312,361,413,464];
Mn$_2$(CO)$_{10}$ [364]; Fe(CO)$_5$ [311,312,560]; Fe(CO)$_4$Br$_2$ [108]; π-pyrrolyl-Mn(CO)$_3$ [557]

Arsines, Stibines (bidentate):

CpMo(CO)$_3$X [211]; (R)CpMn(CO)$_3$ [218,312,361]

Arsenites, Antimonites:

(R)CpMn(CO)$_3$ [413]

Alcohols:

Mo(CO)$_6$ [440]; W(CO)$_6$ [440]

Ketones:

W(CO)$_6$ [440]; CpMn(CO)$_3$ [475]

Ethers:

Cr(CO)$_6$ [457,559,444] [1]; ArCr(CO)$_3$ [205,226,458,559] [1]; Mo(CO)$_6$ [440,559,444] [1];
W(CO)$_6$ [440,559,444] [1]; CpMn(CO)$_3$ [450,36,164,226,460,475,225] [1]

Thioethers:

Cr(CO)$_6$ [455]; ArCr(CO)$_3$ [451]; Mo(CO)$_6$ [455]; W(CO)$_6$ [455]; (R)CpMn(CO)$_3$ [455,459];
Fe(CO)$_5$ [455]

Sulfoxides:

Cr(CO)$_6$ [455]; ArCr(CO)$_3$ [458,476,451]; Mo(CO)$_6$ [455]; W(CO)$_6$ [455];
(R)CpMn(CO)$_3$ [455,459,472]; Fe(CO)$_5$ [455]

SO$_2$:

Cr(CO)$_6$ [455]; ArCr(CO)$_3$ [451]; W(CO)$_6$ [455]; (R)CpMn(CO)$_3$ [455,459];
Dien-Fe(CO)$_3$ [75]

---

[1] Complex with THF utilized for synthesis of other complexes via ligand exchange.

Table 4 (continued)

Dialkylsulfites:

$Cr(CO)_6$ [455]; $ArCr(CO)_3$ [451]; $W(CO)_6$ [455]; $(R)CpMn(CO)_3$ [455,459]

CS_2:

$Fe(CO)_5$ [25]

^{14}CO, ^{13}CO, $C^{18}O$:

$ArCr(CO)_3$ [470]; $ArMo(CO)_3$ [470]; $CpMn(CO)_3$ [470]; $Co(CO)_3NO$ [69]; $Fe(CO)_5$ [568]; $(C_6H_5)_3PFe(CO)_4$ [568]

Other Sulfur Compounds:

$[CpMnCONO]_2 + CF_3—CS—CS—CF_3$ [266]; $[CpFe(CO)_2]_2 + CF_3—CS—CS—CF_3$ [266]; $[CpFe(CO)_2]_2 + CH_3—SS—CH_3$ [271]

The photochemical formation of *(1)* from $Fe_3(CO)_{12}$ and a bidentate arsine provides another interesting preparative example [123].

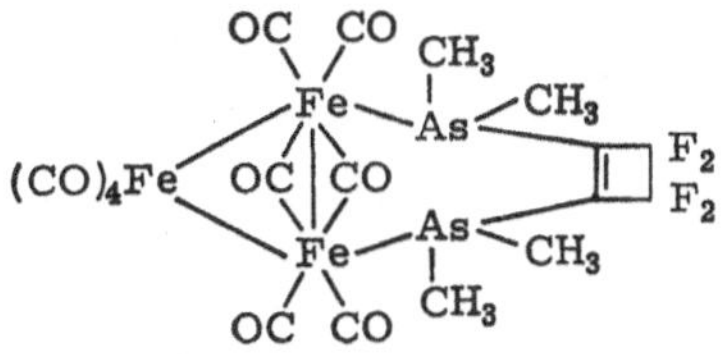

(1)

A valuable method developed by *Strohmeier* [466] for the preparation of photosensitive complexes or complexes of strongly absorbing ligands is that of ligand exchange with photochemically formed labile tetrahydrofurane (THF) complexes (e.g. [133,204]) according to Eq. (29) and (30).

$$M(CO)_6 + THF \xrightarrow{h\nu} M(CO)_5THF + CO \qquad (29)$$

$$M(CO)_5THF + L \longrightarrow M(CO)_5L + THF \qquad (30)$$

Another method comprises ligand transfer from other complexes with or without detachment of the primarily coordinated metal:

$$M(CO)_6 + n\, NiS_4C_4R_4 \xrightarrow{h\nu} M(CO)_{6-2n}(S_2C_2R_2)_n + 2n\, CO + n/2\, (Ni_2S_4C_4R_4)_x\ [405]$$

$M = Mo, W; \quad n = 1,2$

$R = $ alkyl, aryl

$$\pi\text{-}C_5H_5Ti[N(CH_3)_2]_3 + M(CO)_6 \xrightarrow{h\nu} \pi\text{-}C_5H_5Ti[N(CH_3)_2]_3M(CO)_3 + 3\, CO$$

The $N(CH_3)_2$ groups are bridging between Ti and M (Cr, Mo, W); the detailed structure is as yet unknown [72].

Substitution of CO in a cationic complex by the corresponding anion has been reported by *King* [554]:

$$[CpMo(CO)_2(diphos)]^+Cl^- \xrightarrow{h\nu} CpMo(CO)(diphos)\ Cl + CO$$

3. Substitution with π-Donors

One of the first π-donors applied in intermolecular photochemical substitution reactions is acrylonitrile [261], which can function as a π-donor (C=C) and as a *n*-donor ($-C\equiv NI$) [205] as demonstrated with complexes (2)—(4).

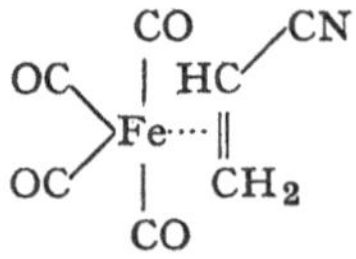

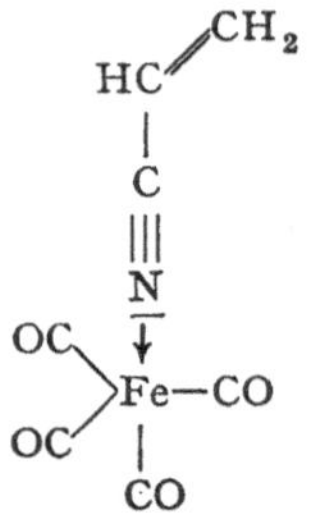

King reported recently the formation of the interesting complex

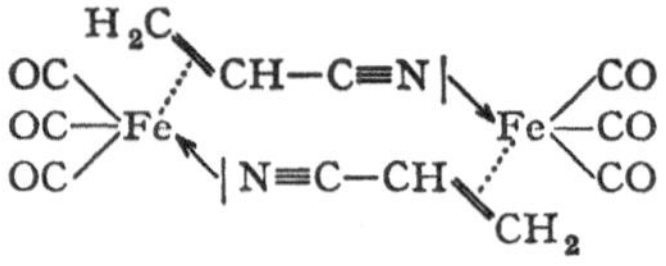

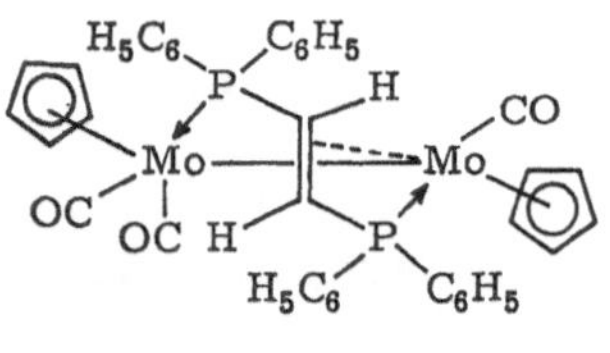

from the irradiation of $[(C_5H_5)Mo(CO)_3]_2$ or $RMo(CO)_3(C_5H_5)$ with trans-$(C_6H_5)_2P-CH=CH-P(C_6H_5)_2$, which functions as a *n*- and π-donor [556].

During the last eight years an impressive list (Table 5) of carbonyl complexes with π-donors has been prepared, using monoolefins, dienes, trienes, acetylenes, and aromatic compounds. Excellent surveys have been given by *Fischer* and *Herberhold* [164,228]. Bonding to the metal can be rationalized as an interaction of a donor π-orbital with an empty hybrid orbital of the $M(CO)_{n-1}$ species combined with backbonding from the highest occupied metal d-orbital into the lowest antibonding π^*-orbital of the ligand [118]. The better the charge built up at the metal through the donor bond is compensated by backdonation into acceptor orbitals of the ligand, the more stable are the complexes of formally zero valent metals [288].

Philodiene iron tetracarbonyls fulfil this postulate very nicely, back-donation works well into the low lying philodiene π^*-orbitals; e.g. maleic anhydride iron tetracarbonyl is stable up to 130 °C and can be dissolved in concentrated H_2SO_4 without decomposition [396].

The low lying π^*-orbitals of conjugated dienes explain the stability of many diene metal carbonyl complexes e.g. butadiene iron tricar-bonyl [382]. Photochemical preparation of the latter [286] is superior to the thermal procedure [272]. Mercury can be used as a sensitizer in the photoreaction of metal carbonyls with dienes [169,170,176,241].

Not all dienes necessarily have to be π-bonded; e.g. perfluorobuta-diene forms stable σ-bonds yielding

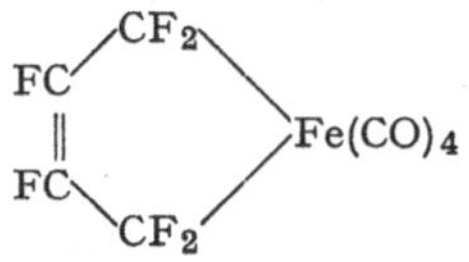

$$\begin{array}{c} FC \diagup CF_2 \\ \| \qquad\qquad \diagup Fe(CO)_4 \qquad [241] \\ FC \diagdown CF_2 \end{array}$$

Cyclooctatetraene (COT) iron carbonyl complexes and ruthenium carbonyl complexes (Table 5) have raised a good deal of interest as flux-ional molecules which display temperature dependent NMR spectra; e.g. $COT\text{-}Fe(CO)_3$ shows only one sharp peak at room temperature, but a pattern conceivable with

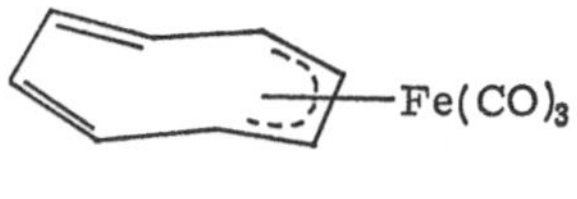

at very low temperatures. This has been attributed to rapid 1.2-shifts [117].

Ligand systems containing several functional groups usually react step-wise with metal carbonyls with successive replacement of CO. An example is the photochemical reaction of $Fe(CO)_5$ with 1.5-cycloocta-diene [284,287].

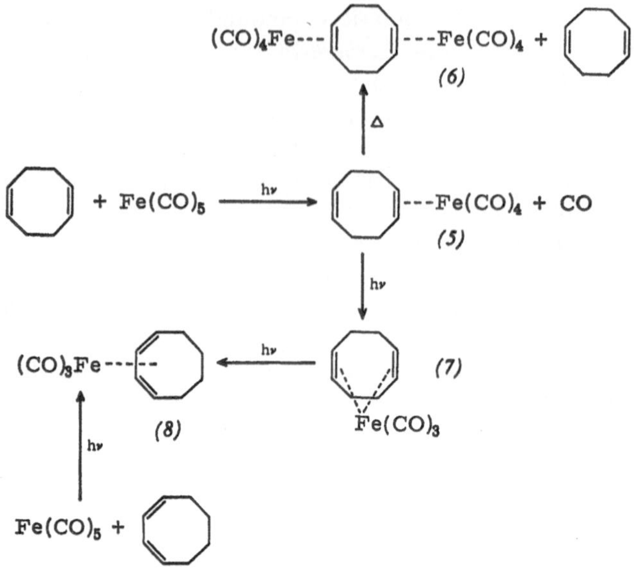

The primary process is the formation of *(5)*, which can be isolated but very easily decomposes to *(6)* and 1.5-cyclooctadiene. Absorption of another light quantum converts *(5)* → *(7)*, in a third photochemical step *(7)* can be isomerized to 1.3-cyclooctadiene iron tricarbonyl *(8)*, which can be prepared directly from 1.3-cyclooctadiene, too. Reaction *(7)* → *(8)* comprises an example of the photochemical rearrangement of a photochemically formed metal carbonyl complex. Such reactions occur quite frequently and can be of preparative value (see section F. 2,3). In *(6)* the diene is acting as a bridging ligand, other cases of this type of coordination are found in [164,228].

Reaction *(5)* → *(7)* is an example of intromolecular substitution reactions [Eq. (31)], as discussed in section E 4.

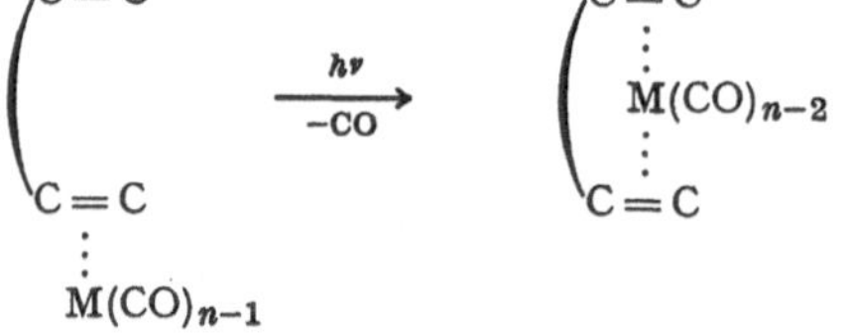

$$(31)$$

The "bis-ylid" $[(C_6H_5)_3P]_2C$ represents a very interesting type of ligand, illumination of $W(CO)_6$ in its presence yields

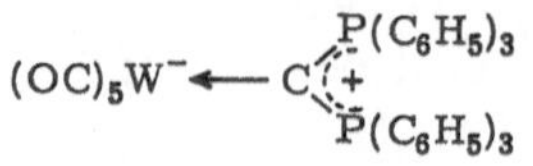

[567)]

Table 5. *Photochemical substitution of metal carbonyl compounds by π-donors*

1. Monoolefins: $M(CO)_{n-1}L$, $M(CO)_{n-2}L_2$

Olefin L	$M(CO)_n$ (References)
Ethylene	$ArCr(CO)_3$ [174,169,456]; $Mo(CO)_6$ [439] [1,2]; $ArMo(CO)_3$ [174]; $W(CO)_6$ [439] [1,2]; $CpMn(CO)_3$ [15,169,164,302]; $Me_3SnMn(CO)_5$ [98]
Propene	$Mo(CO)_6$ [439] [1,2]; $W(CO)_6$ [439] [1]; $CpMn(CO)_3$ [15]; $Fe(CO)_5$ [291]
cis-2-Butene	$W(CO)_6$ [439] [1,2]
trans-2-Butene	$W(CO)_6$ [439] [2]
1-Pentene	$CpMn(CO)_3$ [15] [2]
Styrene	$Fe(CO)_5$ [291]
Vinyl chloride	$Fe(CO)_5$ [291]
Vinyl acetate	$Fe(CO)_5$ [298]
Vinyl ethyl ether	$Fe(CO)_5$ [291]
Acrylonitrile	$Fe(CO)_5$ [261]; $CpMn(CO)_3$ [533];
Methyl methacrylate	$Fe(CO)_5$ [298]
Maleic acid, Fumaric acid	$ArCr(CO)_3$ [12]
Dimethyl maleate, Dimethyl fumarate	$Fe(CO)_5$ [396]
Tetracyanoethylene	$Mo(CO)_6$; $Cr(CO)_6$; $W(CO)_6$; $CpMn(CO)_3$ [227]
Tetrahaloethylenes (C_2ClF_3, $C_2Cl_2F_2$, C_2F_4)	$Fe(CO)_5$ [150]
Hexafluoropropene	$Fe(CO)_5$ [150]
1,2-Dihaloethylenes	$Fe(CO)_5$ [288,285,202]
Cyclopentene	$ArCr(CO)_3$ [456]; $CpMn(CO)_3$ [15,164]
Cycloheptene	$ArCr(CO)_3$ [456]; $CpMn(CO)_3$ [15,164]
cis-Cyclooctene	$CpMn(CO)_3$ [15,164]
cis-Cyclononene	$CpMn(CO)_3$ [164]
Norbornene	$CpMn(CO)_3$ [15]
Maleic anhydride	$ArCr(CO)_3$ [12]; $CpMn(CO)_3$ [225]; $Fe(CO)_5$ [300,396]
Vinylene carbonate, 2.3-Dihydrofuran, 1-Cyclopenten-3-one	$CpMn(CO)_3$ [225]
Endo-cis-bicyclo[2.2.1] -5-heptene-2.3- dicarboxylic anhydride	$ArCr(CO)_3$ [12]; $CpMn(CO)_3$ [15,225]
Citraconic anhydride	$ArCr(CO)_3$ [12]
Cyclobutene-3.4-di- carboxylic anhydride	$Fe(CO)_5$ [202,285]

2. Diolefins: $M(CO)_{n-2}L$, $M(CO)_{n-1}L$[4]), $(CO)_{n-1}M-L-M(CO)_{n-1}$[5]), $M(CO)_{n-4}L_2$

Diolefin L	$M(CO)_n$ (References)
1.3-Butadiene	$CpV(CO)_4$ [176,170]; $[CpCr(CO)_3]_2$ [176] [3]); $Mo(CO)_6$ [176] [1]), [170] [1]), [439] [1,2,4]; $ArMo(CO)_3$ [176] [6]); $W(CO)_6$ [439] [1,4]; $Mn_2(CO)_{10}$ [534];

Table 5 (continued)

Diolefin L	$M(CO)_n$ (References)
1.3 Butadiene	$CpMn(CO)_3$ [176,170,533] [4,5]; $Fe(CO)_5$ [286]; $Co_2(CO)_8$ [175] [7]
Isoprene	$Fe(CO)_5$ [278]
2.3-Dimethyl-1.3-butadiene	$CpV(CO)_4$ [176,170]
Tetraphenylbutadiene	$Fe(CO)_5$ [411]
Cyclobutadiene	$Fe(CO)_5$ [386] [10]
Pentamethylcyclopen-tadiene	$Co_2(CO)_8$ [265] [8]
1.3-Cyclohexadiene	$CpV(CO)_4$ [176]; $CpMn(CO)_3$ [173] [5]; $Fe(CO)_5$ [63] [9]
Norbornadiene	$CpMn(CO)_3$ [164] [4,5]; $Fe(CO)_5$ [371]
1.5-Cyclooctadiene	$CpMn(CO)_3$ [164] [5]; $Fe(CO)_5$ [287,341]
1.3-Cyclooctadiene	$Fe(CO)_5$ [287]
Dicyclopentadiene	$CpMn(CO)_3$ [164] [4]
Methoxy-1.3-cyclohexa-dienes and derivatives	$Fe(CO)_5$ [63] [9]
Pentafluorocyclopen-tadiene	$Fe(CO)_5$ [35] [5]; $CpCo(CO)_2$ [35] [4]
Cyclopentadienone $(C_6H_5)_4$	$Fe(CO)_5$ [411]
α-Pyrone	$Fe(CO)_5$ [386]

3. Cyclic Trienes and Tetraenes

Olefin L	Starting Material	Product (References)
N-Carbethoxyazepine resp. 2.4.6-Trimethyl-	} $Fe(CO)_5$	$Fe(CO)_3L$ [162]
2.7-Dimethyloxepine	$Fe(CO)_5$	$Fe(CO)_3L$ [160]; $Fe_2(CO)_6L$ [160]
1-Benzoxepine	$Fe(CO)_5$	$Fe(CO)_3L$ [160]
1.3.6-Cyclooctatriene	$Fe(CO)_5$	$Fe(CO)_3L$ [339]
Cyclooctatraene	$Fe(CO)_5$	$Fe(CO)_3L$ [342,381] $Fe_2(CO)_6L$ [5] [342,381]
	$CpCo(CO)_2$	$CpCoL$ [338]
	$CpRh(CO)_2$	$CpRhL, (CpRh)_2L$ [5] [77]
	$Os_3(CO)_{12}$	$σ,π-C_8H_8Os(CO)_3$ [83]
	$Ru_3(CO)_{12}$	$Ru(CO)_3L, Ru_2(CO)_5L$ [82]
Phenylcyclooctatetraene	$Fe(CO)_5$	$Fe(CO)_3L$ [340,341] $Fe_2(CO)_6L$ [340]
	$Fe_2(CO)_9$ or $Fe_3(CO)_{12}$	$Fe(CO)_3L$ [340]
Cycloheptatriene	$Co_2(CO)_8$	$C_7H_7Co(CO)_3$ [275]
$(C_7H_7)(C_5H_5)Mo(CO)_2$	$Fe(CO)_5$ or $Fe_2(CO)_9$	$(C_5H_5)(CO)_2Mo(C_7H_7)Fe(CO)_3$ [116]

Table 5 (continued)

4. Acetylenes: $M(CO)_{n-1}L$, $M(CO)_{n-2}L$ [11], $M(CO)_{n-4}L$ [1,11],
$(CO)_{n-1}M-L-M(CO)_{n-1}$ [5]

Acetylene L	$M(CO)_n$ (References)
Acetylene	$CpV(CO)_4$ [209] [11], [511] [11]; $Mo(CO)_6$ [439] [2]
	$W(CO)_6$ [439] [1,2]
Propyl- and t-Butyl-acetylene	$CpV(CO)_4$ [209] [11], [511] [11]
Propyne, Diethylacetylene	$Mo(CO)_6$ [439] [2]; $W(CO)_6$ [439] [2]
Phenylacetylene	$ArCr(CO)_3$ [456]
Tolan	$CpV(CO)_3(PPh_3)$ [209] [11], [511] [11]; $CpNb(CO)_4$ [352] [11], [344] [1,11]; $ArCr(CO)_3$ [456]; $Mo(CO)_6$ [461] [12]; $W(CO)_6$ [461] [12]; $(R)CpMn(CO)_3$ [480,485]
Acetylenedicarboxylic acid	$Co_2(CO)_6L$ [373] [5,13]
Dialkyl Acetylenedicar-boxylates	$ArCr(CO)_3$ [12,456]; $Co_2(CO)_6L$ [372] [5,13]
Hexafluoro-2-butyne	$CpMn(CO)_3$ [70]

5. Aromatic Compounds

Aromatic Ligand L	Starting Material	Product (References)
Benzene	$CpMn(CO)_3$	$CpMnL$ [164]
Indene, Acenaphthene	$CpMn(CO)_3$	$CpMn(CO)_2L$ [225]
Benzene, substituted Benzenes, polycyclic aromatic compounds	$W(CO)_6$	$W(CO)_5L$ [438]

[1] Monosubstitution and some disubstitution.
[2] Complex has not been isolated; identification from spectra or by ligand exchange
[3] Mononuclear product: $C_5H_6CrC_4H_6(CO)_2$; C_5H_5 converted to C_5H_6.
[4] Ligand is monodentate.
[5] Bridging ligand in binuclear complexes.
[6] Substitution of Ar and CO: $Mo(CO)_2(C_4H_6)_2$.
[7] Disubstitution: $[C_4H_6Co(CO)_2]_2$.
[8] Formation of $Me_5C_5Co(CO)_2$.
[9] Isomerization of 1,4-cyclohexadienes to 1,3-cyclohexadienes in the course of reaction.
[10] By decarboxylation of photo-pyrone in the course of reaction.
[11] Acetylene as bidentate ligand.
[12] Composition: $M(CO)(tolan)_3$, $M = Mo$, W.
[13] $HgCo_2(CO)_8$ as starting material.

4. Intromolecular Substitution with n- and π-Donors

Intramolecular substitution is easily brought about with π- or n-donor systems attached to σ-bonded ligands in carbonyl complexes.

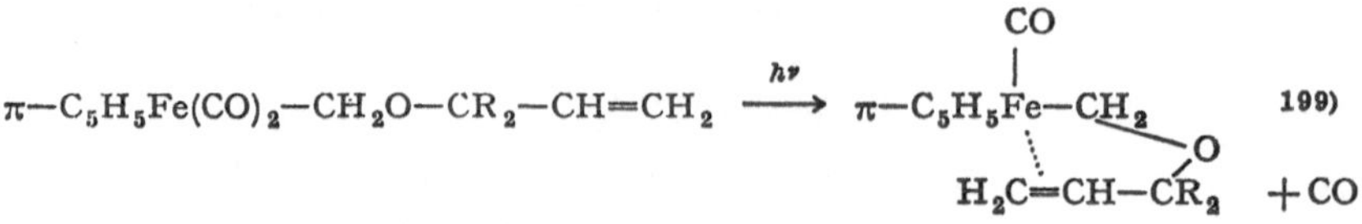

$$\pi-C_5H_5Fe(CO)_2-CH_2O-CR_2-CH=CH_2 \xrightarrow{h\nu} \underset{\substack{\displaystyle | \\ \displaystyle H_2C=CH-CR_2}}{\overset{\displaystyle CO \atop \displaystyle |}{\pi-C_5H_5Fe-CH_2}} \quad +CO \qquad 199)$$

$$\pi\text{-}C_5H_5Fe(CO)_2-S-\overset{\displaystyle SR \atop \displaystyle |}{C}=S \xrightarrow{h\nu} \underset{\substack{\displaystyle | \ \ \displaystyle \| \\ \displaystyle S-C-SR}}{\overset{\displaystyle CO \atop \displaystyle |}{\pi\text{-}C_5H_5Fe\leftarrow S}}+CO \quad 84)$$

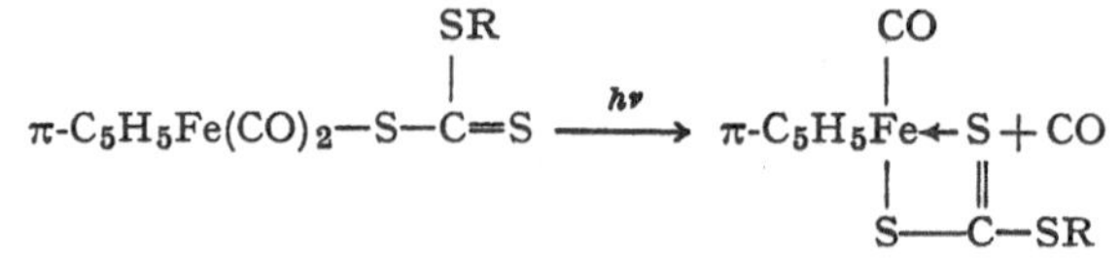

π-allyl complexes result from σ-allyl compounds:

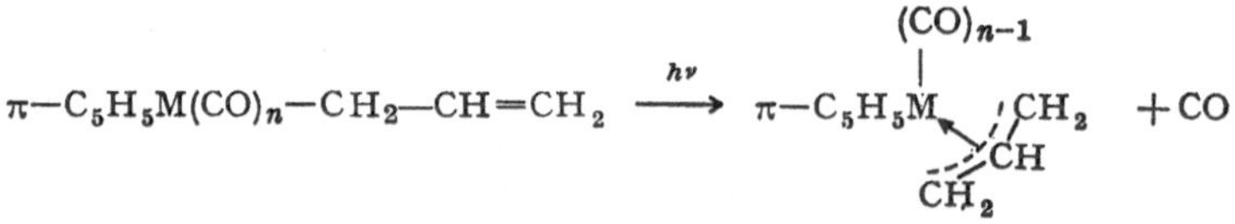

$$\pi-C_5H_5M(CO)_n-CH_2-CH=CH_2 \xrightarrow{h\nu} \overset{\displaystyle (CO)_{n-1} \atop \displaystyle |}{\pi-C_5H_5M} \begin{smallmatrix} CH_2 \\ CH \\ CH_2 \end{smallmatrix} \quad +CO$$

M = Fe, $n = 2$ 201)
M = Mo, $n = 3$ 120)
M = W, $n = 3$ 200)
M = Ru, $n = 2$ 558)

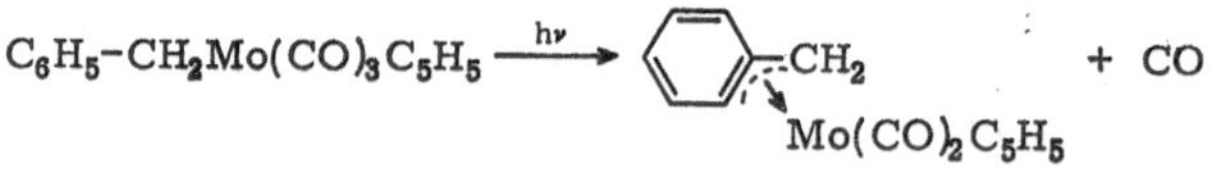

$$C_6H_5-CH_2Mo(CO)_3C_5H_5 \xrightarrow{h\nu} \underset{Mo(CO)_2C_5H_5}{\overset{}{\bigcirc}}{=}CH_2 \quad + CO \qquad 268)$$

$$\overset{}{\underset{S}{\bigcirc}}-CH_2-Mo(CO)_3(C_5H_5) \xrightarrow{h\nu} OC-Mo-CO + CO \quad 556)$$

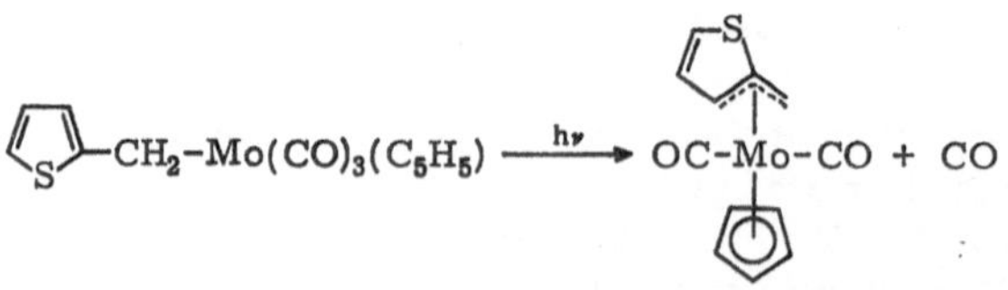

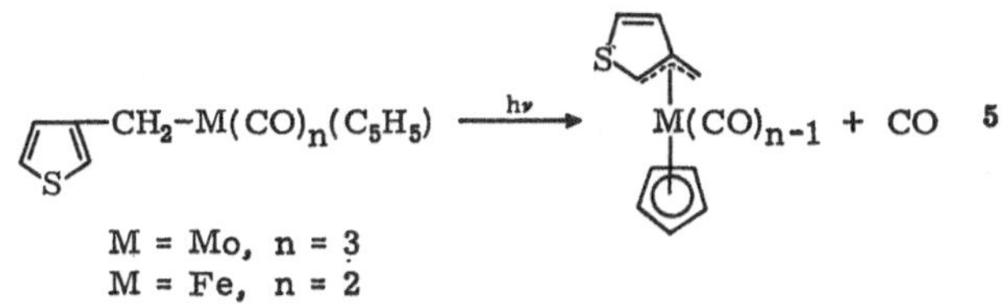

$$\text{(structure)}\ -CH_2-M(CO)_n(C_5H_5) \xrightarrow{h\nu} M(CO)_{n-1} + CO \quad 556)$$

$$M = Mo,\ n = 3$$
$$M = Fe,\ n = 2$$

A variety of reactions is observed with sulfur-containing complexes [270,276,277] of the general formula $CH_3-S-(CH_2)_nM(CO)_x(C_5H_5)_y$, $n = 1,2,3$; $M = Fe, Mo, Mn$; $x = 2,3,5$; $y = 1,1,0$.

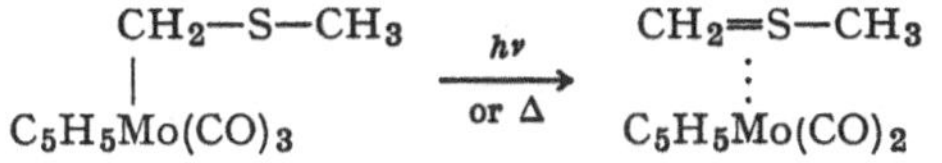

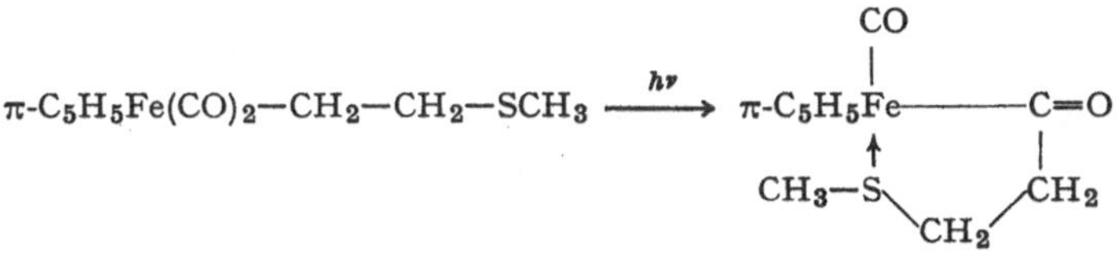

$$\begin{array}{ccc} CH_2-S-CH_3 & & CH_2=S-CH_3 \\ | & \xrightarrow[\text{or } \Delta]{h\nu} & \vdots \\ C_5H_5Mo(CO)_3 & & C_5H_5Mo(CO)_2 \end{array}$$

$$\pi\text{-}C_5H_5Fe(CO)_2-CH_2-CH_2-SCH_3 \xrightarrow{h\nu} \begin{array}{c} CO \\ | \\ \pi\text{-}C_5H_5Fe\text{———}C=O \\ \uparrow \quad | \\ CH_3-S \quad CH_2 \\ CH_2 \end{array}$$

$$+ \pi\text{-}C_5H_5Fe(CO)_2-S-CH_3 + [\pi\text{-}C_5H_5Fe(CO)S-CH_3]_2 + [\pi\text{-}C_5H_5Fe(CO)_2]_2$$

5. Substitution with Substituted Metal Carbonyls

Polynuclear complexes are produced by irradiation of substituted metal carbonyls in analogy to the reaction observed with simple metal carbonyls (see section E 1). In most cases either CO or another ligand attached to the mononuclear complex serve as bridging functions in the polynuclear products.

a) *CO as Bridging Ligand:* Nitrosomanganese tetracarbonyl (isoelectronic with $Fe(CO)_5$) forms a binuclear complex [510] according to:

$$2\ Mn(CO)_4NO \xrightarrow{h\nu} ON(CO)_2Mn(CO)_3Mn(CO)_2NO + CO$$

$C_5H_5Rh(CO)_2$ yields $(C_5H_5)_2Rh_2(CO)_3$ and $[C_5H_5RhCO]_3$ (two isomers) on illumination [331,332,333,367]. Two isomeric complexes with the composition $[C_5H_5CoCO]_3$ are obtained from $C_5H_5Co(CO)_2$ [269].

b) *Other Bridging Ligands:*

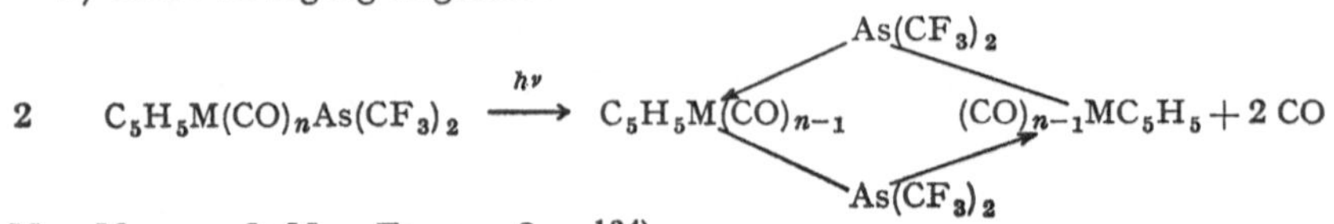

$$2 \quad C_5H_5M(CO)_n As(CF_3)_2 \xrightarrow{h\nu} C_5H_5M(CO)_{n-1} \overset{\displaystyle As(CF_3)_2}{\underset{\displaystyle As(CF_3)_2}{\diamond}} (CO)_{n-1}MC_5H_5 + 2\,CO$$

M = Mo, $n = 3$; M = Fe, $n = 2$ [124]

The products formed from the corresponding perfluorophenyl compounds are polymeric according to their high melting points and low solubilities [111].

$$m(C_6F_5)_2AsM(CO)_nC_5H_5 \rightarrow [(C_6F_5)_2AsM(CO)_{n-1}C_5H_5]_m + mCO$$

M = Mo, $n = 3$; M = Fe, $n = 2$

Binuclear complexes with bridging sulfur compounds result from the following reactions:

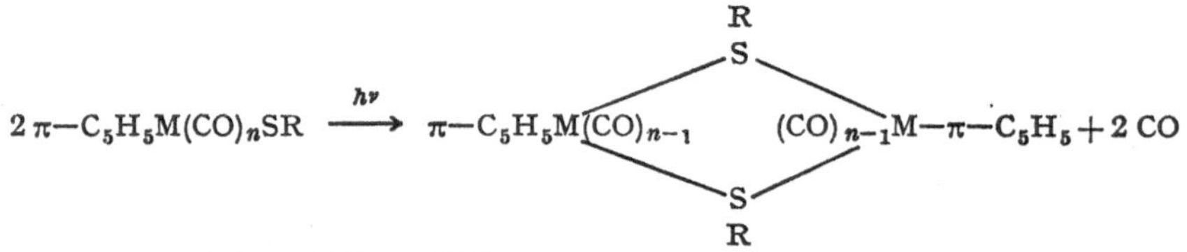

$$2\,\pi-C_5H_5M(CO)_nSR \xrightarrow{h\nu} \pi-C_5H_5M(CO)_{n-1} \overset{\displaystyle \overset{R}{S}}{\underset{\displaystyle \underset{R}{S}}{\diamond}} (CO)_{n-1}M-\pi-C_5H_5 + 2\,CO$$

M = W, $n = 3$, R = C_6H_5 [216]

M = Fe, $n = 2$, R = CH_3, C_2H_5, C_6H_5 [7]

A hexanuclear manganese complex $[(C_5H_5)_6Mn_6(NO)_8]$ free of CO is obtained from $[C_5H_5Mn(CO)NO]_2$ [274]. $RhCl_3(CO)\,[(n\text{-propyl})_3P]_2$ gives another carbonyl free complex $[Rh_2Cl_6([n\text{-propyl}]_3P)_4]$ [78]. Hydrogen is eliminated in the irradiation of $C_5H_5Fe[P(OC_6H_5)_3]_2H$ to yield $\{C_5H_5Fe[P(OC_6H_5)_3]_2\}_2$ [350].

6. Substitution of Ligands Other than CO

Only a few substitution reactions of ligands other than CO in carbonyl complexes have been investigated.

$$(CO)_4Fe[CF_2=CFCl] + C_2F_4 \xrightarrow{h\nu} (CO)_4Fe[CF_2=CF_2] + C_2F_3Cl \quad [150]$$

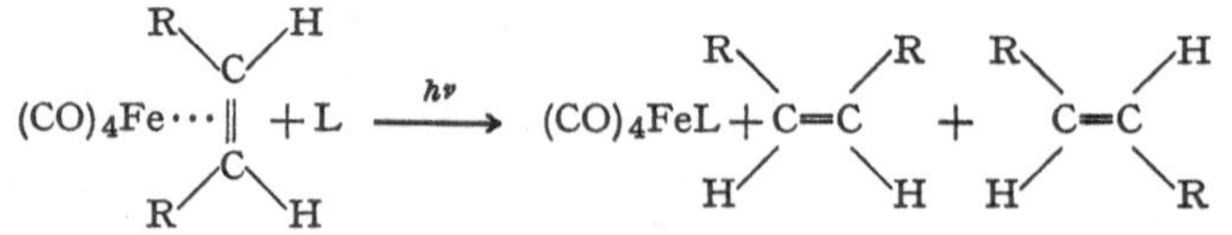

$$(CO)_4Fe\cdots \overset{\displaystyle R\diagdown_C\diagup H}{\underset{\displaystyle R\diagup^C\diagdown H}{\|}} + L \xrightarrow{h\nu} (CO)_4FeL + \overset{R}{\underset{H}{}}C=C\overset{R}{\underset{H}{}} + \overset{R}{\underset{H}{}}C=C\overset{H}{\underset{R}{}}$$

R = $COOCH_3$; L = dimethyl fumarate [282]

The mechanism of this reaction is not as yet quite clear. Since cis/trans isomerization of the replaced ligand is observed in the photochemical but not in the thermal substitution reaction, the mechanisms of these two reactions appear to be different [282]. The photochemical reaction must involve a species, which can rotate about the C=C bond.

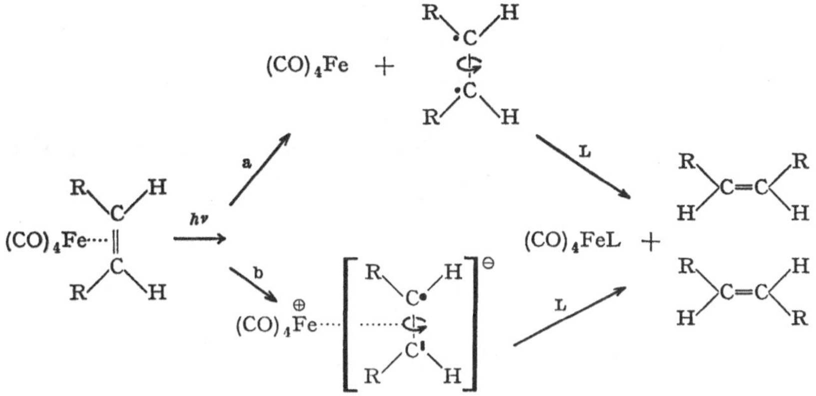

Two different reactions can be supposed to result from excitation of a CTML transition:

a) Dissociation into $Fe(CO)_4$ and an excited olefin, which can produce cis/trans isomers. In a S_N1 reaction $Fe(CO)_4$ can combine with L or with the (isomerized) olefin.

b) CTML in an iron tetracarbonyl olefin complex can lead to a situation resembling the bonding in an olefin radical anion to a metal cation (with an unpaired electron). Even though bonding may still be too strong to allow complete dissociation, rotation could occur in the olefin anion radical. Internal return could lead to the complex of the cis/trans isomerized olefin, attack of a nucleophile on the metal (in the transient) to replacement of the (isomerized) olefin.

Current work in the authors' laboratory is directed towards a clarification of this problem [285]. (See section F 2 for further discussion).

Strohmeier [469] reported the photoinduced exchange of unlabeled benzene chromium tricarbonyl with ^{14}C-labeled benzene (57% in 3 hrs., 366 nm irradiation).

$$(C_6H_6)Cr(CO)_3 + {}^{14}C_6H_6 \xrightarrow{h\nu} ({}^{14}C_6H_6)Cr(CO)_3 + C_6H_6$$

Chlorobenzene, toluene, and cycloheptatriene could be exchanged as well. The reaction worked with $Mo(CO)_3$ compounds, too, but very slowly with $W(CO)_3$ complexes.

Complete dissociation into $M(CO)_3$ and the ligand has been proposed for this process [467,469], which is obviously brought about by $M \to C_6H_6$ charge-transfer as discussed in section B.

It appears worthwhile mentioning in this connexion that the thermal formation of $C_6H_6Mo(CO)_3$ from C_6H_6 and $Mo(CO)_6$ could not be accelerated photochemically [486].

7. $M(CO)_{n-1} + X{-}Y \to (CO)_{n-1} M{<}^{X}_{Y}$

Photochemically produced $M(CO)_{n-1}$ can add to covalent molecules $(X{-}Y)$ or ionic compounds (X^+Y^-) with the formation of

$$(CO)_{n-1}M{<}^{X}_{Y} \quad \text{or} \quad [(CO)_{n-1}MX]^{\oplus}Y^{\ominus}$$

and secondary products derived therefrom.

a) $X = H$: Novel deep red $[Fe(CO)_3X]_2$ ($X = Br, I$) is formed in the reaction of $Fe(CO)_4$ with HX in pentane [284] according to Eq. (32) and (33).

$$Fe(CO)_4 + HX \to (CO)_4Fe{<}^{H}_{X} \tag{32}$$

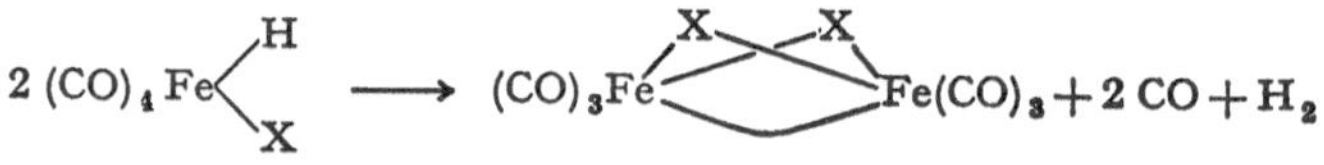

$$2\,(CO)_4Fe{<}^{H}_{X} \longrightarrow (CO)_3Fe{<}\!\!\overset{X\diagdown\diagup X}{\diagdown\!\!\diagup}\!\!{>}Fe(CO)_3 + 2\,CO + H_2 \tag{33}$$

An analogous reaction occurs with RSH [383]:

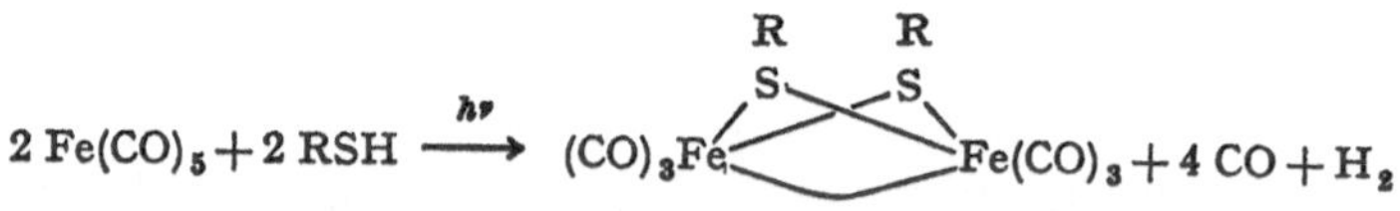

$$2\,Fe(CO)_5 + 2\,RSH \xrightarrow{h\nu} (CO)_3Fe{<}\!\!\overset{\overset{R\quad R}{S\diagdown\diagup S}}{\diagdown\!\!\diagup}\!\!{>}Fe(CO)_3 + 4\,CO + H_2$$

The corresponding products derived from alcohols and $Fe(CO)_4$ appear to be unstable [238,382] and disproportionation is observed [235].

Alcoholates are produced from benzene chromium tricarbonyl [81] and $Cr(CO)_6$ [79] according to Eq. (34).

$$C_6H_6Cr(CO)_3 + 3\,CH_3OH \xrightarrow{h\nu} Cr(OCH_3)_3 + C_6H_6 + 3\,CO + 3/2\,H_2 \tag{34}$$

CH_3OH/THF mixtures yield $Cr(OCH_3)_2 \cdot THF$ and phenol $Cr(OC_6H_5)_3$ with $C_6H_6Cr(CO)_3$ [79].

Other organic active hydrogen compounds such as acetylacetone (AAH), hexafluoroacetylacetone, dibenzoylmethane, and 8-hydroxy-quinoline gave similar photochemical oxidative decarbonylation reactions [188,383]. A preparative application is the preparation of ferric acetyl-acetonate according to Eq. (35):

$$Fe(CO)_5 + 3\ AAH \xrightarrow[\text{in } C_6H_6]{h\nu} Fe(AA)_3 + 5\ CO + 3/2\ H_2 \qquad (35)$$

Attempted photopinacolization of acyl-cyclopentadienylmanganese tricarbonyl with isopropanol caused decomposition [122].

Diphenylphosphinic acid forms coordination polymers with $Fe(CO)_5$ and $Cr(CO)_6$ [376,377]:

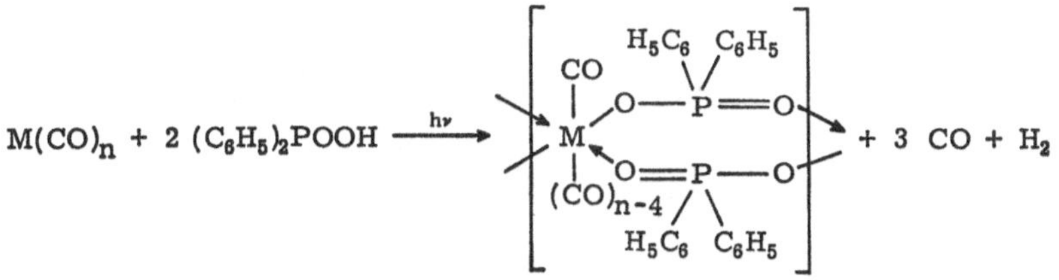

$$M(CO)_n + 2\ (C_6H_5)_2POOH \xrightarrow{h\nu} \left[\begin{array}{c} H_5C_6\ \ C_6H_5 \\ \diagdown / \\ CO \quad \diagdown O \!-\! P \!=\! O \\ | \ \diagup \\ M \\ | \ \diagdown O \!=\! P \!-\! O \\ (CO)_{n-4} \diagup \diagdown \\ H_5C_6\ \ C_6H_5 \end{array} \right] + 3\ CO + H_2$$

The photochemical reaction of trimethylhydrides of group IV elements with osmium and ruthenium carbonyls has been studied by *Stone* [564]:

$$(CH_3)_3MH + Os_3(CO)_{12} \xrightarrow{h\nu} (CH_3)_3MOs(CO)_4H$$

$M = Si, Ge, Sn$

$$(CH_3)_3SiH + Ru_3(CO)_{12} \xrightarrow{h\nu} [(CH_3)_3SiRu(CO)_4]_2 + [(CH_3)_3Si]_2Ru(CO)_4$$

The corresponding reaction of trichlorosilane with a variety of transition metal compounds has been recently reported by *Graham* [573]:

$$Cl_3SiH + C_6H_6Cr(CO)_3 \xrightarrow{h\nu} C_6H_6CrH(CO)_2(SiCl_3) + CO$$

$$Cl_3SiH + C_5H_5Mn(CO)_3 \xrightarrow{h\nu} C_5H_5MnH(CO)_2(SiCl_3) + CO$$

$$Cl_3SiH + C_5H_5Fe(CO)_2(SiCl_3) \xrightarrow{h\nu} C_5H_5FeH(CO)(SiCl_3)_2 + CO$$

$$Cl_3SiH + C_5H_5Co(CO)_2 \xrightarrow{h\nu} C_5H_5CoH(CO)(SiCl_3) + CO$$

The photochemical reaction of $(C_6H_5)_2SiH_2$ with $Re_2(CO)_{10}$ yields the novel $(C_6H_5)_2SiH_2Re_2(CO)_8$ [572].

b) $X-Y=RSSR$:

$$C_6H_6Cr(CO)_3 \xrightarrow[CH_3SSCH_3]{h\nu} Cr(SCH_3)_3+C_6H_6+3\,CO \quad 80)$$

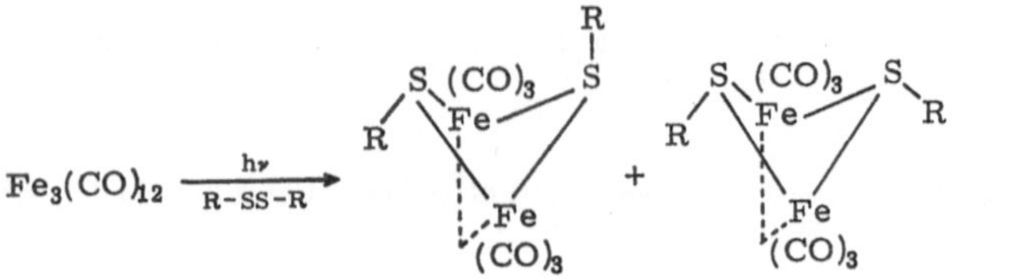

c) $X-Y=R\text{-}Halogen$: Spectroscopic and chemical evidence has been presented in the case of $CHBr_3$ for the intermediacy of $(CO)_4Fe(R)$-halogen in the photochemical reaction of $Fe(CO)_5$ with $CHCl_3$, $CHBr_3$, CH_3COCl, C_6H_5COCl, $C_6H_5CH_2Cl$, and $(CH_3)_3CBr$ [289,295]:

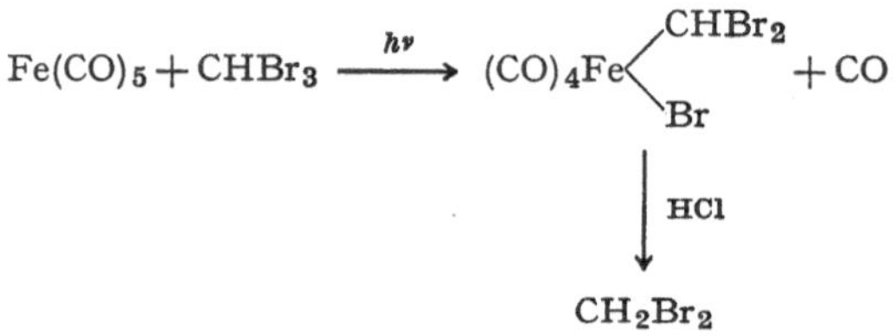

Coupling is observed with tropylium bromide [289]:

$$Fe(CO)_5+2\,C_7H_7^{\oplus}Br^{\ominus} \xrightarrow[CH_3OH]{h\nu} C_7H_7-C_7H_7+FeBr_2+5\,CO$$

Alkyl halides containing easily abstractable β-hydrogen undergo elimination reactions [284]:

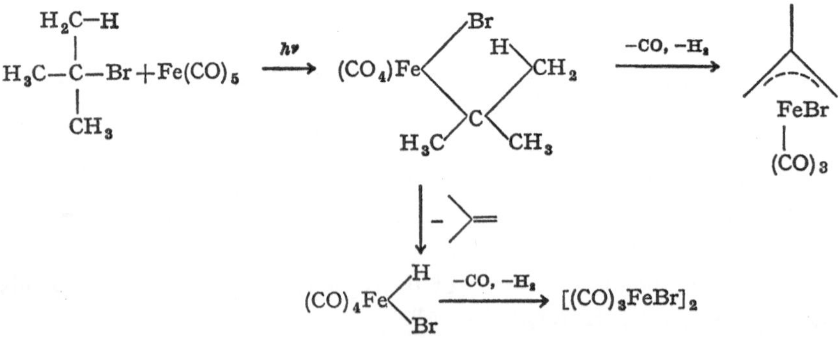

Backbonding in coordinated halo-olefins activates the $=C-X$ bond due to contribution of X-orbitals to the antibonding MO's [202,285].

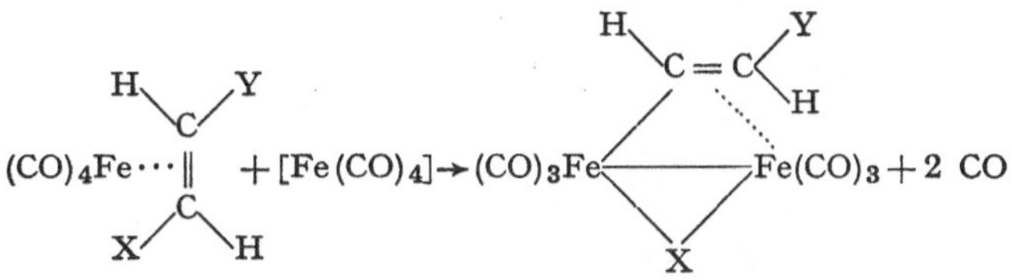

$$X = Cl, Br, I; \ Y = F, Cl, Br, I$$

Allylic halides form π-allyl iron tricarbonyl halides on illumination in the presence of $Fe(CO)_5$ [219,220].

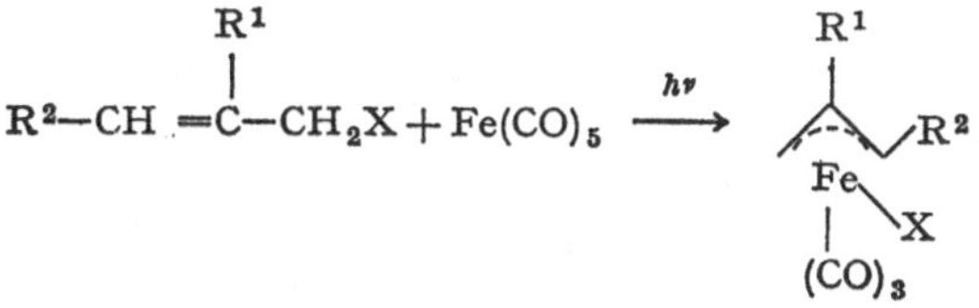

$$X = Cl, Br, I; \ R^1 = H, CH_3; \ R^2 = H, CH_3$$

A cleavage reaction of a $Fe-C$ bond has been observed with C_3F_7I [348]:

$$\pi\text{-}C_5H_5Fe(CO)_2C_6H_5 + C_3F_7I \xrightarrow{h\nu} \pi\text{-}C_5H_5Fe(CO)_2I + \text{products}$$

Photochemical reactions which do not involve cleavage of M(CO)-bonds have been described by *Hieber* [230]:

$$HMn(CO)_3(PR_3)_2 + CCl_4 \xrightarrow{h\nu} ClMn(CO)_3(PR_3)_2 + \text{halogenated hydrocarbons}$$

$$Hg\{Mn(CO)_3[P(OC_6H_5)_3]_2\}_2 \xrightarrow[CCl_4]{h\nu} 2\,ClMn(CO)_3[P(OC_6H_5)_3]_2 + HgCl_2$$

d) $X-Y = O_2$: Dimeric cyclopentadienylmolybdenum tricarbonyl yields compounds *(9—12)* on irradiation in CHCl$_3$ in the presence of O_2 [119].

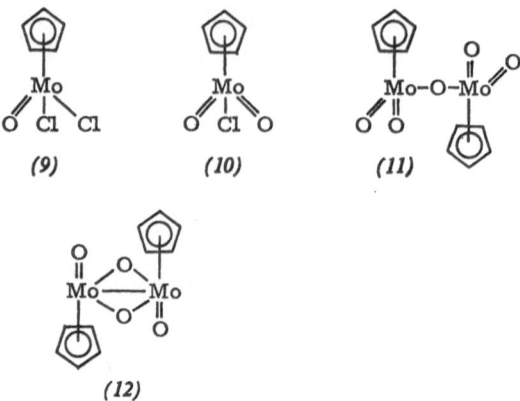

(9)　　　　*(10)*　　　　*(11)*

(12)

e) $X-Y=X^{\oplus}Y^{\ominus}$: A variety of anions have been employed as donor systems in carbonyl complexes, e.g. [1]:

$$Fe(CO)_5 + (C_2H_5)_4N^{\oplus}I^{\ominus} \xrightarrow{h\nu} [(C_2H_5)_4N]^{\oplus}[Fe(CO)_4I]^{\ominus} + CO$$

Similar reactions have been carried out with $Cr(CO)_6$, $Mo(CO)_6$, and $W(CO)_6$ using $[(C_6H_5)_3P]_2NX$ (X = Cl, Br, I) [390], $[(C_6H_5)_3P]_2N^{\oplus}$ $PF_2S_2^{\ominus}$, $[(C_6H_5)_4As]^{\oplus}POSF_2^{\ominus}$ [389], and with $Cr(CO)_6$, $W(CO)_6$, and $Fe(CO)_5$ using $[(C_6H_5)_3P]_2N^{\oplus}X^{\ominus}$ (X = NCS, CN) [388]. In most instances polynuclear complexes are formed, too. Some related reactions have been mentioned in section E 5.

Recently, even "dicarbollide" ions have been coordinated to carbonyl compounds [217] providing an example of a triple "cross-over" between carborane chemistry, coordination chemistry, and photochemistry.

$$Na_2^{++}(3)\text{-}1.2\text{-}B_9C_2H_{11}^{--} + M(CO)_6 \xrightarrow[\text{THF}]{h\nu} Na_2^{++}[\pi\text{-}(3)\text{-}1.2\text{-}B_9C_2H_{11}]M(CO)_3^{--} + 3\,CO$$

M = Mo, W

The complex anions were precipitated as $(CH_3)_4N^+$-salts.

Another very interesting reaction is the coordination of triphenyl-cyclotriphosphine dipotassium *(13)* to metal carbonyls [247].

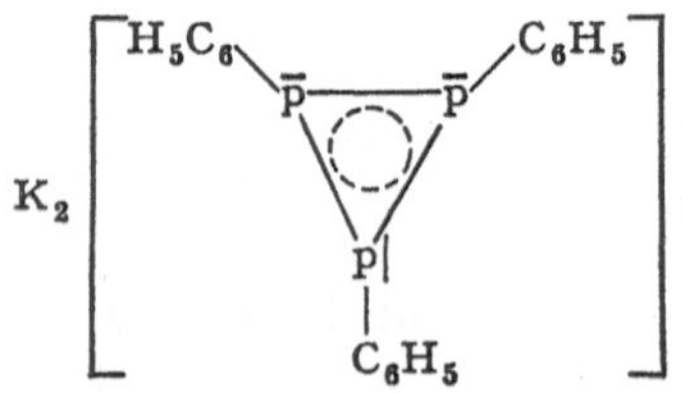

$$K_2 \left[\begin{array}{c} H_5C_6 \diagdown \overline{\underset{}{P}} \diagup \overline{\underset{}{P}} \diagup C_6H_5 \\ \underset{C_6H_5}{\overset{P|}{\diagdown\diagup}} \end{array} \right] + M(CO)_n \xrightarrow[-CO/\text{in THF}]{h\nu} K_2[M(CO)_{n-1}(PC_6H_5)_3] \cdot THF$$

M = Ni, Fe

(13)

Similar reactions have been described with $KP(C_6H_5)_2$ and $Ni(CO)_4$ [246].

The reviewers feel that coordination of silver nitrate to metal carbonyls may play a role in its reduction by a variety of irradiated metal carbonyls, a process which has been claimed for photocopying in a patent [248].

f) $X-Y = M-M'$: $Cr(CO)_6$, $Mo(CO)_6$, $W(CO)_6$, and $Fe(CO)_5$ give $(C_6H_5)_4As-M(CO)_{n-1}M'Cl_3$ (M = Cr, Mo, W, Fe; M' = Ge, Sn) on illumination with $(C_6H_5)_4As-M'Cl_3$ [392], possibly by insertion of the photochemically formed $M(CO)_{n-1}$ species into the As—M' bond.

8. Disproportionation by Substitution

Metal carbonyls of the 7. and 8. group can react with strong n-donors under disproportionation to ionic metal carbonylates in a so-called "base-reaction", e.g. $Fe(CO)_5$ with pyridine [235].

$$5\ Fe(CO)_5 + 6\ C_6H_5N \xrightarrow[85\ °C]{h\nu} [Fe(C_6H_5N)_6]\ [Fe_4(CO)_{13}] + 12\ CO \qquad (36)$$

It has been shown that pyridine and other ligands known to cause the "base-reaction" can yield normal substitution products, e.g. $(CO)_4FeC_6H_5N$, if the illumination is carried out at room temperature in hexane [414] (see section E 2). Since, e.g. pyridine iron tetracarbonyl decomposes at 65 °C (melting point) in the solid state and quickly in solution, it appears possible that Eq. (36) and many other "base-reactions" are thermal decomposition reactions of photochemically formed "normal" complexes, due to the high temperatures employed in the photolyses. Therefore, the reader is referred for further information to existing review articles [234,235] and to a few other references [210,236,237].

9. Chemical Transformations of Ligands

Several cases of desoxygenation of organic compounds have been observed in the course of photochemical reactions with metal carbonyls.

A unique complex is formed in the reaction of $Fe(CO)_5$ with diphenylketene [330,334].

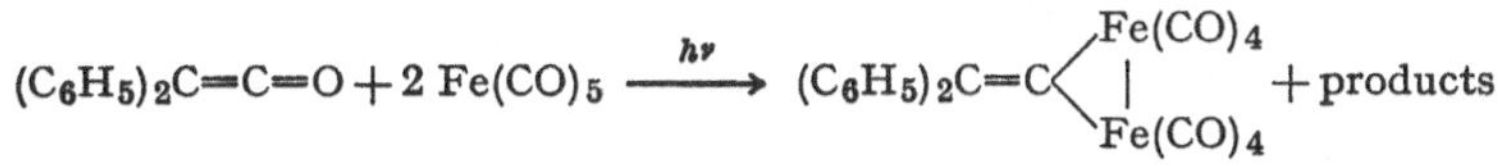

$$(C_6H_5)_2C=C=O + 2\ Fe(CO)_5 \xrightarrow{h\nu} (C_6H_5)_2C=C{\Large\langle}{\small\begin{matrix}Fe(CO)_4\\|\\Fe(CO)_4\end{matrix}} + products$$

The fate of the removed oxygen has not been reported (CO_2?).

Nitrobenzene is reduced to nitrosobenzene with $Fe(CO)_5$ and yields dimeric nitrosobenzene iron tricarbonyl [293].

$$2\,C_6H_5-NO_2 + 2\,Fe(CO)_5 \xrightarrow{h\nu} [(CO)_3FeC_6H_5NO]_2 + 2\,CO + 2\,CO_2$$

In this instance CO is oxidized to CO_2. Substituted aromatic nitro compounds give the same reaction with excellent yields [290]. X-ray structural analysis of these compounds provides final proof of the structure [329].

It is worthwhile mentioning that NO_2^- is desoxygenated to NO in its photochemical reaction with $[C_5H_5Mn(CO)_2NO]PF_6$ yielding $[C_5H_5Mn(NO)_2]_n$, $n > 1$ [267]. A very surprising reaction is the formation of the thionitroso complexes (14 and 15). While (14) results from the photochemical reaction of the sulfur diimide RN=S=NR with $Fe(CO)_5$, (15) is obtained by desoxygenation of RNSO with $Fe_2(CO)_9$ [365].

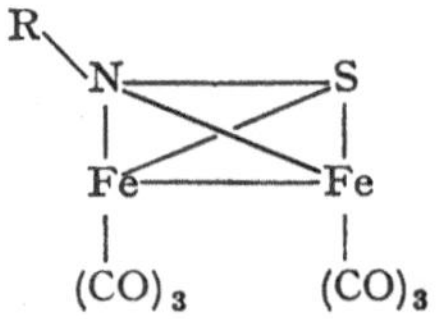

(14): R = t-butyl

(15): R = C_6H_5

Desulfurization of thiophene occurs with $Fe(CO)_5$ [414]:

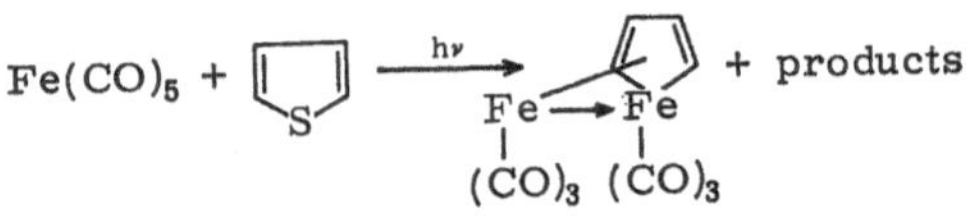

(16)

(16) is also produced by heating of the components [255].

Diphenyldiazomethane forms the complex (17) with $Fe(CO)_5$, the origin of the additional hydrogen is unknown [23].

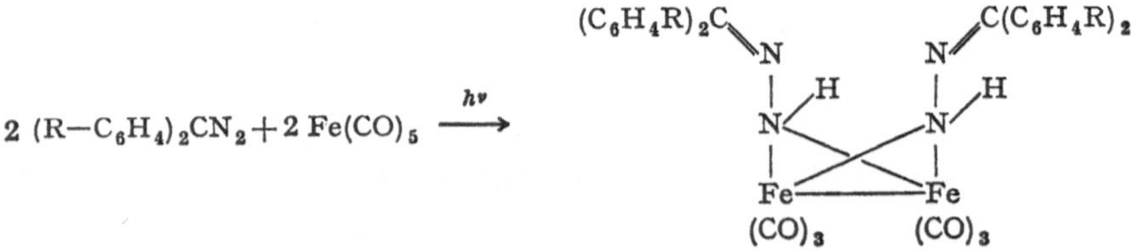

$$2\,(R-C_6H_4)_2CN_2 + 2\,Fe(CO)_5 \xrightarrow{h\nu}$$

(17)

NO_2^- in KOH/water is photochemically reduced by $Fe(CO)_5$ to the complex *(18)* [181].

$$H_2N \diagdown \quad NH_2$$
$$(CO)_3Fe \diagup \diagdown Fe(CO)_3$$

(18)

Rosenblum et al. [386] reported a very elegant synthesis of cyclobutadiene iron tricarbonyl starting from α-pyrone. The corresponding reaction with cyclopentadienylcobalt dicarbonyl produces cyclobutadiene-(cyclopentadienyl)cobalt [385].

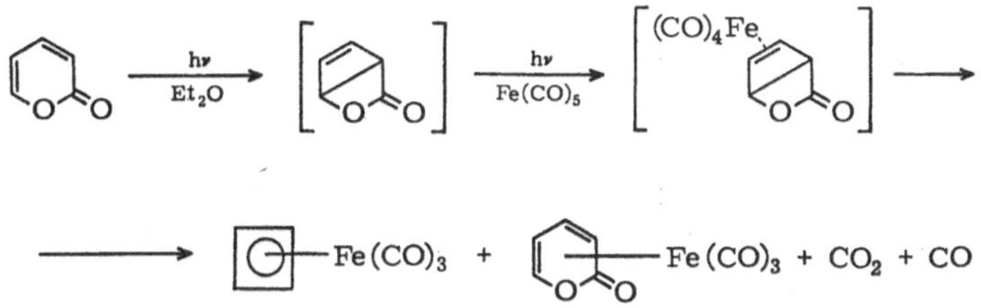

$$\text{α-pyrone} \xrightarrow[Et_2O]{h\nu} \left[\quad\right] \xrightarrow[Fe(CO)_5]{h\nu} \left[(CO)_4Fe \quad \right] \longrightarrow$$

$$\longrightarrow \boxed{\ominus}-Fe(CO)_3 \ + \ \text{(pyrone)}-Fe(CO)_3 \ + \ CO_2 \ + \ CO$$

The catalytic formation of diacetone alcohol has been observed in the irradiation of cyclopentadienyl manganese tricarbonyl in acetone [473,475]. The mechanism of this reaction is unknown: possibly, manganese oxides formed in this reaction function as base catalysts in a condensation reaction.

Illumination of $Fe(CO)_5$ and hexamethyl Dewar benzene produces 0.3% of dimeric hexamethylbenzene iron dicarbonyl [154]. This reaction leads us to a variety of isomerizations in metal carbonyl complexes brought about by illumination, as discussed in the next section.

F. Photochemical Isomerization of Metal Carbonyl Compounds

1. Positional Isomerization

Positional isomerization of the iridium complex *(19)* to *(20)* is brought about by irradiation [78].

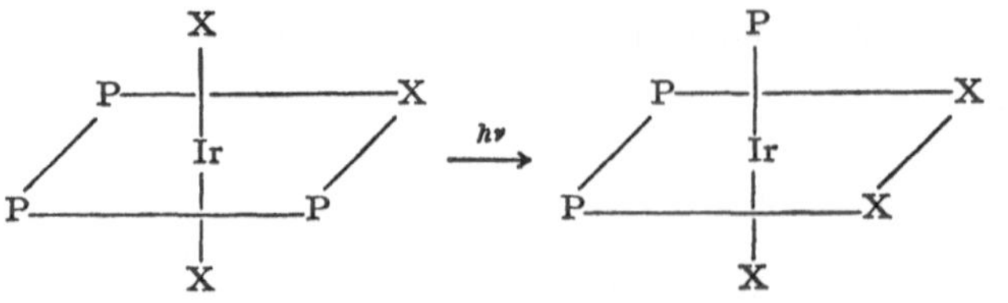

$$P = P(C_2H_5)_3,$$
$$P(C_2H_5)_2C_6H_5$$
$$\text{etc.}$$
$$X = Cl, Br$$

(19)   (20)

Similar isomerizations have been observed with complexes of the general formula $IrCl_3(CO)L_2$, $IrHX_2L_3$, $IrCl_3LL'_2$, and $RuCl_2(CO)_2L_2$ [78]. Irradiation into the long wavelength d—d transition of the square complexes cis- and trans-$[P(C_2H_5)_3]_2PtCl_2$ produces a solvent polarity dependent photostationary state of both isomers [207]. The following path was proposed for this reaction:

Square $\xrightarrow{h\nu}$ Excited $\longrightarrow$ Excited $\longrightarrow$ Tetra- $\longrightarrow$ Square
ground $\rightleftharpoons$ square $\longrightarrow$ square $\longrightarrow$ hedral $\longrightarrow$ ground (cis/trans)
state   singlet   triplet   triplet   state

The solvent effect could also be due to excitation of a solvated ground-state molecule, isomerization proceeding through a trigonal bipyramid containing a solvent molecule.

Many examples of such isomerizations are known with ionic inorganic coordination compounds [518]. However, of simple metal carbonyls only $Fe(CO)_4I_2$ has been investigated: the usually obtained cis-isomer is photochemically rearranged to the unstable trans-isomer [359]. Contradicting claims exist with respect to the reversibility of this process [250, 366].

2. Isomerization of Ligands

Isomerization of the olefinic ligand is observed in the photochemical rearrangement of dimethyl maleate iron tetracarbonyl to dimethyl fumarate iron tetracarbonyl [300,396].

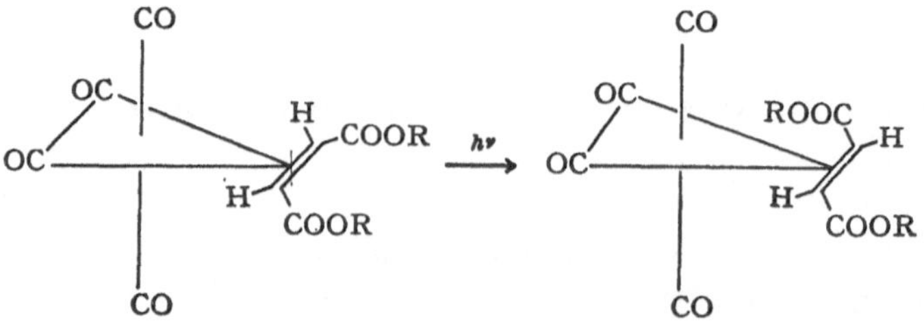

The equatorial position of the olefinic ligands remains unchanged in this cis/trans isomerization [285]. Some mechanistic aspects of this reaction have already been discussed in section E6.

The photochemical isomerization of 1.5-cyclooctadiene iron tricarbonyl *(7)* to 1.3-cyclooctadiene iron tricarbonyl *(8)* mentioned in section E3 is an example of ligand isomerization by formal shift of C=C bonds [284]. No deuterium is incorporated in the complex if the isomerization is carried out in a D_2-atmosphere. It appears not impossible, that the rearrangement occurs by a skeletal reorganization involving cleavage of C—C bonds and not hydrogen shift [284].

Ligand isomerization must also occur in the photochemical transformation of cis- and trans-(1.2-dibromoethylene)iron tetracarbonyl to a binuclear complex (see section E7c), since both iron tetracarbonyls yield the same product with a trans-configurated C=C bond [202,285].

3. Photoinduced Catalytic Isomerization of Olefins with Metal Carbonyls

Two mechanisms have been advanced for the catalytic thermal isomerization of olefins with iron carbonyls [318]:

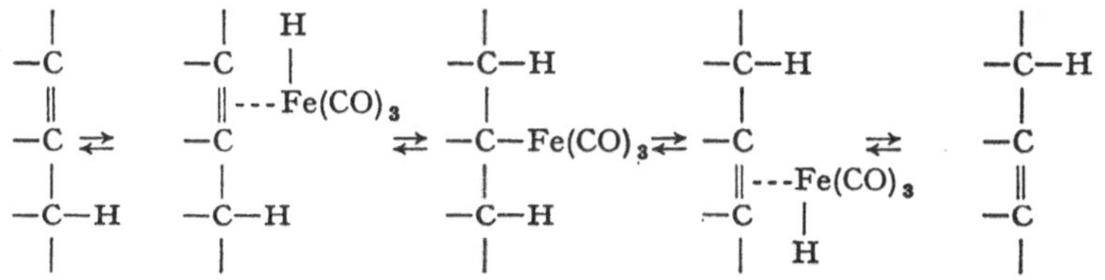

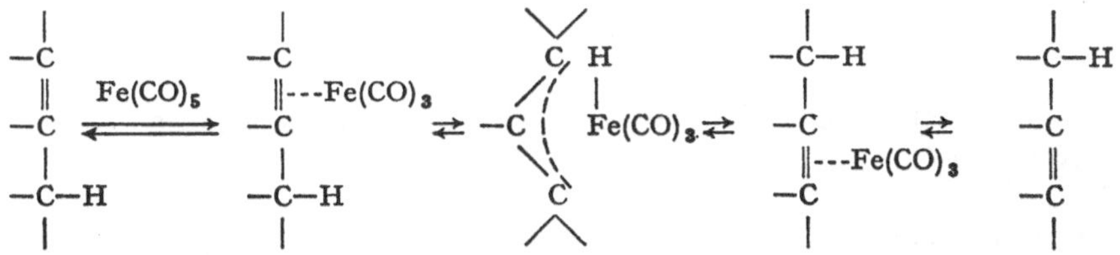

Both mechanisms postulate addition elimination reactions with Fe—H. The basic difference is that a) involves the mutual transformation of σ-and π-bonded species and b) that of π- and π-allyl bonded species. Both pathways need a coordinatively unsaturated iron carbonyl species to start with. Therefore it is not surprising that irradiation facilitates and accelerates the catalytic isomerization so that it can be carried out at room temperature.

1-Octene and 1-undecene are converted to internal olefins by illumination in the presence of $Fe(CO)_5$ or $Fe_3(CO)_{12}$ [20,21,91] in an intramolecular reaction [149]. Cic/trans isomerization of 1.6-cyclodecadiene has been reported [222] (illumination with $Fe(CO)_5$). Shift of the double bond in allyl ethers to the corresponding propenyl ethers [251] or isomerization of unsaturated ethers [126] can be photoinitiated in the presence of $Fe(CO)_5$, under similar conditions ketones and aldehydes can be prepared from unsaturated alcohols [127].

4. Photoinduced H Abstraction from Phenyl Groups

Irradiation of $[(C_6H_5)_2P-CH_2-CH_2P(C_6H_5)_2]_2Fe \cdot CH_2=CH_2$ results in liberation of ethylene and in the formation of $[(C_6H_5)_2P-CH_2-CH_2-P(C_6H_5)_2]Fe(H) [C_6H_4P(C_6H_5)-CH_2-CH_2-P(C_6H_5)_2]$ [215]. Hydrogen is transferred from a phenyl group to the iron. Under ethylene pressure this hydrogen returns to the original ligand.

Hydrogen transfer from a phenyl group to nitrogen is observed in the photochemical formation of (21) from azobenzene and $Fe(CO)_5$ [24]. It appears not impossible, that intermediates with Fe—H bonds are involved in this reaction.

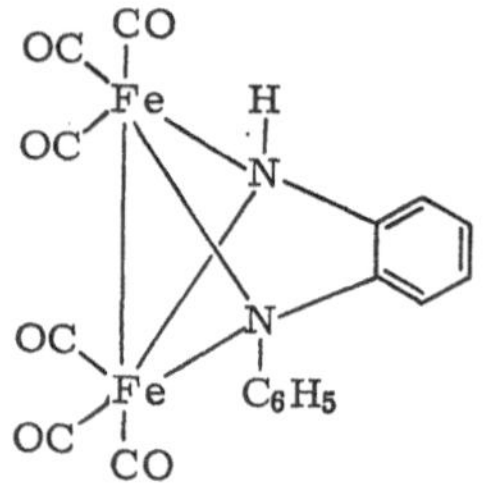

(21)

G. Photochemical Addition and Elimination Reactions of Metal Carbonyl Compounds

1. Insertion into M—M bonds

a) *Insertion of Olefins and Acetylenes:* The photochemical reaction of trimethyltin manganese pentacarbonyl with tetrafluoroethylene yields a variety of products according to Eq. (37) [98,99,100].

$CF_2=CFH$ forms products according to Eq. (38) mainly. $CF_2=CFCl$ behaves similarly. Ethylene replaces CO in the starting material (see Table 5) but does not give any insertion.

$$(CH_3)_3Sn-Mn(CO)_5 + C_2F_4 \xrightarrow{h\nu} (CH_3)_3Sn-CF_2-CF_2-Mn(CO)_5 \qquad (37)$$

$$\downarrow$$

$$CF_2=CF-Mn(CO)_5 + (CH_3)_3SnF \qquad (38)$$

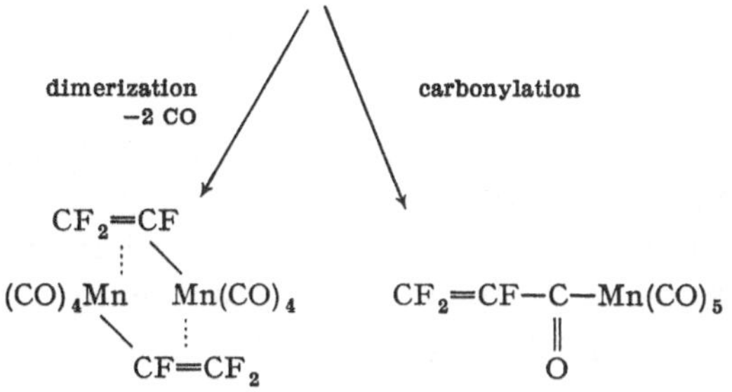

dimerization   carbonylation
−2 CO

$C_4F_5Mn(CO)_5$ is obtained from perfluoro-cyclobutene and $(CH_3)_3$ $Sn-Mn(CO)_5$ [62].

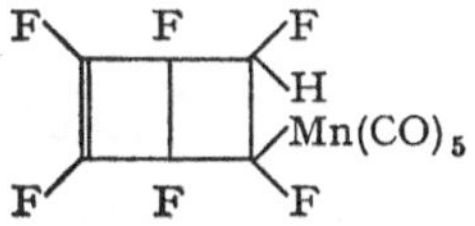

$Mn(CO)_5$ results from the corresponding reaction with Dewar hexafluorobenzene [110]. $(CH_3)_3Ge-Mn(CO)_5$ forms $(CH_3)_3Ge-CF_2-CF_2-Mn(CO)_5$ with $CF_2=CF_2$ but $CHF=CF-Mn(CO)_5$ with trifluoro-ethylene [97]. Tetrafluoroethylene gives insertion with $(CH_3)_3Sn-Co(CO)_4$, trifluoroethylene reacts differently [60,61].

The insertion reactions do not occur in the dark. A free radical process is unlikely and a four center transition state has been proposed [97]. The action of light is not clear, whether activation of the M−M bonds results, remains to be checked.

$CF_2=CF-CF=CF-Mn(CO)_5$ and $(CH_3)_3SnF$ are produced in the irradiation of $(CH_3)_3Sn-Mn(CO)_5$ with perfluorobutadiene [197]. Insertion is observed in the corresponding reaction with $(CH_3)_3Ge-Mn(CO)_5$ combined with 1.3−F-shift to yield $(CH_3)_3Ge-CF_2-CF_2-CF=CF-Mn(CO)_5$ [196], constituting an interesting 1.4-addition reaction. Perfluoro-but-2-yne and $(CH_3)_3Sn-Fe(C_5H_5)(CO)_2$ give $(CH_3)_3Sn-C(CF_3)=C(CF_3)-Fe(C_5H_5)(CO)_2$ [62].

Insertion reactions are also observed with trifluoropropyne [569]:

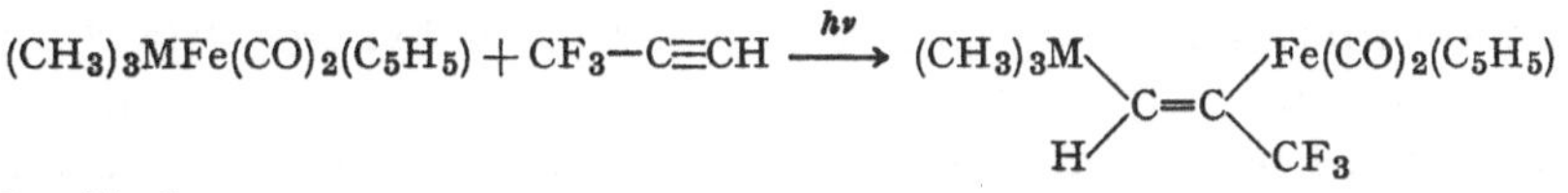

M = Si, Ge

The corresponding tin compound yields

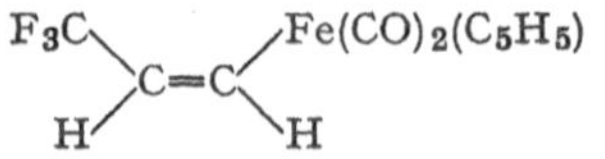

$$\underset{H}{\overset{F_3C}{\diagdown}}C=C\underset{H}{\overset{Fe(CO)_2(C_5H_5)}{\diagup}}$$

as a secondary product.

b) *Insertion of SnX$_2$:* Insertion of SnX$_2$ (X = Cl, Br) into the Co—Co bond in $(CO)_3Co(L)-CO(L)(Co)_3$ (L = P[butyl]$_3$) can be brought about thermally and photochemically. The reactive intermediate in the latter reaction is supposed to be an electronically excited species with an activated Co—Co bond, while a $(CO)_3(L)Co$ radical pair or $\ \overset{\diagup}{\diagdown}Co-CO-Co\overset{\diagdown}{\diagup}$ have been considered for the thermal reactions on the grounds of kinetic data [40].

The suggested activation of the M—M bond in the excited state is supported by a variety of observations:

$$2\ \text{phenanthroline}(CO)_3Re-Mn(CO)_5 \xrightarrow[\text{THF}]{h\nu} [\text{phenanthroline}(CO)_3Re]_2 + Mn_2(CO)_{10}$$
[306,307]

$$[(\pi\text{-}C_5H_5)Mo\,(CO)_2P(OCH_3)_3]_2Hg \xrightarrow{h\nu} [(\pi\text{-}C_5H_5)\,Mo\,(CO)_2P(OCH_3)_3]_2 + Hg \quad [323]$$

$$Hg[Fe(CO)_3NO]_2 \xrightarrow{h\nu} HgFe(CO)_4 + Fe(CO)_2(NO)_2 \quad [232]$$

2. Insertion into M—C- and M—H-Bonds

a) *Insertion of Olefins and Acetylenes:* R—Mn(CO)$_5$ (R = CH$_3$, C$_6$H$_5$) reacts photochemically with tetrafluoroethylene to give R—CF$_2$—CF$_2$—Mn(CO)$_5$, with CF$_2$=CFCl CH$_3$—CF$_2$—CFCl—Mn(CO)$_5$ is produced [523, 524]. With perfluorobutadiene CH$_3$—CF$_2$—CF=CF—CF$_2$—Mn(CO)$_5$ was obtained, which isomerized to CH$_3$—(CF$_2$)$_2$—CF=CF—Mn(CO)$_5$ on standing in solution [121].

Irradiation of trans-[P(C$_2$H$_5$)$_3$]$_2$PtClH and CF$_3$—C≡CH produced cis-[P(C$_2$H$_5$)$_3$]$_2$PtCl—C(CF$_3$)=CH$_2$ [213]. Elimination of C$_2$H$_4$ from —Co—C$_2$H$_5$ in ethyl-cobalamine and the (intermediate) formation of —Co—H can be looked upon as the reversal of this type of reaction [402].

Isobutylene trimers are formed in the irradiation of π-2-methallyl-iron tricarbonyl bromide and of trimethylenemethyl iron tricarbonyl in the presence of isobutylene and t-butyl bromide. The same trimers are obtained photochemically from t-butyl bromide and Fe(CO)$_5$. The following scheme has been proposed for this reaction [284]:

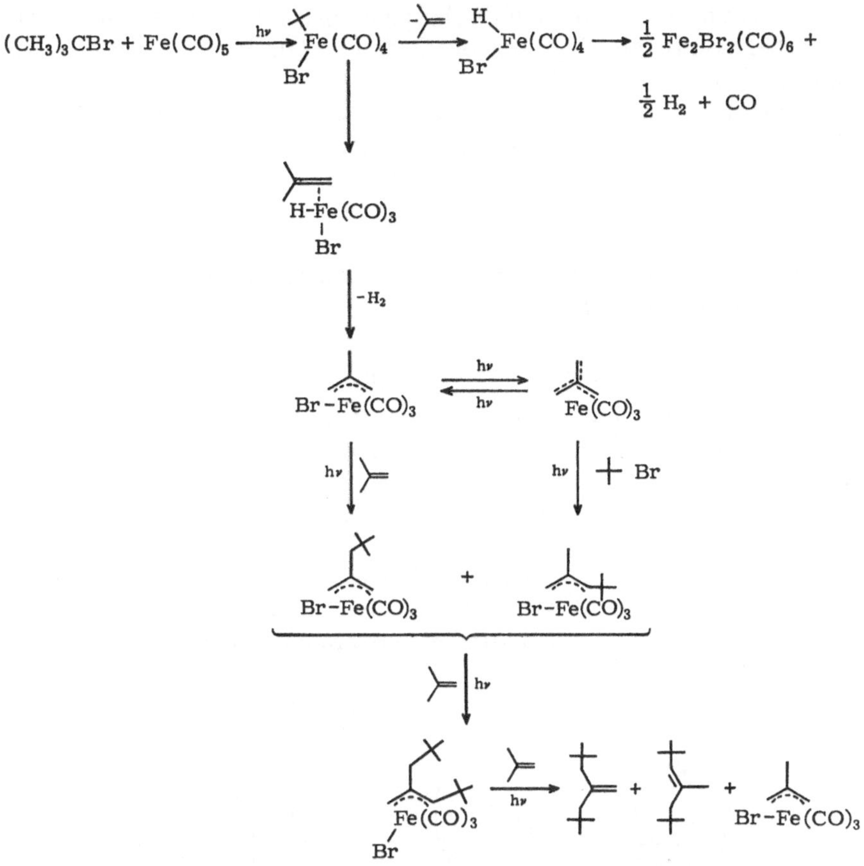

$$(CH_3)_3CBr + Fe(CO)_5 \xrightarrow{h\nu} \underset{Br}{\overset{\displaystyle\times}{Fe(CO)_4}} \xrightarrow{} \underset{Br}{\overset{\displaystyle H}{Fe(CO)_4}} \longrightarrow \tfrac{1}{2} Fe_2Br_2(CO)_6 +$$

$$\tfrac{1}{2} H_2 + CO$$

$$\underset{\underset{Br}{|}}{H\text{-}Fe(CO)_3}$$

$$\downarrow -H_2$$

$$Br-Fe(CO)_3 \underset{h\nu}{\overset{h\nu}{\rightleftarrows}} Fe(CO)_3$$

$$h\nu\downarrow \qquad\qquad h\nu\downarrow + Br$$

$$Br-Fe(CO)_3 \qquad + \qquad Br-Fe(CO)_3$$

$$\underset{Br}{Fe(CO)_3} \xrightarrow{h\nu} \qquad + \qquad + \qquad Br-Fe(CO)_3$$

b) *Insertion (and Elimination) of CO:* Several incidental observations of photochemical CO insertion have been reported, so for $(C_5H_5)Fe(CO)$-$P(C_6H_5)_3R \rightarrow (C_5H_5)Fe(CO)P(C_6H_5)_3-COR$ [37,346,509] (see also [38,113, 114,348]) and section H3 for additional examples).

Actually, the reverse reaction seems to be the preferred one:

$$R-CO-Fe(CO)_2(C_5H_5) \xrightarrow{h\nu} R-Fe(CO)_2(C_5H_5) + CO$$

$$R = CF_3,\ C_2F_5,\ C_3F_7,\ CH_3,\ C_6H_5,\ -CH=CH_2 \text{ [273]}$$

It appears very probable that the CO expelled is one of the metal bonded ones and not that from the acyl group. That means this reaction can be looked upon as an internal substitution.

411

Expulsion of CO from $(C_6H_5)_3PMo(CO)_2(C_5H_5)-COCH_3$ has been reported recently [67]. Expulsion of SO_2 from $(CO)_5Mn-SO_2-CH_2-C_6H_5$ was not achieved [214].

Elimination of the phenyl group results from illumination of $(C_5H_5)Fe(CO)[P(OC_6H_5)_3]C_6H_5$ producing $\{(C_5H_5)Fe(CO)[P(OC_5H_5)_3]\}_2$ [351].

H. Photochemical Cycloaddition Reactions of Olefin and Acetylene Complexes

1. Intermolecular Cycloaddition Reactions

Irradiation of $Fe(CO)_5$ in the presence of norbornadiene affords three major products: norbornadiene iron tricarbonyl (see section E3), dimeric norbornadiene and a ketonic compound [371].

The dimerization of norbornadiene can be brought about photochemically with $Fe(CO)_5$ or in the dark with $Fe_2(CO)_9$. Possibly, the photochemical part of the reaction consists merely in the formation of the $Fe(CO)_4$ species [64,309]. Detailed investigations have shown that several saturated and unsaturated dimers are formed, e.g.

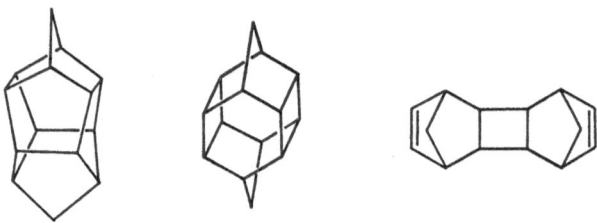

The structural isomers obtained according to different methods have been surveyed [19,258,404].

A thorough study of the photochemical dimerization of norbornadiene with a variety of transition metal carbonyl compounds has revealed some systematic effect of the nature of the transition metal involved on the stereospecifity of the dimerization observed, which is increased going from left to right across the transition metal series [571].

$$L_z(M)_x(CO)_y + \text{[norbornadiene]} \xrightarrow{h\nu} \text{[dimer]}$$

	exo-trans-exo	exo-trans-endo	endo-trans-endo
$Cr(CO)_6$ $R—C_6H_5Cr(CO)_3$	1,8	1	1,4
butadiene-$Fe(CO)_3$ 2,3-dimethylbutadiene-$Fe(CO)_3$ 1,1'-dicyclohexenyl-$Fe(CO)_3$ 1,4-diphenylbutadiene-$Fe(CO)_3$	1,4	—	1
$Fe(CO)_2(NO)_2$ $[(C_6H_5)_3]PFe(CO)(NO)_2$	1	traces	traces
$Co(CO)_3NO$ $[(C_6H_5)_3P]Co(CO)_2(NO)$ $[(C_6H_5)_3P]_2Co(CO)(NO)$	1	traces	traces
$[(C_6H_5)_3P]_2Ni(CO)_2$	1	traces	traces

In the case of the chromium carbonyl compounds the intermediacy of norbornadiene-$Cr(CO)_4$ has been demonstrated. 1,4-diphenylbutadiene-$Fe(CO)_3$ and $[(C_6H_5)_3P]_2Ni(CO)_2$ have been recovered to >90% showing their catalytic function. UV irradiation of norbornadiene in the presence of $Ni(CO)_4$ gives the following product resulting from a formal $(2+2+2)$ cycloaddition reaction [571]:

Photochemical dimerization of norbornene to *(22)* (97%) and *(23)* (3%) is achieved in the presence of catalytic amounts of copper(I) halides, due to the formation of a complex $(C_7H_{10})_n(CuX)_n$ which absorbs strongly at 41 800 cm^{-1} [507,508].

(22)   *(23)*

Dimeric norbornene is formed with a quantum yield of ~0.1. The concentration dependence of the quantum yield is in accord with the interaction of two ground-state norbornene molecules with the excited complex to yield the dimer. "This may be reconciled in terms of a new transient complex, existing only in the excited state, which comprises three olefins coordinated to a single cuprous halide" [507].

Possible details of this mechanism could be pictured as follows [282]:

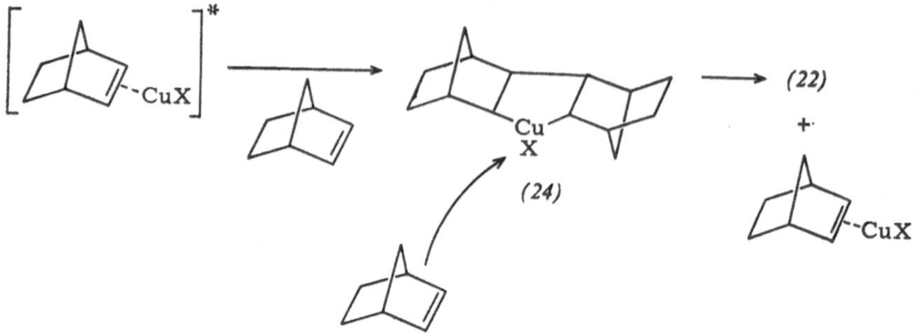

Nucleophilic attack of a ground-state olefin on the (CT) excited complex leads to a σ-bonded intermediate (24), which with a second ground-state olefin as "activating ligand" [221] is converted to (22) and a ground state complex molecule. The inspection of molecular models shows no detectable steric hindrance for a trans-exo approach of nor-bornene towards the exo-coordinated [22] norbornene · CuX leading to (24). A cis-approach appears extremely unlikely for steric reasons. The trans-endo approach has to overcome some steric interaction, in agreement therewith only 3% (23) is observed besides (22).

The following arguments can be advanced for the stepwise bond formation via (24), which contradicts the concept of "π-complex multi-center processes" [404]:

a) (25) is formed in the photochemical reaction of $Fe(CO)_5$ with tetrafluoro-ethylene [150].

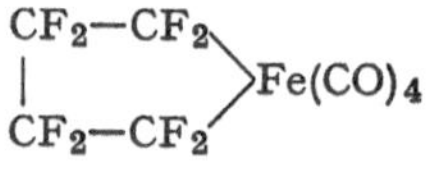

b) (26) has been isolated from the reaction of iron carbonyls with diphenyl acetylene [240].

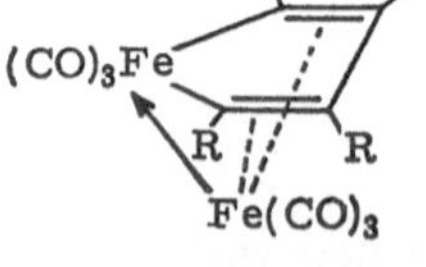

Decafluorotolan yields the corresponding complex on illumination in the presence of $Fe(CO)_5$ [577].

c) Cyclobutene-(3.4)-dimethyl-dicarboxylate forms *(27)* with $Fe(CO)_4$, which is converted to *(28)* using CO or $P(C_6H_5)_3$ as activating ligand [202, 285].

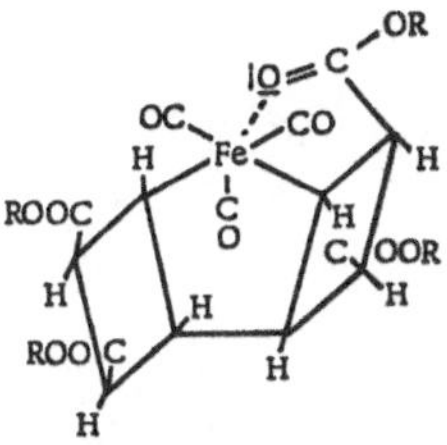

.*(27)* 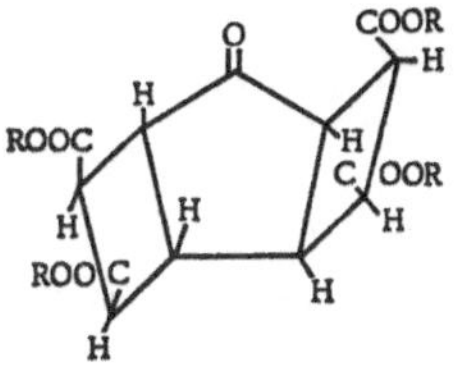   *(28)*

Two $Fe(CO)_3$ complexes of dimeric cyclooctatetraenes *(29* and *30)* are formed by irradiation of cyclooctatetraene iron tricarbonyl in the presence of free cyclooctatetraene, posing a formidable problem with respect to the mechanism of this cycloaddition reaction [244,403,407,410].

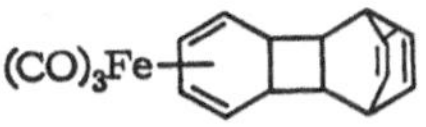

(29) 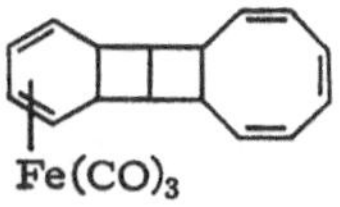   *(30)*

The catalytic dimerization at $\sim 100\ °C$ of butadiene (and isoprene) with dicarbonyldinitrosyliron and dicarbonylnitrosyl(π-allyl)iron [1—2 wt. %] to 4-vinylcyclohex-1-ene (CH_3-substituted vinylcyclohexenes) can be carried out photochemically at $-10\ °C$ to $25\ °C$. It has been suggested, that intermediates formed by UV decomposition of the complexes may be the active catalysts, and that these intermediates could be the same as those produced thermally. π-Allyl iron tricarbonyl iodide was found to be ineffective as a thermal dimerization catalyst [90]. However, π-methallyl iron tricarbonyl bromide has been shown to be effective in the trimerization of isobutylene [284] [see section G2a].

Photochemical dimerization of perfluorobut-2-yne with $(CH_3)_3Sn—Mn(CO)_5$ affords *(31)* [62].

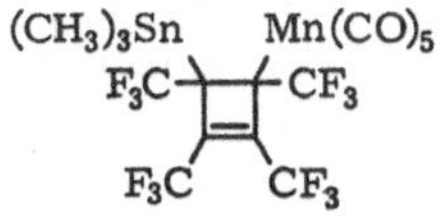

(31)

415

Cycloaddition reactions with cyclobutadiene [513], trimethylene-methyl [131], and benzyne [187] can be accomplished by photolysis of complexes containing the coordinated species or its precursor.

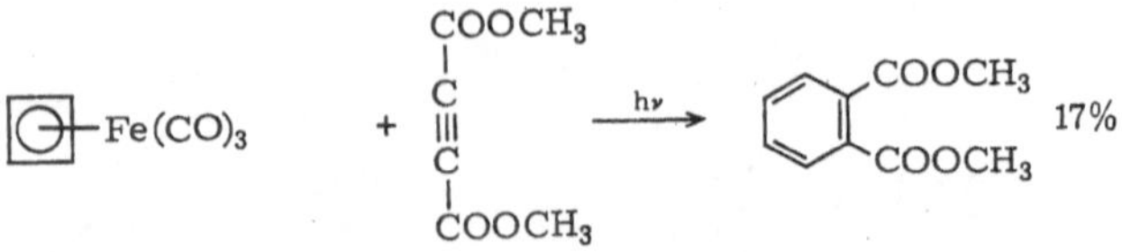

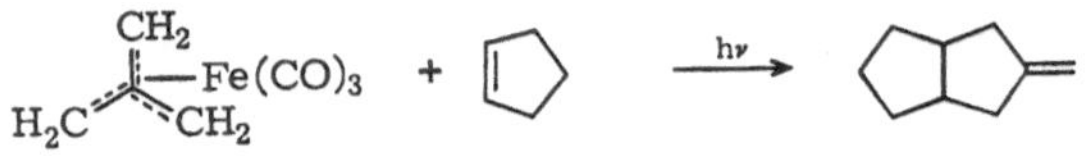

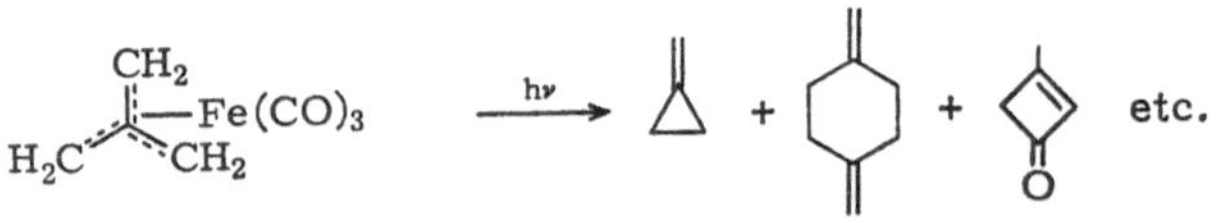

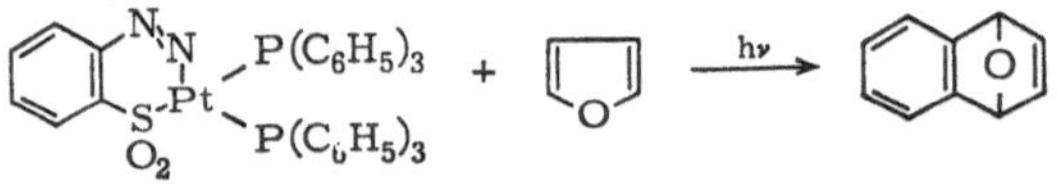

2. Intramolecular Cycloaddition Reactions

cis,cis-1.5-Cyclooctadiene yields $\sim 28\%$ tricyclo-[3.3.0.0$^{2.6}$]-octane *(32)* on illumination (in ether) with 254 nm in the presence of Cu_2X_2 (Rh_2Cl_2 gives different products) [431,433].

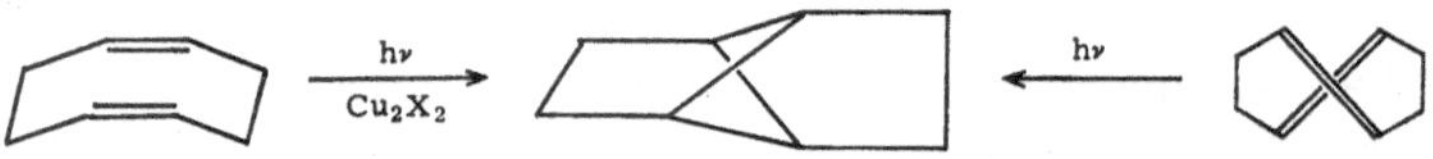

Using deuterated cyclooctadiene it has been demonstrated that no hydrogen abstraction is involved. The reaction has to be considered as an intramolecular process [212] and not as a free radical reaction [26].

Decomposition of the Cu_2X_2 complexes left as residues from such illuminations produced cis,cis-1.5-cyclooctadiene, cis,trans-1.5-cycloocta-diene, and small amounts of trans,trans-1.5-cyclooctadiene *(33)*. *(33)* gives *(32)* on illumination [522]. However, is it not yet clear whether free *(33)* is formed in cyclooctadiene photolysis, and then undergoes rearrangement to *(32)* due to its intense absorption at $40\,700\ cm^{-1}$ ($\varepsilon \sim 1500$) [caused by the proximity of the π-orbitals]. Cis, anti, cis-tricyclo-[3.3.0.0$^{2.4}$]-octane has been reported as an additional product ($\sim 10\%$ yield) from the illumination of cis, cis-1,5-cyclooctadiene and cyprous chloride [579].

Cu$_2$X$_2$ increases the yield of bicyclobutane *(34)* in the solution photolysis of butadiene [432].

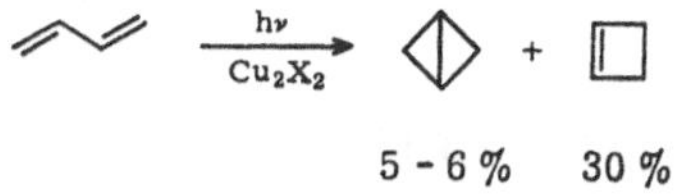

$$5 - 6\% \qquad 30\%$$
(34)

Another type of cyclooctadiene rearrangement is observed with a palladium complex [10]:

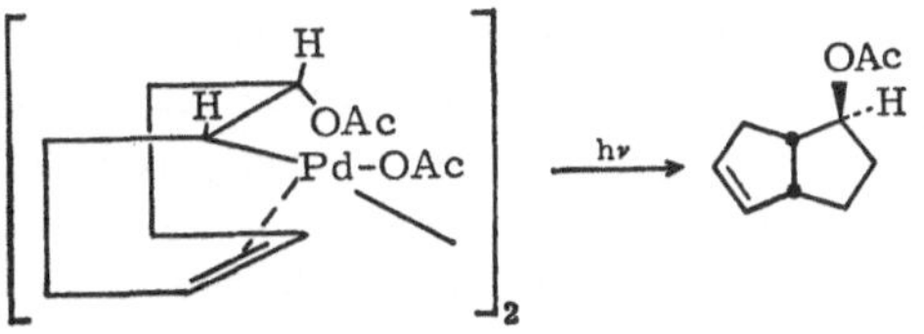

3. Cycloaddition with CO-Insertion

In many cycloaddition reactions brought about with metal carbonyls, insertion of CO is observed. The probable mechanism of such reactions has already been depicted with the transformation *(27)* → *(28)* (see section H 1).

The ketone obtained by *Pettit* [371] from the illumination of Fe(CO)$_5$ and norbornadiene has the exo-trans-exo structure *(35)* [198].

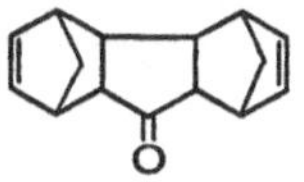

(35)

A unique synthesis of duroquinone results from the photoreaction of but-2-yne and Fe(CO)$_5$ [437]:

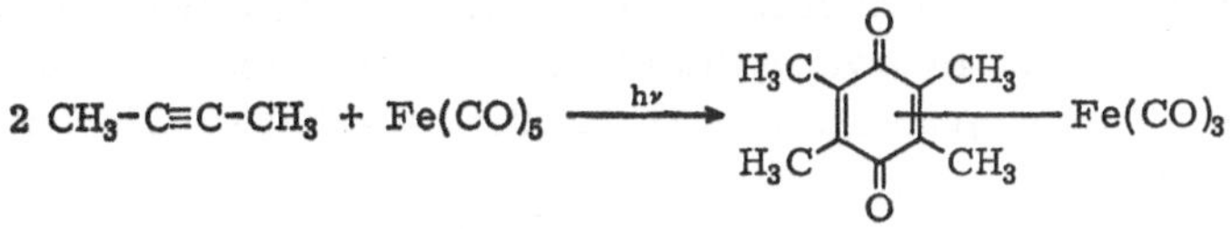

Other quinones can be produced by similar reactions.

The formation of cyclopentadienone complexes is a general phenomenon in photoreactions of acetylenes and metal carbonyls compounds
[240,319,411,480], e. g.

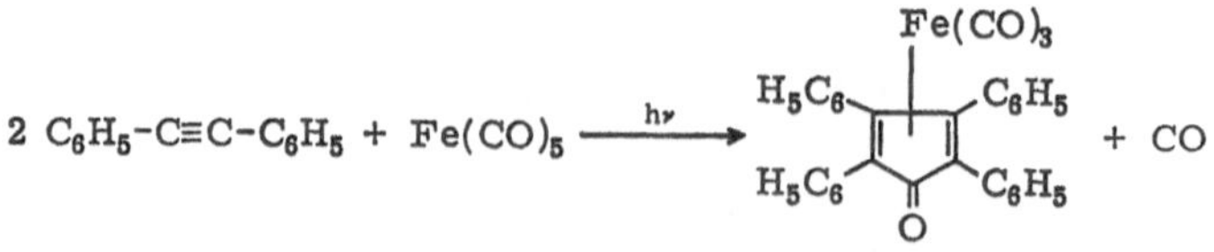

$$2\ C_6H_5-C\equiv C-C_6H_5\ +\ Fe(CO)_5\ \xrightarrow{h\nu}\quad + \ CO$$

Compounds like *(26)* appear to be intermediates in this reaction.

4. Application of the Woodward-Hoffmann-Rules to Photochemical
Reactions of Transition Metal Complexes, in particular to Cycloaddition
Reactions

It has become clear from the Woodward-Hoffmann-rules how "orbital
symmetry controles in an easily discernible manner the feasibility and
stereochemical consequences of every concerted reaction" [239]. For
cycloaddition reactions of a m-π-electron system to a n-π-electron
molecule the following stereochemical selection rules have been established ($q = 0,1,2, \ldots$):

$m+n$	allowed thermal cycloaddition	allowed photochemical cycloaddition
$4\,q$	cis-trans trans-cis	cis-cis trans-trans
$4\,q+2$	cis-cis trans-trans	cis-trans trans-cis

These rules state which cycloaddition reactions are allowed to
proceed in a concerted (one step) fashion and which are not; e.g. a
thermal cis-cis $2\,\pi + 2\,\pi$ cycloaddition reaction (cyclobutane formation)
is forbidden as a concerted process. The rules do not state that every
symmetry allowed reaction has to proceed in a concerted fashion. If a
reaction can occur by two pathways (concerted or disconcerted), then,
other factors being equal, the allowed pathway will have the lower activation energy [521].

An enormous wealth of experimental facts demonstrates how strictly
these rules are followed.

Many symmetry forbidden reactions (e.g. $2\pi + 2\pi$ cycloaddition), which can not be carried out thermally, occur under mild conditions in the presence of a suitable transition metal complex. It has been argued that this might be due to the presence in the complexes of occupied molecular orbitals with symmetries that make these reactions allowed concerted electrocyclic processes which proceed in a stereospecific manner [317,325,517].

Recently, some doubt has been raised as to whether these reactions involving metal complexes really are concerted electrocyclic reactions, since products and intermediates have been observed "which have no place in the previous mechanistic scheme" [258] [see also the discussion of the norbornene dimerization in section H 1].

It may very well turn out that electrocyclic reactions with transition metal complexes containing metal carbon bonds actually proceed as a sequence of several discrete steps, involving mutual changes from σ- to π-bonding [221]. The steps of such reactions can be stereoelectronically controlled in accord with the Woodward-Hoffmann rules, as shown by recent SCCC–MO calculations of nickel catalyzed cycloaddition reactions [505,506], e.g.:

Obviously, the appropriate application of the Woodward-Hoffmann rules to transition metal complex reactions involving metal carbon bonds requires information about the mechanistic and stereochemical details which is not easy to come by and definitely not yet sufficiently available for photochemical cycloaddition reactions of complexed olefinic systems.

However, for such well-known reactions [45] as positional isomerization and bimolecular substitution in simple square planar coordination compounds, very recently promising attempts have been reported to derive selection rules on the basis of the orbital symmetry conservation principle [139] and state correlation diagrams [521]; e.g. the cis-trans isomerization of square planar complexes has been predicted as a thermally forbidden and photochemically allowed process, in accordance with experiments [28,31,370] [see also section F 1].

I. Photopolymerization with Metal Carbonyls

Polymerization of methyl methacrylate (MMA) is brought about at room temperature by irradiation in the presence of $Fe(CO)_5$ and organic halides [289]. Addition of the halogen component in the dark to the irradiated MMA–Fe(CO)$_5$ mixture causes polymerization, too. In the latter case, CO set free in the course of the reaction causes simultaneous foaming of the polymer, yielding products with density <0.2.

Irradiation of Fe(CO)$_5$ in MMA is known to produce MMA · Fe(CO)$_4$ [298], whose reaction with organic halides (e.g. CCl$_4$) has been shown to initiate the polymerization [289,291,295,299].

The attack of, e.g. CHBr$_3$, can be envisaged as follows:

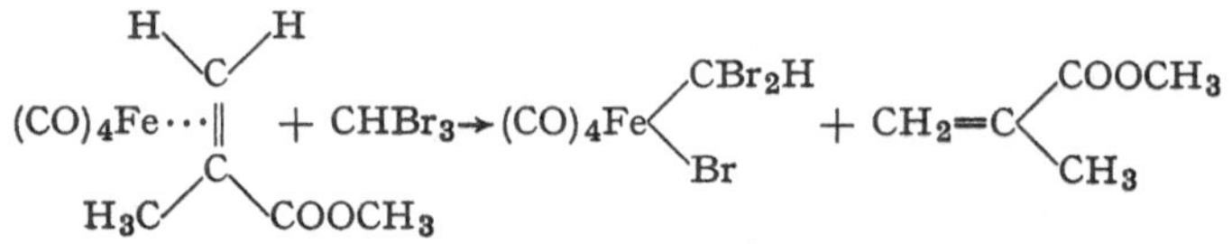

$$(CO)_4Fe\cdots\| \begin{array}{c} H \quad H \\ C \\ \\ C \\ H_3C \quad COOCH_3 \end{array} + CHBr_3 \rightarrow (CO)_4Fe \begin{array}{c} CBr_2H \\ \\ Br \end{array} + CH_2{=}C \begin{array}{c} COOCH_3 \\ \\ CH_3 \end{array}$$

$$(36)$$

Replacement of the olefinic ligand yields dibromomethyl iron tetra-carbonyl bromide. This type of reaction has been directly observed in the following system [282]:

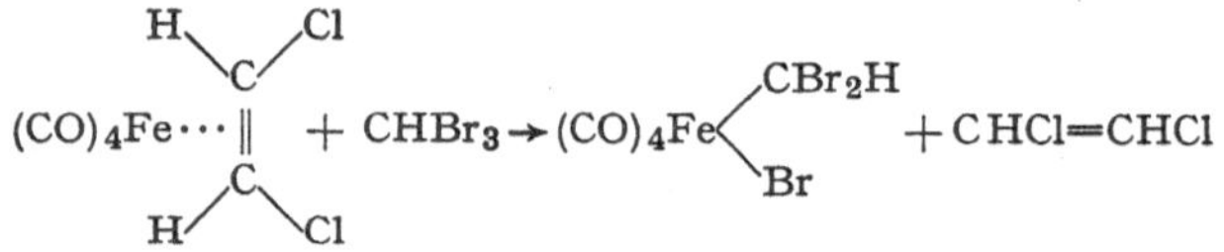

$$(CO)_4Fe\cdots\| \begin{array}{c} H \quad Cl \\ C \\ \\ C \\ H \quad Cl \end{array} + CHBr_3 \rightarrow (CO)_4Fe \begin{array}{c} CBr_2H \\ \\ Br \end{array} + CHCl{=}CHCl$$

$$(36)$$

(36) (see also section E. 7) is an effective polymerization catalyst [289]. One could imagine (36) producing free radicals according to Eq. (39) as the polymerization initiating species:

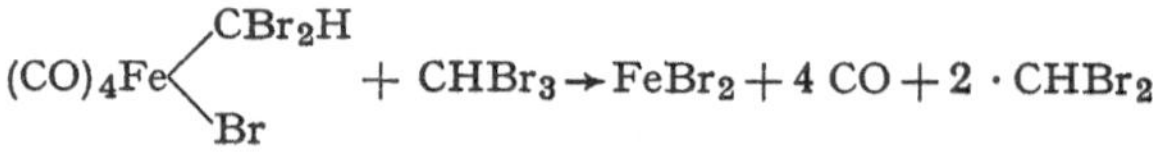

$$(CO)_4Fe \begin{array}{c} CBr_2H \\ \\ Br \end{array} + CHBr_3 \rightarrow FeBr_2 + 4\,CO + 2 \cdot CHBr_2 \qquad (39)$$

However, nucleophilic attack of MMA on (36) [Eq. (40)], as well as the attack of CHBr$_3$ on MMA · Fe(CO)$_4$, could yield (37), which (as a typical intermediate of transition metal catalyzed polymerization reactions [107]) should undergo insertion of the coordinated C=C group into the Fe–CHBr$_2$ σ-bond [Eq. (41)] and polymerize MMA.

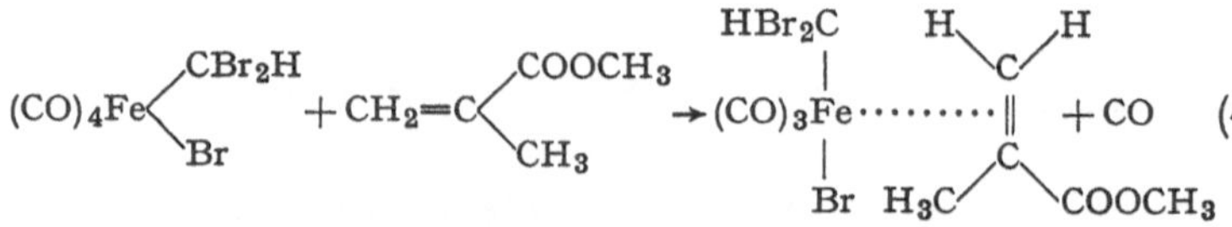

$$(CO)_4Fe \begin{array}{c} CBr_2H \\ \\ Br \end{array} + CH_2{=}C \begin{array}{c} COOCH_3 \\ \\ CH_3 \end{array} \rightarrow (CO)_3Fe \begin{array}{c} HBr_2C \\ | \\ \cdots\cdots\| \\ | \\ Br \quad H_3C \end{array} \begin{array}{c} H \quad H \\ C \\ \\ C \\ COOCH_3 \end{array} + CO \qquad (40)$$

$$(37)$$

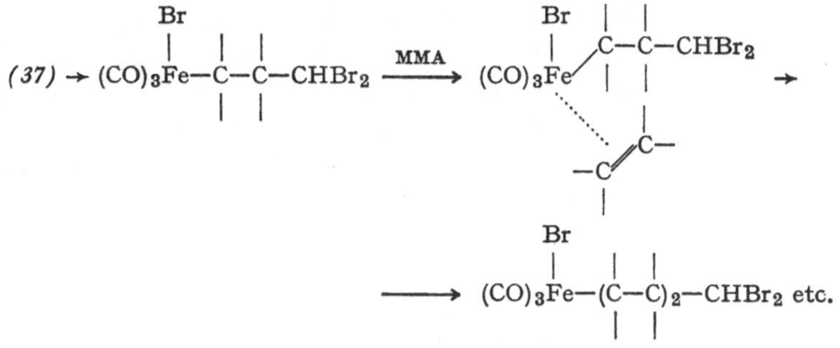

$$(37) \rightarrow (CO)_3Fe-\overset{\displaystyle Br}{\underset{\displaystyle |}{\underset{|}{C}}}-\overset{|}{\underset{|}{C}}-CHBr_2 \xrightarrow{\text{MMA}} (CO)_3Fe \begin{matrix} \overset{Br}{|} \\ C-C-CHBr_2 \\ | \ \ | \\ | \\ C- \\ \parallel \\ -C \\ | \end{matrix} \rightarrow$$

$$\xrightarrow{\hspace{2cm}} (CO)_3Fe-\overset{\displaystyle Br}{\underset{\displaystyle |}{(\underset{|}{C}-\underset{|}{C})_2}}-CHBr_2 \text{ etc.} \qquad (41)$$

Bamford, who extensively investigated the kinetics of the thermal MMA polymerization with metal carbonyls and organic halides [537,538], recently advanced the following free radical mechanism for the photo-polymerization of MMA with $Mn_2(CO)_{10}$ and CCl_4 [33,34]:

$$\text{metal carbonyl} \xrightarrow{h\nu} F$$
$$F + CCl_4 \xrightarrow{\hspace{1.5cm}} (I)$$
$$(I) \xrightarrow{\hspace{1.5cm}} \cdot CCl_3$$
$$(I) + F \xrightarrow{\hspace{1.5cm}} \text{inactive products}$$

Possibly, a transition metal catalyzed polymerization reaction as outlined above could also fit the kinetic data.

Strohmeier has shown that, with a variety of metal carbonyls and organic halides, not only MMA [462,463] but also vinyl chloride [stereo-specific!] [445,446,447] can be photopolymerized.

The photodegradation product of $Mn_2(CO)_{10}$ in propylene oxide was found to be a stereospecific catalyst for the thermal polymerization of propylene oxide [448,449,452,453].

Electric discharge can be used instead of light to activate $M(CO)_m$ and CX_4 systems in MMA polymerization [454].

Networks are formed by irradiation of $Mn_2(CO)_{10}$ in MMA if poly-(vinyl trichloroacetate) is used as a polyfunctional halide [32].

J. Synthesis of Olefin Metal Complexes

A versatile method for the preparation of olefin metal complexes has been developed by *E. O. Fischer* and his coworkers: a Grignard reagent (preferably isopropyl magnesium bromide) is reacted with a mixture of a transition metal halide and an olefin in ether. Subsequent irradiation increases the yield and accelerates the formation of the corresponding metal complexes.

$$MCl_x + x\ i\text{-}C_3H_7MgBr + n\ \text{olefin} \xrightarrow[\text{ether}]{h\nu} M(\text{olefin})_n + x/2\ C_3H_8 + x/2\ C_3H_6 + x\ MgBrCl$$

The mechanism of this reaction has not been as yet completely established. Obviously, metal alkyls formed initially, are decomposed (by illumination) to coordinatively unsaturated species (possibly via metal hydride intermediates), which then form the olefin complexes. The known [539] Ni(1.5-cyclooctadiene)$_2$ is obtained from NiCl$_2$, i-C$_3$H$_7$MgBr and 1.5-cyclooctadiene [336]. The corresponding reaction with azulene and FeCl$_3$ yields Fe(azulene)$_2$ [163]:

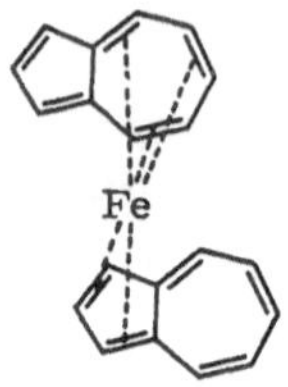

Oxidative work-up of the reaction of CoCl$_2$, RhCl$_3$ and IrCl$_3$ with 6.6-diphenyl fulvene [159,161] and Grignard reagent produces the complexes *(38)*; with 6.6-dimethyl fulvene *(39)* is obtained [159].

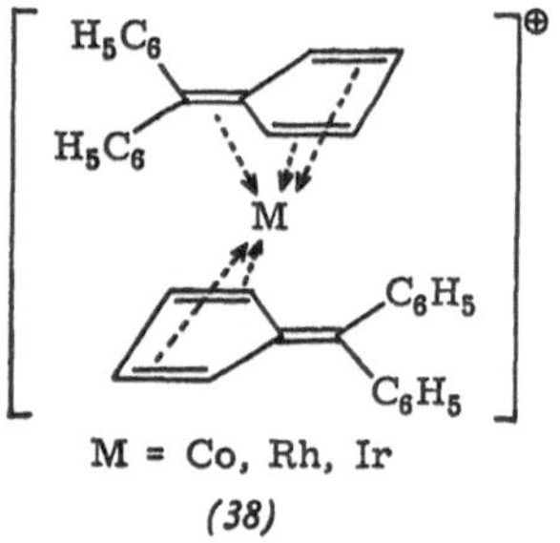

M = Co, Rh, Ir
(38)

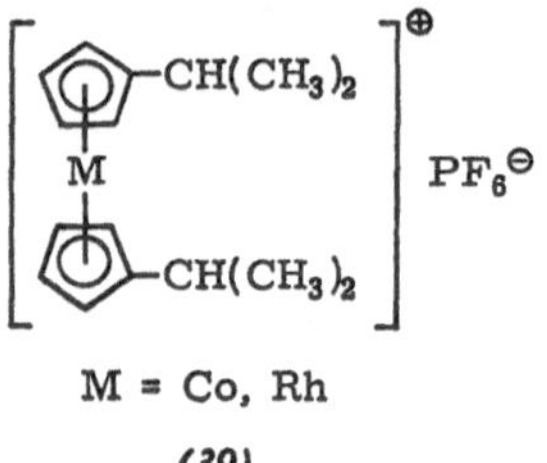

M = Co, Rh
(39)

The preparation of complexes containing two different olefins can be accomplished by the following methods:

a) by reacting MCl$_x$ with a mixture of two olefins

MCl$_x$	product	Ref.
RuCl$_3$	(1.3.5-cyclooctatriene)Ru(1.5-cyclooctadiene)	[165]
FeCl$_3$	(1.3.5-cycloheptatriene)Fe(1.3-cycloheptadiene)	[168]
FeCl$_3$	(1.3.5-cyclooctatriene)Fe(1.5-cyclooctadiene)	[167]

b) by reacting an olefin metal halide in the presence of another olefin

Olefin metal halide	product	Ref.
(1.5-cyclooctadiene)PtCl$_2$	(1.5-cyclooctadiene)$_2$Pt	[336]
(1.5-cyclooctadiene)RuCl$_2$	(1.5-cyclooctadiene)Ru-(1.3.5-cycloheptatriene)	[337]
(1.5-cyclooctadiene)RuCl$_2$	(1.5-cyclooctadiene)Ru-(1.3.5-cyclooctatriene)	[337]
C$_5$H$_5$CrCl$_2$ · THF	C$_5$H$_5$Cr(1.3.5-cycloheptatriene)	[166]
C$_5$H$_5$CrCl$_2$ · THF	C$_5$H$_5$Cr(1.3.5-cyclooctatriene)	[166]

(1.5-cyclooctadiene)Pt(i—C$_3$H$_7$)$_2$ has been isolated from the reaction of (1.5-cyclooctadiene)PtCl$_2$ with (i—C$_3$H$_7$)MgBr, and was converted to (1.5-cyclooctadiene)$_2$Pt by irradiation in the presence of 1.5-cyclooctadiene [336]. The attempted photochemical transformation of (norbornadiene)Pt(i—C$_3$H$_7$)$_2$ to (norbornadiene)$_2$Pt failed [336].

H$_2$ is eliminated in several reactions with 1.3-cyclohexadiene:

$$MCl_3 + 3\ i\text{-}C_3H_7MgBr + 2\ 1.3\text{-}C_6H_8 \xrightarrow[\text{ether}]{h\nu} (C_6H_6)M(1.3\text{-}C_6H_8) + H_2 + 3/2\ C_3H_8$$

$$+ 3/2\ C_3H_6$$

M=Fe [171], Ru, Os [165]

Dibenzene chromium is formed in the corresponding reaction of 1.3-C$_6$H$_8$ with CrCl$_3$ [172]. (1.5-cyclooctadiene)RuCl$_2$ yields a mixture of (1.5-cyclooctadiene)Ru(C$_6$H$_6$) and [by ligand exchange] (1.3-C$_6$H$_8$)Ru(C$_6$H$_6$) in the reaction with 1.3-C$_6$H$_8$ and Grignard reagent [337].

(C$_6$H$_6$)Ru(1.3-C$_6$H$_8$) results from (norbornadiene)RuCl$_2$ and 1.3-C$_6$H$_8$ [337]. (1.5-cyclooctadiene)Fe(1.3.5-cyclooctatriene) mentioned above can also be prepared under concomitant dehydrogenation from 1.5-cyclooctadiene, FeCl$_3$ and i-C$_3$H$_7$MgBr [167].

(40) results from CrCl$_3$, i-C$_3$H$_7$ MgBr and azulene [163].

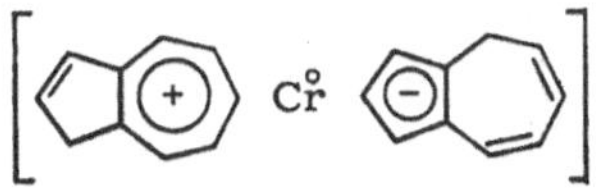

(40)

Simultaneous hydrogenation and dehydrogenation occur in the reaction of cycloheptatriene with CrCl$_3$ yielding (1.3.5-cycloheptatrienyl)-Cr(1.3-cycloheptadiene) [158].

Bonding of both valence tautomers of cyclooctatriene has been reported for (1.3.5-cyclooctatriene)Fe(bicyclo[4.2.0]octa-2.4-diene) and the corresponding Ru complex [337].

The diversity of the complexes described reflects the versatility of the isopropyl Grignard method.

The preparation of PtCl$_2$(CH=CH−CN)$_2$ from CH$_2$=CH−CN and triethylammonium chloroplatinate(II) may illustrate a different method for the photochemical synthesis of olefin metal complexes [308].

K. Photochemistry of Ferrocene

Despite the enormous interest in the chemistry of ferrocene [387] reports on photochemical studies are relatively scarce. This is all the more surprising as there is plenty of information about spectral data (section B), and the ferrocenes are very stable compounds and easily handled.

1. Photochemical Degradation of Ferrocenes

Solutions of ferrocene in cyclohexene, acetone, and 2-propanol [301] as well as in decalin, methanol, and 1-propanol [501] remain unchanged on irradiation. A photoreaction is observed with CCl$_4$, CBr$_4$, CHCl$_3$, CH$_2$Cl$_2$, CCl$_3$COOH etc. as discussed in the next section.

Substituted ferrocenes display a markedly increased photosensitivity: 1.1'-dibenzoylferrocene decomposes or shows colour changes on irradiation in alcohols, ethers, esters, acetone, acetonitrile, pyridine, dimethylsulfoxide etc. [501]. The products of these reactions have not been investigated.

An interesting reaction for the preparation of trialkylammonium cyclopentadienylides is the photolysis of trialkylferrocenyl ammonium hydroxide [347,355]:

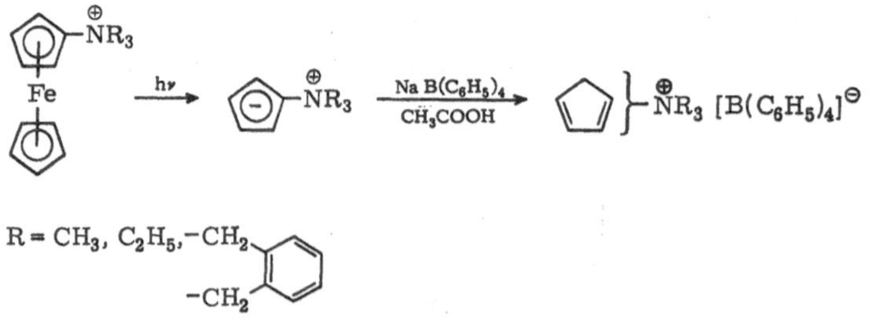

$$R = CH_3, C_2H_5, -CH_2 \cdots -CH_2$$

A similar reaction is observed with α-pyridylferrocene methiodide [357] and α-quinolylferrocene methiodide [355] in alkaline solution:

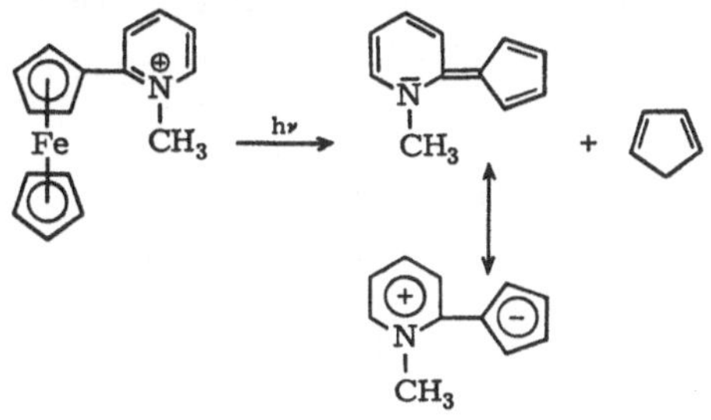

and

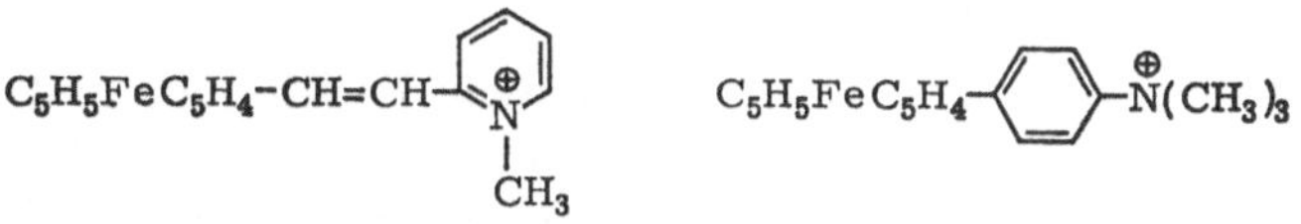

show analogous reactions [347].

Dimeric cyclopentadiene carboxylic acid results from the photolysis of ferrocene-1.1'-dicarboxylic acid in alkaline solution [353]. Ferrocene-1.1'-disulfonic acid forms (via its iron salt) $[(C_5H_5SO_3)_2Fe]_x$ on irradiation in water [355].

The decomposition of ferrocenyl amines in oxalic acid solution gives iron oxalate [356].

All reactions reported by *Nesmeyanov* and his coworkers [347,353,355, 356,357] can be explained by nucleophilic displacement reactions facilitated by a $M \rightarrow MR_\pi$ transition (see section B).

2. Photoreaction of Ferrocenes with Organic Halides

Solutions of ferrocene (Fn) in CCl_4 show a new band in the electronic spectrum due to charge transfer to the solvent [73] [see section B].

$$(C_5H_5)_2Fe + CCl_4 \xrightarrow{h\nu} [(C_5H_5)_2Fe]^+Cl^- + \cdot CCl_3 \qquad (42)$$

Irradiation into this transition causes the almost quantitative formation of ferricenium tetrachloroferrate $[(C_5H_5)_2Fe]^+[FeCl_4]^-$ [296,301]. This product can also be obtained radiation-chemically from Fn and CCl_4 [294,299,300,301] [see section L2], by reacting Cl_2 with excess Fn [301, 540], by treating Fn with $FeCl_3$ [541], by heating Fn with hexachlorocyclopentadiene to 90—120 °C [542] or by illuminating this mixture [430].

425

The corresponding tetrabromoferrate is produced in the irradiation of Fn and CBr_4 [296,299,300,301)] or by treating Fn with Br_2 [540,543)].

The structure of the ferricenium tetrachloroferrate has been a matter for discussion, since the Mössbauer spectrum consisting of a singlet [544)] and the IR data [542)] lend support to *(41)*.

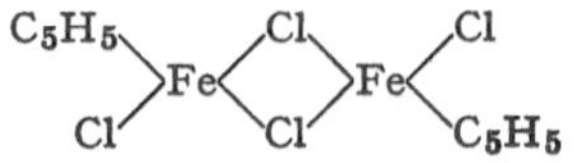

(41)

However, a different interpretation of the IR data has been proposed [545)] and the quantitative reduction of $Fn^+FeCl_4^-$ to ferrocene appears incompatible with *(41)*.

$Fn^+FeCl_4^-$, produced from $FeCl_3$ enriched with ^{57}Fe and ferrocene with natural ^{57}Fe content gave no ferrocene enriched in ^{57}Fe [544)]. This observation as well as the 100% increase of Fn^+ absorption by adding Fn to $Fn^+FeCl_4^-$ in methanol [282)], and the absence of $(R-C_5H_4)_2$ Fe in the reduction of the photochemical reaction product of (C_5H_5)-$Fe(C_5H_4-R)$ with CCl_4, are unequivocal evidence for the $Fn^+FeCl_4^-$ structure [282)].

The mechanism of the photochemical production of $Fn^+FeX_4^-$ from CCl_4, CBr_4, C_2Cl_6, CCl_3COOH [296,301)] and C_5Cl_6 [430)] in high yields is not yet quite clear with respect to the formation of FeX_3 from Fn^+. The known high susceptibility of Fn^+ (as compared with Fn) towards free radical attack [546)] allows the assumption, that the reaction [Eq. (43)] could be responsible for the FeX_3 formation.

$$Fn^+X^- + n \cdot CX_3 \rightarrow FeX_3 + \text{products} \qquad (43)$$

However, photoreactions of Fn^+ according to Eq. (44 and 45) cannot be ruled out and would perhaps explain the low quantum yield of ~ 0.01 for the formation of $Fn^+FeCl_4^-$ [301)].

$$(C_5H_5)_2Fe^+Cl^- + 2\ CCl_4 \xrightarrow{h\nu} FeCl_3 + 2\ C_5H_5CCl_3 \qquad (44)$$

$$2\ (C_5H_5)_2Fe^+Cl^- \xrightarrow{h\nu} (C_5H_5)_2Fe + FeCl_2 + 2 \cdot C_5H_5 \qquad (45)$$

It was demonstrated [282)] that the equilibrium [Eq. (46)] is shifted to the left if $Fn^+FeCl_4^-$ is soluble in the system and no Fe^{3+} is detected.

If $Fn^+FeCl_4^-$ precipitates, no Fe^{2+} is detected in the presence of excess Fn^+Cl^- (in benzene, CCl_4), see [Eq. (47)].

$$(C_5H_5)_2Fe^+Cl^- + FeCl_2 \rightleftharpoons FeCl_3 + (C_5H_5)_2Fe \qquad (46)$$

$$(C_5H_5)_2Fe^+Cl^- + FeCl_3 \rightleftharpoons [(C_5H_5)_2Fe]^+[FeCl_4]^- \qquad (47)$$

Solutions of $Fn^+FeX_4^-$ react with ferrocene according to [Eq. (48)] [282]

$$Fn^+FeX_4^- + Fn \rightarrow 2\ Fn^+X^- + FeX_2 \qquad (48)$$

$Fn^+FeX_4^-$ is reduced photochemically in CH_3OH according to Eq. (49) [282,296].

$$Fn^+FeX_4^- + 2\ R{-}H \xrightarrow{h\nu} Fn + FeX_2 + 2\ H{-}X + 2 \cdot R \qquad (49)$$

In CCl_4/CH_3OH mixtures this reaction can be combined with Eq. (42) thus providing a simple source of free radicals, which can be useful for the photopolymerization of vinyl monomers [296].

Substituted ferricenium tetrachloroferrates have been obtained photochemically from 1.1'-dimethyl-ferrocene and 1.1'-di-t-butyl-ferrocene with CCl_4 [282] and from 1.1'-bis(triphenylsilyl)ferrocene with C_5Cl_6 [430].

Mixtures of ferrocene and organic halides have been claimed as a material for photocopying in a patent [256] based on [296].

A few remarks on photoreactions of the bis-arene chromium(I) cation, which is isoelectronic with Fn^+, appear appropriate in this connexion.

Photolysis of $[(C_6H_6)_2Cr]^+$ in water has been proposed [395] to proceed according to [Eq. (50)].

$$[(C_6H_6)_2Cr]^+ \xrightarrow{h\nu} 2\ C_6H_6 + Cr^+ \xrightarrow{H_2O} Cr_{aq.}^{3+} + 2\ e^- \qquad (50)$$

Disproportionation of the bis(biphenyl) chromium cation is observed in the photoreaction with dipyridyl (Dipy.) in water or methanol [223]:

$$2\ [(C_6H_5{-}C_6H_5)_2Cr]^+Cl^- + 3\ Dipy. \xrightarrow{h\nu} (C_6H_5{-}C_6H_5)_2Cr + 2\ C_6H_5{-}C_6H_5$$
$$+ [Cr(Dipy.)_3]^{2+}Cl_2^{2-}$$

The photoreaction without additives is analogous:

$$2\ [(C_6H_5{-}C_6H_5)_2Cr]^+Cl^- \xrightarrow[H_2O]{h\nu} (C_6H_5{-}C_6H_5)_2Cr + CrCl_2 + 2\ C_6H_5{-}C_6H_5$$

The ratio of $(C_6H_5-C_6H_5)_2Cr$ to "naked" chromium ions is 1.38 in this case, showing some contribution of reaction [Eq. (51)]:

$$[(C_6H_5-C_6H_5)_2Cr]^+ + Cr^{2+} \rightarrow (C_6H_5-C_6H_5)_2Cr + Cr^{3+} \qquad (51)$$

3. Photoreactions of Substituted Ferrocenes Involving the Substituents

The photorearrangement of phenylferrocenoate (Photo-*Fries* reaction) [Eq. (52)] has been reported [151].

$$C_5H_5FeC_5H_4COOC_6H_5 \xrightarrow{h\nu} C_5H_5FeC_5H_4-CO-\!\!\left\langle\!\bigcirc\!\right\rangle\!\!-OH \qquad (52)$$

No ortho product is observed. If the para position is blocked with a methyl group the only identified products are ferrocenoic acid and p-tolyl-ferrocene.

$C_5H_5FeC_5H_4-CH(R)-CH_2-CO-C_6H_5$ $(R = C_6H_5$, cyclohexyl) is converted to benzoyl ferrocene on irradiation [71].

Permanent images can be produced by irradiation of cinnamoyl ferrocene (and substitution products thereof) on a photographic support. An insoluble yellow compound is formed in the areas of exposure, which remains after tieatment with a solvent (dissolving the unreacted starting material). Iodoform is a sensitizer of this reaction [138].

The tosylhydrazone of α-keto-1.1'-trimethylene ferrocene failed to give any well-defined product on irradiation in the presence of sodium methoxide, while heating produced the starting ketone and the α-methoxy compound [16].

Photolysis of ferrocenylsulfonyl azide *(42)* in benzene gives the novel ferrocenophane-thiazine-1,1-dioxide *(44)* as the main product (67%) and some (14%) ferrocenyl sulfonamide *(45)*. No *(44)* is detected in the corresponding thermal reaction, however, *(45)* becomes the major product. The multiplicity of the intermediate ferrocenylsulfonyl nitrene *(43)* has not yet been clarified [565].

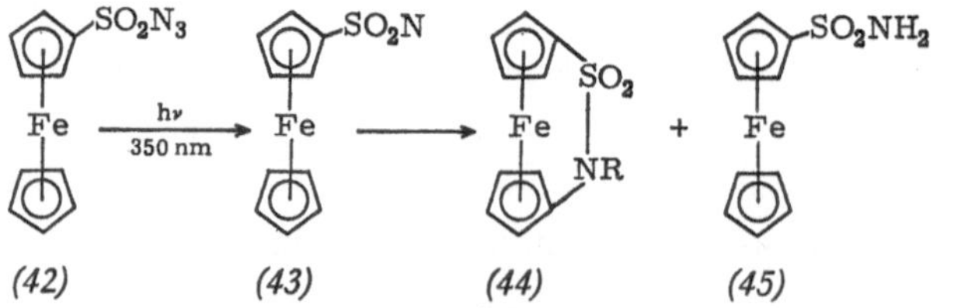

(42)   *(43)*   *(44)*   *(45)*

a) R = H
b) R = CH₃

4. Photosensitization with Ferrocene

Dannenberg and *Richards* were the first to report photosensitization with ferrocene of cis-trans isomerization of piperylene as well as of its dimerization [128,129]. In benzene solution ferrocene behaved like a high-energy triplet sensitizer: the photostationary state of the cis-trans isomerization contained 54—55% trans-piperylene, and the dimers were composed of 92% cyclobutanes and cyclooctadienes and of 8% cyclohexenes. Excitation of the long wavelength absorption was ineffective, at least the 324 nm transition had to be excited. On the basis of non-additive electronic absorption data of ferrocene-piperylene mixtures (only at <300 nm) and on the observation of a NMR shift of the piperylene methyl doublet on addition of ferrocene (in CCl_4) a mechanism involving a ferrocene-piperylene complex was proposed for the energy transfer [129]:

$$\text{ferrocene} + \text{diene} \rightleftharpoons \text{complex} \tag{53}$$

$$\text{complex} \xrightarrow{h\nu} \text{complexS}_n \tag{54}$$

$$\text{complexS}_n \longrightarrow \text{ferroceneT}_n + \text{dieneT}_n \tag{55}$$

$$\text{dieneT}_n \longrightarrow \text{dieneT}_1 \tag{56}$$

$$\text{dieneT}_1 \longrightarrow \text{products} \tag{57}$$

A detailed discussion of other mechanistic possibilities is given in [128].

However, it was pointed out by another group [203] that reaction [Eq. (55)] would require the lowest ferrocene triplet to be ∼ 32 kcal/mole [excitational energy of 313 nm = 91.2 kcal/mole; lowest triplet state of trans-piperylene = 58.8 kcal/mole], a figure not in accord with the actual value of 40.5 kcal/mole [in ethanol] [419]. These workers were unable to detect any ferrocene-piperylene complex by NMR, but observed small shifts in the electronic spectra below 310 nm. Therefore, possible transitions of such a complex must be isoenergetic with those of ferrocene at >310 nm. The conclusion was reached [203], that "if an appreciable amount of complex is formed, the energy-transfer mechanism may be one where the complex in its singlet state (isoenergetic with ferrocene's $^1E_{1g}{}^+$ state) intersystem-crosses to its triplet state (ferrocene's $^3E_{1g}{}^+$ state) and then dissociates to ferrocene in its ground-state and to a triplet piperylene which isomerizes".

This mechanism corresponds to one considered earlier by *Dannenberg* [128]. The consideration of a simple triplet energy transfer from the ferrocene T_2 (54 kcal/mole) [see section B] to piperylene [$T_1 = 56.9$ for the cis isomer and 58.8 kcal/mole for the trans isomer] has been rejected, since ferrocene was thought to behave like a high-energy sensitizer [128,129]. However, in trans-1.2-dimethylcyclohexane it was shown to be a low-energy sensitizer and the quantum yield for photosensitized cis-trans isomerization of piperylene was found to be $2.6 \cdot 10^{-3}$ at 313 nm [203].

Ferrocene is a very efficient triplet quencher, its quenching constants showing practically no difference for triplets between 66.6 (triphenylene) and 42.6 kcal/mole (anthracene) [185]. Two conclusions are possible: either energy is transferred to ferrocene T_1 (40.5 kcal/mole) or the quenching does not involve energy-transfer at all, and weak electronic interaction on contact between ferrocene and the triplet molecule effects rapid intersystem crossing of the latter by spin-orbit coupling.

Ferrocene does not affect the sensitized and the unsensitized photo-reaction of diazomethane with cyclohexene [520].

L. Radiation Chemistry of Transition Metal Complexes

"Radiation chemistry may be defined as the study of the chemical effects produced in a system by the absorption of ionizing radiation" [547].

Absorption of γ-rays in a system will cause the formation of fast electrons by *Compton* scattering, by pair production, and by photoelectric absorption. Therefore, looking upon chemical effects brought about by β- and γ-rays we will have to consider similar primary processes. Differences are mainly due to different absorption characteristics (half-thickness values) of organic matter for the β- and γ-rays [548,549,562].

The following interactions of electrons with matter can be visualized [551]:

Ionization:

$$M + e \rightarrow M^+ + e' + e''$$
$$N + e_{slow} \rightarrow N^-$$

Electronic Excitation:

a) optically allowed transitions: $\quad M + e_{fast} \rightarrow M_S^* + e'$

b) optically forbidden transitions: $M + e_{slow} \rightarrow M_T^* + e'$

c) highly excited states by capture of subexcitational electrons: $\qquad M^+ + e_{sub} \rightarrow M^{**}$

The ions produced can undergo the following reactions:

Interaction of charged particles: $M^+ + N^- \rightarrow M^{**} + N^{**}$

Ion molecule reactions: $\qquad\qquad M^+ + NL \rightarrow MN^+ + L$

Charge transfer: $\qquad\qquad\qquad M^+ + N \rightarrow M + N^+$

Neutralization: $\qquad\qquad\qquad\quad M^+ + e \rightarrow M^{**}$

The final fate of ions in many instances is the formation of electronically excited species. Therefore, similarities of radiation-chemically induced processes with those induced photochemically can be frequently expected. However, this point should be made with some "reservatio mentalis" only.

One major difference of photo- and radiation-chemical reactions is in the factors controlling energy absorption: the absorption of light is controlled by the *Lambert-Beer* law, whereas the absorption of ionizing radiation (in a crude approximation) depends on the electron density of a compound (i. e. on its electron fraction in a mixture). This means that in cases, where the high extinction coefficient of a colored product affects the possible degree of photochemical conversion or causes secondary photochemical reactions (internal light filter), the application of ionizing radiation can be of preparative advantage. On the other side only a few radiation-chemical reactions can be expected to show the same selectivity as photochemical reactions very often do. However, appropriate energy transfer in radiation processes [550] can provide interesting preparative possibilities, as has been shown, e.g. for cycloaddition reactions in aromatic solvents [553].

The range of electrons with energies < 100 eV in liquids or solids is relatively small and they will produce unusually high concentrations of excited or ionized species in a small cluster or "spur". Therefore, reactions are possible in these "spurs", which are usually very improbable and not observed e.g. in photochemical reactions.

For detailed information about recent trends in radiation chemistry the reader is referred to [552].

1. Radiation Chemistry of Metal Carbonyls

$Fe_2(CO)_9$ and $[Fe(CO)_4]_3$ are formed in the ^{60}Co-γ-irradiation of neat $Fe(CO)_5$ with G-values (number of formed or converted molecules per 100 eV absorbed energy) of 6.7 and 1.65 [43,44]. The same products are obtained in benzene and cyclohexane [43,297,299]. Energy transfer from these solvents to $Fe(CO)_5$ has been demonstrated [41,43]. The ratio $G_{Fe_2(CO)_9}$ to $G_{[Fe(CO)_4]_3}$ is 12.8 in cyclohexane ($G_{[Fe(CO)_4]_3} = 0.22$) and 152 in benzene ($G_{[Fe(CO)_4]_3} = 0.028$) for a solution of 0.2 mole/l $Fe(CO)_5$ [41].

Charge transfer has been proposed for the energy transfer from cyclohexane ions to $Fe(CO)_5$ yielding $Fe_2(CO)_9$, while in benzene excited solvent molecules have been considered for this process.

The decrease of $G_{Fe_2(CO)_9}$ in benzene in the presence of stilbene suggests that benzene triplets are involved [41]. However, the known reactivity of stilbene (and of the other additives studied, e.g. diethyl ether, benzophenone) with iron carbonyls [see section E] complicates the interpretation of this effect.

Ferrocene does not react with iron carbonyls under irradiation. Its low ionization potential [$\sim$ 7—7.5 eV [387]] and its high triplet quenching efficiency [185] would cause a strong decrease of $G_{Fe_2(CO)_9}$ and $G_{[Fe(CO)_4]_3}$ if benzene triplets or any ions were involved in their formation. This is not the case. 0.54 mole/l ferrocene causes $G_{Fe_2(CO)_9}$ to decrease by only 20%, and has no effect on $G_{[Fe(CO)_4]_3}$ [0.21 mole Fe(CO)$_5$/1 benzene] [289]. This observation suggests that energy transfer from excited benzene singlets could play a role.

The formation of [Fe(CO)$_4$]$_3$ is obviously not due to ions, since N$_2$O, diethyl ether [41] and ferrocene do not decrease $G_{[Fe(CO)_4]_3}$. No Fe$_2$(CO)$_9$ is observed in the ^{60}Co-γ-irradiation of Fe(CO)$_5$ [0.84 mole/l] in neat Ni(CO)$_4$, $G_{[Fe(CO)_4]_3}$ is still $\sim$ 1/3 of its value in benzene. The radiation-chemical precursor of Fe$_2$(CO)$_9$ appears to have the same properties as that in its photochemical formation (see section C), and probably is Fe(CO)$_4$. The formation of [Fe(CO)$_4$]$_3$ in a "spur" reaction [as suggested earlier [44]] would appear to be the best explanation why it is still observed in Ni(CO)$_4$.

Ni(CO)$_4$ is not affected by intense ^{60}Co-γ-irradiation and can be looked upon as one of the most radiation stable compounds [44]. A study of the radiolysis of mixtures of cyclohexane and/or benzene and Ni(CO)$_4$ has demonstrated its protective properties. Probably, energy transfer to Ni(CO)$_4$ and the reversible dissociation

$$Ni(CO)_4^* \rightarrow Ni(CO)_3 + CO \rightarrow Ni(CO)_4$$

provides a very efficient "energy sink" [42,43].

[60]Co-γ-irradiation of Fe(CO)$_5$ in benzene in the presence of π-donors produces the corresponding π-donor iron tetracarbonyls [289,295,297,299,300]. This is not very surprising, if the same transient, namely Fe(CO)$_4$, results from Fe(CO)$_5$ by illumination or ionizing irradiation. However, a comparison of the yields shows the interesting fact that the selectivity appears to be very similar under both conditions [289,297]:

Complex	Yield [1]	
	$h\nu$	^{60}Co-γ
(Maleic anhydride)Fe(CO)$_4$	52%	37%
(Dimethyl fumarate)Fe(CO)$_4$	53%	61%
(Vinyl acetate)Fe(CO)$_4$	48%	46%
(Methyl methacrylate)Fe(CO)$_4$	57%	31%

[1] On the basis of CO formation.

Also the reduction of aromatic nitro compounds with $Fe(CO)_5$ to complexes of the corresponding nitroso compounds (see section E9) can be brought about in good yields by ^{60}Co-γ-irradiation [290,292,293].

It appears worthwhile mentioning in this connexion, that the reaction of active nitrogen (produced by electrodeless discharge) with $Ni(CO)_4$, $Fe(CO)_5$, $Cr(CO)_6$, $W(CO)_6$, $Mn_2(CO)_{10}$ and $Co(NO)(CO)_3$ and the resulting emission has been studied. Stepwise degradation according to

$$M(CO)_n + N \rightarrow M(CO)_{n-1} + NCO$$

has been proposed [76].

2. Radiation Chemistry of Ferrocene

Ferrocene is very stable towards high-energy electron irradiation in the solid state [304]:

$$G_{H_2} \qquad : 0.01$$
$$G_{hydrocarbons}: 0.004$$
$$G_{inorganic\ Fe} : 0.04$$

Ferrocene exhibits a substantial protective effect (similar to that of naphthalene) in the ^{60}Co-γ-irradiation of cyclohexane, $G_{cyclohexene}$ and $G_{dicyclohexyl}$ being reduced to $\sim 50\%$ by 0.1 mole/l ferrocene [65]. Ferricenium sulfate in water is reduced to ferrocene with $G_{Fn} = 5.46$ [with exclusion of O_2] [65].

The radiation sensitivity of ferrocene in CCl_4 [109,301] is due to the formation of chlorine atoms which produce ferricenium chloride. The latter is attacked by the irradiation products of CCl_4 and yields ferricenium tetrachloroferrate (see section K.2) as the final product [296,299, 300,301], and not ferricenium chloride [109]. $Fn^+FeCl_4^-$ precipitates and can be easily determined gravimetrically. $G_{Fn^+FeCl_4^-}$ is independent of dose, dose-rate, and temperature and the system has been proposed for chemical dosimetry [294]. For a detailed discussion of the mechanism and its possible implication on the determination of the radical yield from CCl_4, the reader is referred to [299].

M. Synthesis of Labeled Metal Carbonyls and Metallocenes by Nuclear Processes

The reviewers feel that it could be of interest for future developments of transition metal complex photochemistry and its possible application in biologically important systems to make use of labeled metals. Therefore, a brief survey is given of the wealth of information available concerning the direct preparation of radioactive metal carbonyls and metallocenes by nuclear processes. The chemical reactions due to the *Szilard-Chalmers* effect [182] provide additional information about preferential cleavage reactions of these complexes.

Thermal neutron irradiation [$^{50}Cr(n,\gamma)^{51}Cr$] of $(C_6H_6)_2{}^{50}Cr$ [12%] [55,56], $(C_6H_6)^{50}Cr(CO)_3$ [10%] [49,56], and $^{50}Cr(CO)_6$ [60%] [56] gives the ^{51}Cr labeled parent compounds (retention) in reasonable or good yields [in square bracketts]. 13.5% $^{51}Cr(CO)_6$ and small amounts of $^{51}Cr(C_6H_6)_2$ were identified as by-products in the irradiation of $^{50}Cr(C_6H_6)(CO)_3$.

21% $^{59}Fe(C_5H_5)_2$ can be obtained from neutron irradiation of ferrocene [498].

7.6% retention (^{56}Mn) has been observed in the neutron irradiation of methylcyclopentadienyl manganese tricarbonyl; further products identified were $CH_3{}^{56}Mn(CO)_5$ and $^{56}Mn_2(CO)_{10}$ [115,253,434,528,530].

The central metal in metal carbonyls and metallocenes can be substituted by uranium fission products if intimate mixtures of U_3O_8 and the complex are neutron irradiated. In this way $^{103}Ru(C_5H_5)_2$ has been obtained from $Fe(C_5H_5)_2$, and $^{99}Mo(CO)_6$ from $Cr(CO)_6$ [48,53,54].

Neutron capture and subsequent β-emission convert an element into its next right neighbour in the periodic table. According to this principle the following conversions have been substantiated:

$$^{104}Ru(C_5H_5)_2 \rightarrow {}^{105}Rh(C_5H_5)_2 \quad [57]$$

$$Mo(C_6H_6)_2 \rightarrow [^{99m}Tc(C_6H_6)_2]^+ \quad [51,52]$$

$$[C_5H_5Mo(CO)_3]_2 \rightarrow C_5H_5{}^{99m}Tc(CO)_3 \quad [50]$$

$$U(C_5H_5)_3X \rightarrow {}^{239}Np(C_5H_5)_3X \quad [47]$$

N. Conclusions

Photochemical methods are very useful for the synthesis of transition metal complexes, expecially of thermally sensitive compounds.

The assumption of a free coordination site produced by dissociation of a metal-CO bond can explain many photoreactions of metal carbonyl compounds.

Photoreactions of specific interest for the organic synthetic chemist are photochemical isomerization and addition (especially cycloaddition) reactions of coordinated organic systems.

The metallocenes constitute an interesting class of compounds for the physical organic photochemist due to the diversity of excited states possible.

The applications of the described reactions in photopolymerization and optical fixation of information (e.g. photocopying, data processing) could be of industrial interest; in many of these reactions color changes are accompanied by changes in the electric properties of the material.

To learn more about the factors governing the photochemical behaviour of organic systems coordinated to transition metals systematic studies appear necessary, e.g. the comparison of olefin photochemistry with that of olefin metal complexes (e.g. see [428]).

An impressive example of the possibilities of metal complex photochemistry is the substantiation of the critical step in corrin synthesis reported by *Eschenmoser* [145]: Ring closure is brought about by antarafacial cycloisomerization of a secocorrinoidic palladium complex by photochemical 1.16-hydrogen shift.

Acknowledgements

Our thanks are due to Mrs. *Z. Pfajfer* and to Miss *R. Wagner* for their help in collecting the references.

Valuable discussions with Dr. *M. Herberhold*, Munich, who also provided us with a number of references, are gratefully appreciated Prof. Dr. *W. Strohmeier*, Würzburg, most kindly supported our efforts with a collection of his reprints.

We are thankful to Drs. *R. B. King*, Athens, Georgia, *P. W. Jennings*, Bozeman, Montana, and *R. G. Sutherland*, Saskatoon, Canada, for personal communications and preprints of recent work.

The assistance of Miss *I. Heuer* in typing the manuscript is most gratefully acknowledged.

O. References

[1] *Abel, E. W., I. S. Butler*, and *C. R. Jenkins:* J. Organometal. Chem. (Amsterdam) *8*, 382 (1967).

[2] — Quart. Rev. (London) *17*, 133 (1963).

[3] *Adamson, A. W.:* In: Proc. 11th Int. Conf. Coord. Chem., Haifa-Jerusalem (1968), p. 63 Ed. *M. Cais.* Amsterdam: Elsevier 1968.

[4] —, *W. L. Waltz, E. Zinato, D. W. Watts, P. D. Fleischhauer*, and *R. D. Lindholm:* Chem. Rev. *68*, 541 (1968).

[5] — Rec. Chem. Progr. *29*, 191 (1968).

[6] *Agron, P. A., R. D. Ellison*, and *H. A. Levy:* Acta Cryst. *23*, 1079 (1967).

[7] *Ahmad, M., R. Bruce*, and *G. R. Knox:* J. Organometal Chem. (Amsterdam) *6*, 1 (1966).

[8] *Alderdice, D. S.:* J. Mol. Spectry. *15*, 509 (1965).

[9] *Amiet, R. G., P. C. Reeves*, and *R. Pettit:* Chem. Commun. 1208 (1967).

[10] *Anderson, C. B.*, and *B. J. Burreson:* Chem. Ind. (London) 620 (1967).

11) *Angelici, R. J.:* Organometal. Chem. Rev. *3*, 173 (1968).
12) —, and *L. Busetto:* Inorg. Chem. *7*, 1935 (1968).
13) —, and *M. D. Malone:* Inorg. Chem. *6*, 1731 (1967).
14) —, and *J. R. Graham:* Inorg. Chem. *6*, 988 (1967).
15) —, and *W. Loewen:* Inorg. Chem. *6*, 682 (1967).
16) *Applebaum, M. N., R. W. Fish,* and *M. Rosenblum:* J. Org. Chem. *29*, 2452 (1964).
17) *Armstrong, A. T., D. G. Carroll,* and *S. P. McGlynn:* J. Chem. Phys. *47*, 1104 (1967).
18) —, *F. Smith, E. Elder,* and *S. P. McGlynn:* J. Chem. Phys. *46*, 4321 (1967).
19) *Arnold, D. R., D. J. Trecker,* and *E. B. Whipple:* J. Am. Chem. Soc. *87*, 2596 (1965).
20) *Asinger, F., B. Fell* u. *K. Schrage:* Chem. Ber. *98*, 381 (1965).
21) — — — Chem. Ber. *98*, 372 (1965); *F. Asinger, B. Fell, G. Collin:* Chem. Ber. *96*, 716 (1963).
22) *Baenziger, N. C., H. L. Haight,* and *J. R. Doyle:* Inorg. Chem. *3*, 1535 (1964).
23) *Bagga, M. M., P. E. Baikie, O. S. Mills,* and *P. L. Pauson:* Chem. Commun. 1106 (1967).
24) —, *P. L. Pauson, F. J. Preston,* and *R. I. Reed:* Chem. Commun. 543 (1965); *P. E. Baikie, O. S. Mills:* Chem. Commun. 707 (1966).
25) *Baird, M. C., G. Hartwell,* and *G. Wilkinson:* J. Chem. Soc. A 2037 (1967).
26) *Baldwin, J. E.,* and *R. H. Greeley:* J. Am. Chem. Soc. *87*, 4514 (1965).
27) *Ballhausen, C. J.,* and *H. B. Gray:* Molecular Orbital Theory. New York: W. A. Benjamin, Inc. 1965.
28) *Balzani, V.,* and *V. Carassiti:* J. Phys. Chem. *72*, 383 (1968).
29) —, *L. Moggi* u. *V. Carassiti:* Ber. Bunsenges. Physik. Chem. *72*, 288 (1968).
30) — —, *F. Scandola,* and *V. Carassiti:* Inorg. Chim. Acta Rev. *1*, 7 (1967).
31) —, *V. Carassiti, L. Moggi,* and *F. Scandola:* Inorg. Chem. *4*, 1243 (1965).
32) *Bamford, C. H., R. W. Dyson,* and *G. C. Eastmond:* J. Polymer Sci. Part C, *16*, 2425 (1967).
33) —, *P. A. Crowe,* and *R. P. Wayne:* Proc. Roy. Soc. (London) A *284*, 455 (1965); *C. H. Bamford,* and *R. P. Wayne:* In: Photochemistry and Reaction Kinetics. *P. G. Ashmoore, F. S. Dainton, T. M. Sugden* (Editors). Cambrigde University Press 1967.
34) —, *J. Hobbs,* and *R. P. Wayne:* Chem. Commun. 469 (1965).
35) *Banks, R. E., T. Harrison, R. N. Haszeldine, A. B. P. Lever, T. F. Smith,* and *J. B. Walton:* Chem. Commun. 30 (1965).
36) *Barbeau, C.:* Can. J. Chem. *45*, 161 (1967).
37) *Barnett, K. W.:* Ph. D. Thesis, The University of Wisconsin, 1967, Diss. Abstr. B *28*, 3203 (1968).
38) —, and *P. M. Treichel:* Inorg. Chem. *6*, 294 (1967).
39) *Barr, T. H.,* and *W. E. Watts:* J. Organometal. Chem. (Amsterdam) *15*, 177 (1968).
40) *Barrett, P. F.,* and *A. J. Poe:* J. Chem. Soc. A, 429 (1968).
41) *Barzynski, H.:* Z. Physik. Chem. (Frankfurt) *60*, 185 (1968).
42) —, *R. R. Hentz,* and *M. Burton:* J. Phys. Chem. *69*, 2034 (1965).
43) —, u. *D. Hummel:* Z. Physik. Chem. (Frankfurt) *39*, 148 (1963).
44) — — Z. Physik. Chem. (Frankfurt) *38*, 103 (1963).
45) *Basolo, F.,* and *R. G. Pearson:* Mechanisms of Inorganic Reactions, A Study of Metal Complexes in Solution. New York: Wiley 1968.
46) —, and *A. Wojcicki:* J. Am. Chem. Soc. *83*, 520 and 525 (1961).

47) *Baumgärtner, F., E. O. Fischer* u. *P. Laubereau:* Naturwissenschaften *52*, 560 (1965).
48) —, and *P. Reichold:* Ger. Patent 1156387 (1963); Chem. Abstr. *60*, 5018c (1964).
49) —, and *U. Zahn:* Radiochim. Acta *1*, 51 (1963).
50) —, *E. O. Fischer* u. *U. Zahn:* Naturwissenschaften *49*, 156 (1962).
51) — — — Naturwissenschaften *48*, 478 (1961).
52) — — — Chem. Ber. *94*, 2198 (1961).
53) —, u. *P. Reichold:* Z. Naturforsch. A *16*, 945 (1961).
54) — — Z. Naturforsch. A *16*, 374 (1961).
55) —, *U. Zahn* u. *J. Seeholzer:* Z. Naturforsch. A *15*, 1086 (1960).
56) — Annual Meeting German Chem. Soc., Stuttgart 1960.
57) —, *E. O. Fischer* u. *U. Zahn:* Chem. Ber. *91*, 2336 (1958).
58) *Beach, N. A.,* and *H. B. Gray:* J. Am. Chem. Soc. *90*, 5713 (1968).
59) *Berry, R. S.:* J. Chem. Phys. *35*, 2025 (1961).
60) *Beveridge, A. D.,* and *H. C. Clark:* J. Organometal. Chem. (Amsterdam) *11*, 601 (1968).
61) — — Inorg. Nucl. Chem. Letters *3*, 95 (1967).
62) *Bichler, R. E. J., M. R. Booth,* and *H. C. Clark:* Inorg. Nucl. Chem. Letters *3*, 71 (1967).
63) *Birch, A. J., P. E. Cross, J. Lewis, D. A. White,* and *S. B. Wild:* J. Chem. Soc. A 332 (1968).
64) *Bird, C. W., D. L. Colinese, R. C. Cookson, J. Hudec,* and *R. O. Williams:* Tetrahedron Letters 373 (1961).
65) *Blackburn, R.,* and *A. Kabi:* Chem. Commun. 862 (1966).
66) *Bock, H.,* u. *H. Tom Dieck:* Z. Anorg. Allgem. Chem. *345*, 9 (1966).
67) *Bolton, E. S., M. Dekker, G. R. Knox,* and *C. G. Robertson:* Chem. Ind. (London) 327 (1969).
68) *Bor, G.:* J. Organometal. Chem. (Amsterdam) *11*, 195 (1968).
69) — J. Organometal. Chem. (Amsterdam) *10*, 343 (1967).
70) *Boston, J. L., S. O. Grimm,* and *G. Wilkinson:* J. Chem. Soc. 3468 (1963).
71) *Bozak, R. E.:* Chem. Ind. (London) 24 (1969).
72) *Bradley, D. C.,* and *A. S. Kasenally:* Chem. Commun. 1430 (1968).
73) *Brand, J. C. D.,* and *W. Snedden:* Trans. Faraday Soc. *53*, 894 (1957).
74) *Braye, E. H.,* and *W. Hübel:* In: Inorganic Syntheses, Vol. VIII, pp. 178, Ed. *H. F. Holtzclaw, jr.* New York: McGraw Hill 1966.
75) — — Angew. Chem. *75*, 345 (1963).
76) *Brennen, W. R.,* and *G. B. Kistiakowsky:* J. Chem. Phys. *44*, 2695 (1966).
77) *Brenner, K. S., E. O. Fischer, H. P. Fritz* u. *C. G. Kreiter:* Chem. Ber. *96*, 2632 (1963).
78) *Brookes, P. R.,* and *B. L. Shaw:* Chem. Commun. 919 (1968).
79) *Brown, D. A., D. Cunningham,* and *W. K. Glass:* J. Chem. Soc. A, 1563 (1968).
80) —, *W. K. Glass,* and *B. Kumar:* Chem. Commun. 736 (1967).
81) —, *D. Cunningham,* and *W. K. Glass:* Chem. Commun. 306 (1966).
82) *Bruce, M. I., M. Cooke,* and *M. Green:* J. Organometal. Chem. (Amsterdam) *13*, 227 (1968).
83) — — — Angew. Chem. *80*, 662 (1968).
84) *Bruce, R.,* and *G. R. Knox:* J. Organometal. Chem. (Amsterdam) *6*, 67 (1966).
85) *Calderazzo, F.,* and *F. l'Eplattenier:* Inorg. Chem. *6*, 1220 (1967).
86) *Callear, A. B.,* and *R. J. Oldman:* Nature *210*, 730 (1966); Trans. Faraday Soc. *63*, 2888 (1967).
87) — Proc. Roy. Soc. (London) A *265*, 88 (1962).
88) — Proc. Roy. Soc. (London) A *265*, 71 (1962).

89) *Calvert, J. G.*, and *J. N. Pitts, jr.*: Photochemistry. New York: Wiley 1966.
90) *Candlin, J. P.*, and *W. H. Janes*: J. Chem. Soc. C 1856 (1968).
91) *Carr, M. D., V. V. Kane*, and *M. C. Whiting*: Proc. Chem. Soc. (London) 408 (1964).
92) *Carroll, D. G.*, and *S. P. McGlynn*: Inorg. Chem. *7*, 1285 (1968).
93) *Caulton, K. G.*, and *R. F. Fenske*: Inorg. Chem. *7*, 1273 (1968).
94) *Chatt, J.*, and *H. R. Watson*: J. Chem. Soc. 4980 (1961).
95) *Chaudhari, F. M.*, and *P. L. Pauson*: J. Organometal. Chem. (Amsterdam) *5*, 73 (1966).
96) *Chien, J. C. W.*: J. Phys. Chem. *67*, 2477 (1963).
97) *Clark, H. C., J. D. Cotton*, and *J. H. Tsai*: Inorg. Chem. *5*, 1582 (1966).
98) —, and *J. H. Tsai*: Inorg. Chem. *5*, 1407 (1966).
99) — — , and *W. S. Tsang*: Chem. Commun. 171 (1965).
100) — — Chem. Commun. 111 (1965).
101) *Clark, R. J., J. P. Hargaden, H. Haas*, and *R. K. Sheline*: Inorg. Chem. *7*, 673 (1968).
102) — Inorg. Chem. *6*, 299 (1967).
103) —, and *P. I. Hoberman*: Inorg. Chem. *4*, 1771 (1965).
104) — Chem. Eng. News *42*, No. 43, 52 (1964).
105) — Inorg. Chem. *3*, 1395 (1964).
106) *Coates, G. E., M. L. H. Green, P. Powell*, and *K. Wade*: Principles of Organometallic Chemistry. London: Methuen 1968.
107) — —, and *K. Wade*: Organometallic Compounds, Vol. I and II. London: Methuen 1967/1968.
108) *Cohen, I. A.*, and *F. Basolo*: J. Inorg. Nucl. Chem. *28*, 511 (1965).
109) *Collinson, E., F.S. Dainton*, and *H. Gillis*: J. Phys. Chem. *65*, 695 (1961).
110) *Cook, D. J., M. Green, N. Mayne*, and *F. G. A. Stone*: J. Chem. Soc. A 1771 (1968).
111) *Cooke, M., M. Green*, and *D. Kirkpatrick*: J. Chem. Soc. A 1507 (1968).
112) *Corey, E. R.*, and *L. F. Dahl*: J. Am. Chem. Soc. *83*, 2203 (1961).
113) *Costa, G., G. Mestroni*, and *G. Pellizer*: J. Organometal. Chem. (Amsterdam) *15*, 187 (1968).
114) — — Tetrahedron Letters 1781 (1967).
115) *Costea, T., I. Negoescu, P. Vasudev*, and *D. R. Wiles*: Can. J. Chem. *44*, 885 (1966).
116) *Cotton, F. A.*, and *C. R. Reich*: J. Am. Chem. Soc. *91*, 847 (1969).
117) — Accounts Chem. Research *1*, 257 (1968).
118) —, and *G. Wilkinson*: Advanced Inorganic Chemistry. New York: Interscience 1966.
119) *Cousins, M.*, and *M. L. H. Green*: J. Chem. Soc. 1567 (1964).
120) — — J. Chem. Soc. *889* (1963).
121) *Craig, P. J., M. Green, A. J. Rest*, and *F. G. A. Stone*: J. Organometal. Chem. (Amsterdam) *12*, 548 (1968).
122) *Cuingnet, E.*, et *M. Adalberon*: Compt. Rend. *257*, 461 (1963).
123) *Cullen, W. R., D. A. Harbourne, B. V. Liengme*, and *J. R. Sams*: J. Am. Chem. Soc. *90*, 3293 (1968).
124) —, and *R. G. Hayter*: J. Am. Chem. Soc. *86*, 1030 (1964).
125) *Dahl, J. P.*, and *C. J. Ballhausen*: Kgl. Danske Videnskab. Selskab., Mat.-Fys. Medd. *33*, 3 (1961).
126) *Damico, R.*: J. Org. Chem. *33*, 1550 (1968).
127) —, and *T. J. Logan*: J. Org. Chem. *32*, 2356 (1967).

128) *Dannenberg, J.:* Ph. D. Thesis, California Institute of Technology 1967; Diss. Abstr. B 27, 3456 (1967).

129) —, and *J. H. Richards:* J. Am. Chem. Soc. 87, 1626 (1965).

130) *Darensbourg, D. J.,* and *T. L. Brown:* Inorg. Chem. 7, 959 (1968).

131) *Day, A. C.,* and *J. T. Powell:* Chem. Commun. 1241 (1968).

132) *Denning, R. G., F. R. Hartley,* and *L. M. Venanzi:* J. Chem. Soc. A 1322 (1967).

133) *Deubzer, B.,* and *H. D. Kaesz:* J. Am. Chem. Soc. 90, 3276 (1968).

134) *Dewar, J.,* and *H. O. Jones:* Proc. Roy. Soc. (London) A 77, 66 (1906).

135) — — Proc. Roy. Soc. (London) A 76, 558 (1905).

136) *Dobson, G. R.:* J. Phys. Chem. 69, 677 (1965).

137) —, *M. F. A. El-Sayed, I. W. Stolz,* and *R. K. Sheline:* Inorg. Chem. 1, 526 (1962).

138) *Dubosç, J. P. C. G.:* (Eastman Kodak Co.), U.S. Patent 3335008 (1967); Chem. Abstr. 67, 77884g (1967).

139) *Eaton, D. R.:* J. Am. Chem. Soc. 90, 4272 (1968).

140) *Egger, H.,* u. *K. Schlögl:* Monatsh. Chem. 95, 1750 (1964).

141) *Eisch, J. J.:* The Chemistry of Organometallic Compounds, The Main Group Elements. New York: Macmillan 1967.

142) *El-Sayed, M. F. A.:* J. Phys. Chem. 68, 433 (1964).

143) *L'Eplattenier, F.,* and *F. Calderazzo:* Inorg. Chem. 7, 1200 (1968).

144) *Erhard, K.:* Z. Physik. Chem. (Frankfurt) 36, 126 (1963).

145) *Eschenmoser, A., Y. Yamada, D. Miljkovic, P. Wehrli, B. Golding, P. Löliger, R. Keese* u. *K. Müller:* Angew. Chem. 81, 301 (1969).

146) *Evans, G. O.,* and *R. K. Sheline:* J. Inorg. Nucl. Chem. 30, 2862 (1968).

147) —, *J. P. Hargaden,* and *R. K. Sheline:* Chem. Commun. 186 (1967).

148) *Eyber, G.:* Z. Physik. Chem. 144, 1 (1929).

149) *Fell, B., P. Krings,* u. *F. Asinger:* Chem. Ber. 99, 3688 (1966).

150) *Fields, R., M. M. Germain, R. N. Haszeldine,* and *P. W. Wiggans:* Chem. Commun. 243 (1967).

151) *Finnegan, R. A.,* and *J. J. Mattice:* Tetrahedron 21, 1015 (1965).

152) *Fischer, E. O., W. Bathelt, M. Herberhold* u. *J. Müller:* Angew. Chem. 80, 625 (1968).

153) —, *E. Louis,* and *R. J. J. Schneider:* Angew. Chem. 80, 122 (1968).

154) —, *W. Berngruber,* and *C. G. Kreiter:* J. Organometal. Chem. (Amsterdam) 14, P25 (1968).

155) —, *E. Louis, W. Bathelt, E. Moser,* and *J. Müller:* J. Organometal. Chem. (Amsterdam) 14, P9 (1968).

156) —, *R. J. J. Schneider,* and *J. Müller:* J. Organometal. Chem. (Amsterdam) 14, 4 (1968).

157) — — J. Organometal. Chem. (Amsterdam) 12, P27 (1968).

158) —, *A. Reckziegel, J. Müller,* and *P. Göser:* J. Organometal. Chem. (Amsterdam) 11, P13 (1968).

159) —, and *B. J. Weimann:* J. Organometal. Chem. (Amsterdam) 8, 535 (1967).

160) —, *C. G. Kreiter, H. Rühle* u. *K. E. Schwarzhans:* Chem. Ber. 100, 1905 (1967).

161) —, u. *B. J. Weimann:* Z. Naturforsch. B 21, 84 (1966).

162) —, u. *H. Rühle:* Z. Anorg. Allgem. Chem. 341, 137 (1965).

163) —, and *J. Müller:* J. Organometal. Chem. (Amsterdam) 1, 464 (1964).

164) —, and *M. Herberhold:* In: Essays in Coordination Chemistry, Experientia Suppl. 9, 259—305. Basel: Birkhäuser 1964.

165) —, u. *J. Müller:* Chem. Ber. 96, 3217 (1963).

166) — — Z. Naturforsch. B 18, 1137 (1963).

167) — — Z. Naturforsch. B 18, 413 (1963).

168) — — J. Organometal. Chem. (Amsterdam) *1*, 89 (1963).

169) —, and *H. P. Kögler:* Ger. Patent 1146053 (1963); Chem. Abstr. *59*, 8788f (1963).

170) — — Ger. Patent 1146054 (1963); Chem. Abstr. *59*, 10118d (1963).

171) —, u. *J. Müller:* Z. Naturforsch. B *17*, 776 (1962).

172) — —, and *P. Kuzel:* Rev. Chim. (Bucharest) 7, 827 (1962).

173) —, u. *M. Herberhold:* Z. Naturforsch. B *16*, 841 (1961).

174) —, u. *P. Kuzel:* Z. Naturforsch. B *16*, 475 (1961)

175) — — u. *H. P. Fritz:* Z. Naturforsch. B *16*, 138 (1961).

176) —, *H. P. Kögler* u. *P. Kuzel:* Chem. Ber. *93*, 3006 (1960).

177) —, and *H. P. Fritz:* In: Adv. Inorg. Chem. Radiochem., Vol. I, p. 55, Ed. *H. J. Emeléus* and *A. G. Sharpe.* New York: Academic Press 1959.

178) —, and *H. Werner:* Metal-π-Complexes, Vol. I, Complexes with Di- and Oligo-Olefinic Ligands. Amsterdam: Elsevier 1966.

179) *Fischer, R. D.:* Theoret. Chim. Acta *1*, 418 (1963).

180) *Flitcroft, N.,* and *D. Sutton:* Chem. Ind. (London) 201 (1969).

181) *Frey, V., W. Hieber,* u. *O. S. Mills:* Z. Naturforsch. B *23*, 105 (1968).

182) *Friedlander, G.,* and *J. W. Kennedy:* Nuclear and Radiochemistry. New York: Wiley 1964.

183) *Fritz, H. P., H. J. Keller* u. *K. E. Schwarzhans:* Z. Naturforsch. B *23*, 298 (1968).

184) — Z. Naturforsch. B *16*, 415 (1961).

185) *Fry, A. J., R. S. H. Liu,* and *G. S. Hammond:* J. Am. Chem. Soc. *88*, 4781 (1966).

186) *Garratt, A. P.,* and *H. W. Thompson:* J. Chem. Soc. 1817 (1934).

187) *Gilchrist, T. L., F. J. Graveling,* and *C. W. Rees:* Chem. Commun. 821 (1968).

188) *Goan, J. C., C. H. Huether,* and *H. E. Podall:* Inorg. Chem. *2*, 1078 (1963).

189) *Graham, J. R.,* and *R. J. Angelici:* J. Am. Chem. Soc. *87*, 5590 (1965).

190) *Grandberg, K. I.,* and *S. P. Gubin:* Izv. Akad. Nauk SSSR, Ser. Khim. 551 (1966); Bull. Acad. Sci. USSR, Chem. Sci. 519 (1966).

191) — —, and *E. G. Perevalova:* Izv. Akad. Nauk SSSR, Ser. Khim. 549 (1966); Bull. Acad. Sci. USSR, Chem. Sci. 516 (1966).

192) *Grant, G. F.,* and *P. L. Pauson:* J. Organometal. Chem. (Amsterdam) *9*, 553 (1967).

193) *Gray, H. B., E. Billig, A. Wojcicki,* and *M. Farona:* Can. J. Chem. *41*, 1281 (1963).

194) —, and *N. A. Beach:* J. Am. Chem. Soc. *85*, 2922 (1963).

195) —, *I. Bernal,* and *E. Billig:* J. Am. Chem. Soc. *84*, 3404 (1962).

196) *Green, M., N. Mayne,* and *F. G. A. Stone:* J. Chem. Soc. A 902 (1968).

197) — — — Chem. Commun. 755 (1966).

198) —, u. *E. A. C. Lucken:* Helv. Chim. Acta *45*, 1870 (1962).

199) *Green, M. L. H., M. Ishaq,* and *R. N. Whiteley:* J. Chem. Soc. A 1508 (1967).

200) —, and *A. N. Stear:* J. Organometal. Chem. (Amsterdam) *1*, 230 (1964).

201) —, and *P. L. I. Nagy:* J. Chem. Soc. 189 (1963).

202) *Grevels, F.-W.:* Diploma Thesis, Technical University Aachen, 1968.

203) *Guillory, J. P., C. F. Cook,* and *D. R. Scott:* J. Am. Chem. Soc. *89*, 6776 (1967).

204) *Guttenberger, J. F.:* Chem. Ber. *101*, 403 (1968).

205) —, u. *W. Strohmeier:* Chem. Ber. *100*, 2807 (1967).

206) — Angew. Chem. Intern. Ed. Engl. *6*, 1081 (1967); Angew. Chem. *79*, 1071 (1967).

207) *Haake, P.,* and *T. A. Hylton:* J. Am. Chem. Soc. *84*, 3774 (1962).

208) *Haas, H.,* and *R. K. Sheline:* J. Am. Chem. Soc. *88*, 3219 (1966).

209) *Hagihara, N.:* Ann. N. Y. Acad. Sci. *125*, 98 (1965).
210) *Haines, R. J., R. S. Nyholm,* and *M. H. B. Stiddard:* J. Chem. Soc. A 43 (1968).
211) — — — J. Chem. Soc. A 94 (1967).
212) *Haller, I.,* and *R. Srinivasan:* J. Am. Chem. Soc. *88*, 5084 (1966).
213) *Harbourne, D. A.,* and *F. A. G. Stone:* J. Chem. Soc. A 1765 (1968).
214) *Hartmann, F. A.,* and *A. Wojcicki:* Inorg. Chem. *7*, 1504 (1968).
215) *Hata, G., H. Kondo,* and *A. Miyake:* J. Am. Chem. Soc. *90*, 2278 (1968).
216) *Havlin, R.,* u. *G. R. Knox:* Z. Naturforsch. B *21*, 1108 (1966).
217) *Hawthorne, M. F., D. C. Young, T. D. Andrews, D. V. Howe, R. L. Pilling, A. D. Pitts, M. Reintjes, L. F. Warren, jr.,* and *P. A. Wegner:* J. Am. Chem. Soc. *90*, 879 (1968).
218) *Hayter, R. G.,* and *L. F. Williams:* J. Inorg. Nucl. Chem. *26*, 1977 (1964).
219) *Heck, R. F.:* (Hercules Inc.), US Patent 3338936 (1967); Chem. Abstr. *68*, 49788f (1968).
220) —, and *C. R. Boss:* J. Am. Chem. Soc. *86*, 2580 (1964).
221) *Heimbach, P.,* u. *R. Traunmüller:* Chemie der Metall-Olefin-Komplexe. Weinheim/Bergstr.: Verlag Chemie, in press.
222) — Angew. Chem. *78*, 604 (1966).
223) *Hein, F.,* u. *H. Scheel:* Z. Anorg. Allgem. Chem. *312*, 264 (1961).
224) *Hennig, H.,* and *O. Gürtler:* J. Organometal. Chem. (Amsterdam) *11*, 307 (1968).
225) *Herberhold, M.,* u. *C. R. Jablonski:* Chem. Ber. *102*, 767 (1969).
226) — — J. Organometal. Chem. (Amsterdam) *14*, 457 (1968).
227) — Angew. Chem. *80*, 314 (1968).
228) — Doctoral Dissertation, Munich University 1963.
229) *Hey, E.,* and *K. Gollnick:* Ber. Bunsenges. Physik. Chem. *72*, 263 (1968); Preprints of the International Conf. on Photochemistry, Munich 1967, Part II, p. 465.
230) *Hieber, W., M. Höfler* u. *J. Muschi:* Chem. Ber. *98*, 311 (1965).
231) —, u. *E. H. Schubert:* Z. Anorg. Allgem. Chem. *338*, 32 (1965).
232) —, u. *H. Beutner:* Z. Anorg. Allgem. Chem. *320*, 101 (1963).
233) — — Z. Naturforsch. B *17*, 211 (1962).
234) —, *W. Beck* u. *G. Zeitler:* Angew. Chem. *73*, 364 (1961).
235) — — u. *G. Braun:* Angew. Chem. *72*, 795 (1960).
236) — u. *A. Lipp:* Chem. Ber. *92*, 2075 and 2085 (1959).
237) —, u. *N. Kahlen:* Chem. Ber. *91*, 2234 (1958).
238) —, u. *E. Becker:* Chem. Ber. *63B*, 1405 (1930).
239) *Hoffmann, R.,* and *R. B. Woodward:* Accounts Chem. Res. *1*, 17 (1968), and references therein.
240) *Hübel, W.,* and *E. H. Braye:* J. Inorg. Nucl. Chem. *10*, 250 (1959).
241) *Hunt, R. L., D. M. Roundhill,* and *G. Wilkinson:* J. Chem. Soc. A 982 (1967).
242) *Ibekwe, S. D.,* and *M. J. Newlands:* J. Chem. Soc. A 1783 (1967).
243) — — Chem. Commun. 114 (1965).
244) *Ihrmann, K. G.,* and *T. H. Coffield:* U. S. Patent 3077489 (1963); Chem. Abstr. *59*, 5199b (1963).
245) *Ingemanson, C. M.,* and *R. J. Angelici:* Inorg. Chem. *7*, 2646 (1968).
246) *Issleib, K.,* u. *W. Rettkowski:* Z. Naturforsch. B *21*, 999 (1966).
247) —, u. *E. Fluck:* Angew. Chem. *78*, 597 (1966).
248) Itek Corporation: Neth. Patent Appl. 6604423 (1966); Chem. Abstr. *66*, 33531q (1967).
249) *Jaffé, H. H.,* and *M. Orchin:* Theory and Applications of Ultraviolet Spectroscopy. New York: Wiley 1962.

441

E. Koerner von Gustorf and F.-W. Grevels

250) *Johnson, B. F. G., J. Lewis, P. W. Robinson,* and *J. R. Miller:* J. Chem. Soc. A 1043 (1968).
251) *Jolly, P. W., F. G. A. Stone,* and *K. Mackenzie:* J. Chem. Soc. 6416 (1965).
252) — — J. Chem. Soc. 5259 (1965).
253) *De Jong, I. G.,* and *D. R. Wiles:* Chem. Commun. 519 (1968).
254) *Jørgensen, C. K.:* Absorption Spectra and Chemical Bonding in Complexes. Oxford: Pergamon 1962.
255) *Kaesz, H. D., R. B. King, T. A. Manuel, L. D. Nichols,* and *F. G. A. Stone:* J. Am. Chem. Soc. *82,* 4749 (1960).
256) *Kalle, A. G.:* Neth. Appl. Patent 6504126 (1965); Chem. Abstr. *64,* 10642e (1966).
257) *Kan, R. O.:* Organic Photochemistry. New York: McGraw-Hill 1966.
258) *Katz, T. J.,* and *S. Cerefice:* J. Am. Chem. Soc. *91,* 2405 (1969).
259) *Kealy, T. J.,* and *P. L. Pauson:* Nature *168,* 1039 (1951).
260) *Keeley, D. F.,* and *R. E. Johnson:* J. Inorg. Nucl. Chem. *11,* 33 (1959).
261) *Kettle, S. F. A.,* and *L. E. Orgel:* Chem. Ind. (London) 49 (1960).
262) *King, R. B.,* and *K. H. Pannell:* J. Am. Chem. Soc. *90,* 3984 (1968).
263) — — Inorg. Chem. *7,* 2356 (1968).
264) — — Inorg. Chem. *7,* 1510 (1968).
265) —, and *M. Bisnette:* J. Organometal. Chem. (Amsterdam) *8,* 287 (1967).
266) — — Inorg. Chem. *6,* 469 (1967).
267) — Inorg. Chem. *6,* 30 (1967).
268) —, and *A. Fronzaglia:* J. Am. Chem. Soc. *88,* 709 (1966).
269) — Inorg. Chem. *5,* 2227 (1966).
270) —, and *M. B. Bisnette:* Inorg. Chem. *4,* 486 (1965).
271) — — Inorg. Chem. *4,* 482 (1965).
272) — Organometallic Syntheses, Vol. I, Transition-Metal Compounds, Ed. *J. J. Eisch* and *R. B. King.* New York: Academic Press 1965.
273) —, and *M. B. Bisnette:* J. Organometal. Chem. (Amsterdam) *2,* 15 (1964).
274) — — Inorg. Chem. *3,* 791 (1964).
275) — — Inorg. Chem. *3,* 785 (1964).
276) — — J. Am. Chem. Soc. *86,* 1267 (1964).
277) — — Proc. 8th Int. Conf. Coord. Chem., Vienna (1964), p. 264, Ed. *V. Gutmann.* Berlin–Heidelberg–New York: Springer 1964.
278) —, *T. A. Manuel,* and *F. G. A. Stone:* J. Inorg. Nucl. Chem. *16,* 233 (1961).
279) *Kling, O., E. Nikolaiski* u. *H. L. Schläfer:* Ber. Bunsenges. Physik. Chem. *67,* 883 (1963).
280) *Koch, E.:* Chem.-Ingr.-Tech., *41,* 916 (1969).
281) —, u. *G. O. Schenck:* Chem. Ber. *99,* 1984 (1966).
282) *Koerner von Gustorf, E.:* unpublished results.
283) —, and *M. E. Langmuir:* to be published.
284) —, and *J. C. Hogan:* unpublished results; *J. C. Hogan, Doctoral Dissertation,* Boston College 1969.
285) —, and *F.-W. Grevels:* unpublished results.
286) —, and *Z. Pfajfer:* unpublished results.
287) —, and *J. C. Hogan:* Tetrahedron Letters 3191 (1968).
288) —, *M. C. Henry* u. *D. J. McAdoo:* Ann. Chem. *707,* 190 (1967), and references therein.
289) —, and *M.-J. Jun:* unpublished results; *M.-J. Jun,* Doctoral Dissertation, Technical University Aachen 1966.
290) —, *M. C. Henry, R. E. Sacher* u. *C. Di Pietro:* Z. Naturforsch. B *21,* 1152 (1966).
291) — — u. *C. Di Pietro:* Z. Naturforsch. B *21,* 42 (1966).

292) — Angew. Chem. *78*, 780 (1966); Angew. Chem. Intern. Ed. Engl. *5*, 739 (1966).
293) —, u. *M.-J. Jun:* Z. Naturforsch. B *20*, 521 (1965).
294) —, *H. Krönert, H. Köller, G. Tillmanns* u. *G. O. Schenck:* Z. Naturforsch. B *19*, 170 (1964).
295) — —, H. *Huhn* u. *G. O. Schenck:* Angew. Chem. *75*, 1120 (1963).
296) —, *H. Köller, M.-J. Jun* and *G. O. Schenck:* Chem.-Ingr.-Tech. *35*, 591 (1963).
297) —, *M.-J. Jun* u. *G. O. Schenck:* Z. Naturforsch. B *18*, 767 (1963).
298) — — — Z. Naturforsch. B *18*, 503 (1963).
299) — —, *H. Köller,* and *G. O. Schenck:* Ind. Uses of Large Radiation Sources, Proc. Conf., Salzburg (Austria) *2*, 73 (1963), Internat. At. En. Agency, Vienna 1963.
300) —, *H. Köller, M.-J. Jun,* and *G. O. Schenck:* Proc. 2. Wiss.-Techn. Tagung des Deutschen Atomforums, Atomstrahlung in Medizin und Technik (Jan. 1963, München), p. 259. München: Thiemig 1964.
301) — — unpublished results; *H. Köller,* Doctoral Dissertation, Göttingen University 1962.
302) *Kögler, H. P.,* u. *E. O. Fischer:* Z. Naturforsch. B *15*, 676 (1960).
303) *Koller, L. R.:* Ultraviolet Radiation. New York: Wiley 1965.
304) *Krotoszynski, B. K.:* J. Chem. Phys. *41*, 2220 (1964).
305) *Kruck, Th.:* Angew. Chem. *79*, 27 (1967).
306) —, *M. Höfler* u. *M. Noack:* Chem. Ber. *99*, 1153 (1966).
307) — — Angew. Chem. *76*, 786 (1964).
308) *Kukushkin, Yu. N., A. A. Lipovskii* u. *Yu. E. Vyaz'menskii:* Zh. Neorg. Khim. *12*, 1090 (1967); Russ. J. Inorg. Chem. *12*, 573 (1967).
309) *Lemal, D. M.,* and *K. S. Shim:* Tetrahedron Letters 368 (1961).
310) *Levy, D. A.,* and *L. E. Orgel:* Mol. Phys. *4*, 93 (1961).
311) *Lewis, J., R. S. Nyholm, S. S. Sandhu,* and *M. H. B. Stiddard:* J. Chem. Soc. 2825 (1964).
312) — —, *A. G. Osborne, S. S. Sandhu,* and *M. H. B. Stiddard:* Chem. Ind. (London) 1398 (1963).
313) *Lindauer, M. W., G. O. Evans,* and *R. K. Sheline:* Inorg. Chem. *7*, 1249 (1968).
314) *Lundquist, R. T.,* and *M. Cais:* J. Org. Chem. *27*, 1167 (1962).
315) *Mais, M. H. B., P. G. Owston,* and *D. T. Thompson:* J. Chem. Soc. A 1735 (1967).
316) *Manchot, W.,* u. *W. J. Manchot:* Z. Anorg. Allgem. Chem. *226*, 385 (1936).
317) *Mango, F. D.,* and *J. H. Schachtschneider:* J. Am. Chem. Soc. *89*, 2484 (1967).
318) *Manuel, T. A.:* J. Org. Chem. *27*, 3941 (1962).
319) *Markby, R., H. W. Sternberg,* and *I. Wender:* Chem. Ind. (London) 1381 (1959).
320) *Mason, C. R., V. Boekelheide,* and *W. A. Noyes, jr.:* In: Technique of Organic Chemistry, Vol. II, p. 257, Ed. *A. Weißberger.* New York: Interscience 1956.
321) *Massey, A. G.:* J. Inorg. Nucl. Chem. *24*, 1172 (1962).
322) —, and *L. E. Orgel:* Nature *191*, 1387 (1961).
323) *Mays, M. J.,* and *S. M. Pearson:* J. Chem. Soc. A 2291 (1968).
324) *Mazlovskii, A. A.,* u. *N. A. Fuks:* Kolloid Zh. *29*, 120 (1967); Colloid J. USSR *29*, 97 (1967).
325) *Merk, W.,* and *R. Pettit:* J. Am. Chem. Soc. *89*, 4788 (1967).
326) *Milazzo, G.,* u. *G. Schiebe:* Z. Physik. Chem. B *31*, 431 (1936).
327) *Miles, jr., W. J.,* and *R. J. Clark:* Inorg. Chem. *7*, 1801 (1968).
328) *Miller, S. A., J. A. Tebboth,* and *J. F. Tremaine:* J. Chem. Soc. 632 (1952).
329) *Mills, O. S.,* and *M. J. Barrow:* Angew. Chem. *81*, 898 (1969).
330) —, and *A. D. Redhouse:* J. Chem. Soc. A 1282 (1968).
331) —, and *J. P. Nice:* J. Organometal. Chem. (Amsterdam) *10*, 337 (1967).

332) —, and *E. F. Paulus:* J. Organometal. Chem. (Amsterdam) *10*, 331 (1967).

333) — — Chem. Commun. 815 (1966).

334) —, and *A. D. Redhouse:* Chem. Commun. 444 (1966).

335) *Moore, J. W.:* Acta Chem. Scand. *20*, 1154 (1966).

336) *Müller, J.,* u. *P. Göser:* Angew. Chem. *79*, 380 (1967); Angew. Chem. Intern. Ed. Engl. *6*, 364 (1967).

337) —, and *E. O. Fischer:* J. Organometal. Chem. (Amsterdam) *5*, 275 (1966).

338) *Nakamura, A.,* and *N. Hagihara:* Nippon Kagaku Zasshi *82*, 1392 (1961); Bull. Chem. Soc. Japan *33*, 425 (1960).

339) — — Nippon Kagaku Zasshi *82*, 1389 (1961); Chem. Abstr. *59*, 2855 h (1963).

340) — — Nippon Kagaku Zasshi *82*, 1387 (1961); Chem. Abstr. *59*, 2855 a (1963).

341) — — Mem. Inst. Sci. Ind. Res. Osaka Univ. *17*, 187 (1960); Chem. Abstr. *55*, 6457 f (1961).

342) — — Bull. Chem. Soc. Japan *32*, 880 (1959).

343) *Neckers, D. C.:* Mechanistic Organic Photochemistry. New York: Reinhold Publ. Corp. 1967.

344) *Nesmeyanov, A. N., A. I. Gusev, A. A. Pasynskii, K. N. Amisimov, N. E. Kolobova,* and *Yu. T. Struchkov:* Chem. Commun. 277 (1969).

345) —, *Yu. A. Chapovsky,* and *Yu. A. Ustynyuk:* J. Organometal. Chem. (Amsterdam) *9*, 345 (1967).

346) — —, *I. V. Polovyanyuk,* and *L. G. Makarova:* J. Organometal. Chem. (Amsterdam) *7*, 329 (1967).

347) —, *V. A. Sazonova, V. I. Romanenko, V. N. Postnov, G. N. Zol'nikova, V. A. Blinova* u. *R. M. Kalyanova:* Dokl. Akad. Nauk SSSR *173*, 589 (1967); Dokl. Chem. *173*, 289 (1967).

348) —, *Yu. A. Chapovskii, B. V. Lokshin, I. V. Polovyanyuk* u. *L. G. Makarova:* Dokl. Akad. Nauk SSSR *166*, 1125 (1966); Dokl. Chem. *166*, 213 (1966).

349) —, *K. N. Anisimov, N. E. Kolobova* u. *A. A. Pasynskii:* Izv. Akad. Nauk SSSR, Ser. Khim. 2231 (1966); Bull. Acad. Sci. USSR Chem. Sci. 2164 (1966).

350) —, *Yu. A. Chapovskii* u. *Yu. A. Ustynyuk:* Izv. Akad. Nauk. SSSR, Ser. Khim. 1871 (1966); Bull. Acad. Sci. USSR Chem. Sci. 1815 (1966).

351) — — — Izv. Akad. Nauk. SSSR, Ser. Khim. 1870 (1966); Bull. Acad. Sci. USSR Chem. Sci. 1814 (1966).

352) —, *K. N. Anisimov, N. E. Kolobova* u. *A. A. Pasynskii:* Izv. Akad. Nauk SSSR, Ser. Khim. 774 (1966); Bull. Acad. Sci. USSR Chem. Sci. 746 (1966).

353) —, *V. A. Sazonova, V. I. Romanenko* u. *G. P. Zol'nikova:* Izv. Akad. Nauk SSSR, Ser. Khim. 1694 (1965); Bull. Acad. Sci. USSR Chem. Sci. 1660 (1965).

354) —, *B. M. Yavorskii, G. B. Zaslavskaya* u. *N. S. Kochetkova:* Dokl. Akad. Nauk. SSSR *160*, 837 (1965); Dokl. Chem. *160*, 131 (1965).

355) —, *V. A. Sazonova, V. I. Romanenko, N. A. Rodionova* u. *G. P. Zol'nikova:* Dokl. Akad. Nauk SSSR *155*, 1130 (1964); Dokl. Chem. *155*, 382 (1964).

356) — — — Dokl. Akad. Nauk SSSR *152*, 1358 (1963); Proc. Akad. Sci. USSR, Sect. Chem. *152*, 835 (1963).

357) — —, *A. V. Gerasimenko* u. *N. S. Sazonova:* Dokl. Akad. Nauk SSSR *149* 1354 (1963); Proc. Acad. Sci. USSR, Sect. Chem. *149*, 376 (1963).

358) *Neuse, E. W.:* J. Organometal. Chem. (Amsterdam) *7*, 349 (1967).

359) *Noack, K.:* Helv. Chim. Acta *45*, 1847 (1962).

360) Norrish, *R. G. W.:* Angew. Chem. *80*, 868 (1968).

361) *Nyholm, R. S., S. S. Sandhu,* and *M. H. B. Stiddard:* J. Chem. Soc. 5916 (1963).

362) *Nyman, F.:* Chem. Ind. (London) 604 (1965).

363) *Orgel, L. E.:* An Introduction to Transition-Metal Chemistry, Ligand Field Theory. London: Methuen 1960.

364) *Osborne, A. G.,* and *M. H. B. Stiddard:* J. Chem. Soc. 634 (1964).
365) *Otsuka, S., T. Yoshida,* and *A. Nakamura:* Inorg. Chem. *7,* 1833 (1968).
366) *Pankowski, M.,* and *M. Bigorgne:* Compt. Rend. Ser. C *263,* 239 (1966).
367) *Paulus, E. F., E. O. Fischer, H. P. Fritz,* and *H. Schuster-Woldan:* J. Organometal. Chem. (Amsterdam) *10,* P3 (1967).
368) *Pauson, P. L.:* Organometallic Chemistry. London: Arnold 1967.
369) *Pavlik, I.,* u. *J. Klikorka:* Z. Chem. *5,* 345 (1965).
370) *Perumareddi, J. R.,* and *A. W. Adamson:* J. Phys. Chem. *72,* 414 (1968).
371) *Pettit, R.:* J. Am. Chem. Soc. *81,* 1266 (1959).
372) *Peyronel, G., A. Ragni* e *E. F. Trogu:* Gazz. Chim. Ital. *92,* 738 (1962).
373) — — — Atti Soc. Nat. Mat. Modena *91,* 100 (1960).
374) *Pitts, jr., J. N., F. Wilkinson,* and *G. S. Hammond:* In: Advances in Photochemistry, Vol. I, p. 1, Ed. *W. A. Noyes, jr., G. S. Hammond,* and *J. N. Pitts, jr.* New York: Interscience 1963.
375) *Plotnikow, J.:* Allgemeine Photochemie. Berlin: de Gruyter 1920.
376) *Podall, H. E.:* U.S. Patent 3 349 019 (1967); Chem. Abstr. *68,* 3403 y (1968).
377) —, and *T. L. Iapalucci:* J. Polymer Sci., Part B, *1,* 457 (1963).
378) *Polovyanyuk, I. V., Yu. A. Chapovskii* u. *L. G. Makarova:* Izv. Akad. Nauk SSSR 387 (1966); Bull. Acad. Sci. USSR Chem. Sci. 368 (1966).
379) *Porter, G.:* Angew. Chem. *80,* 882 (1968).
380) *Pruett, R. L.,* and *J. E. Wyman:* Chem. Ind. (London) 119 (1960).
381) *Rausch, M. D.,* and *G. N. Schrauzer:* Chem. Ind. (London) 957 (1959).
382) *Reihlen, H., A. Gruhl, G. v. Heßling* u. *O. Pfrengle:* Ann. Chem. *482,* 161 (1930).
383) — — — Ann. Chem. *472,* 268 (1929).
384) *Rochow, E. G., D. T. Hurd,* and *R. N. Lewis:* The Chemistry of Organometallic Compounds. New York: Wiley 1957.
385) *Rosenblum, M.,* and *B. North:* J. Am. Chem. Soc. *90,* 1060 (1968).
386) —, and *C. Gatsonis:* J. Am. Chem. Soc. *89,* 5074 (1967).
387) — Chemistry of the Iron Group Metallocenes, Vol. I. New York: Interscience 1965.
388) *Ruff, J. K.:* Inorg. Chem. *8,* 86 (1969).
389) —, and *M. Lustig:* Inorg. Chem. *7,* 2171 (1968).
390) — Inorg. Chem. *7,* 1821 (1968).
391) — Inorg. Chem. *7,* 1818 (1968).
392) — Inorg. Chem. *6,* 1502 (1967).
393) *Saito, H., J. Fujita,* and *K. Saito:* Bull. Chem. Soc. Japan *41,* 863 (1968).
394) — — — Bull. Chem. Soc. Japan *41,* 359 (1968).
395) *Scandola, F., O. Traverso,* and *V. Carassiti:* Proc. 11th Int. Conf. Coord. Chem., Haifa-Jerusalem, p. 88 (1968), Ed. *M. Cais.* Amsterdam: Elsevier 1968.
396) *Schenck, G. O., E. Koerner v. Gustorf,* and *M.-J. Jun:* Tetrahedron Letters 1059 (1962).
397) — Dechema Monograph. *24,* 105 (1955).
398) *Schläfer, H. L.,* u. *G. Gliemann:* Einführung in die Ligandenfeldtheorie. Frankfurt/Main: Akademische Verlagsges. 1967.
399) *Schlögl, K.,* and *W. Steyrer:* J. Organometal. Chem. (Amsterdam) *6,* 399 (1966).
400) —, u. *H. Egger:* Ann. Chem. *676,* 88 (1964).
401) *Schönberg, A., G. O. Schenck,* and *O.-A. Neumüller:* Preparative Organic Photochemistry. Berlin–Heidelberg–New York: Springer 1968.
402) *Schrauzer, G. N., J. W. Sibert,* and *R. J. Windgassen:* J. Am. Chem. Soc. *90,* 6681 (1968).
403) —, and *P. W. Glockner:* J. Am. Chem. Soc. *90,* 2800 (1968).
404) — Advan. Catalysis *18,* 373 (1968).

405) —, V. P. Mayweg, and W. Heinrich: J. Am. Chem. Soc. 88, 5174 (1966).
406) — Inorg. Chem. 4, 264 (1965).
407) —, P. Glockner u. R. Merényi: Angew. Chem. 76, 498 (1964).
408) —, and G. Kratel: J. Organometal. Chem. (Amsterdam) 2, 336 (1964).
409) —, and H. Thyret: Theoret. Chim. Acta 1, 172 (1963).
410) —, u. S. Eichler: Angew. Chem. 74, 585 (1962).
411) — J. Am. Chem. Soc. 81, 5307 (1959).
412) Schreiner, A. F., and T. L. Brown: J. Am. Chem. Soc. 90, 3366 (1968).
413) Schroll, G. E.: (Ethyl Corp.), U.S. Patent 3130215 (1964); Chem. Abstr. 61, 4396h (1964).
414) Schubert, E. H., and R. K. Sheline: Inorg. Chem. 5, 1071 (1966).
415) — — Z. Naturforsch. B 20, 1306 (1965).
416) Schumann, H., O. Stelzer, W. Gick, and R. Weis: Lecture on the Chemie-Dozen-ten-Tagung, Karlsruhe 1969.
417) — — Angew. Chem. 80, 318 (1968).
418) — — J. Organometal. Chem. (Amsterdam) 13, P25 (1968).
419) Scott, D. R., and R. S. Becker: J. Organometal Chem. (Amsterdam) 4, 409 (1965).
420) — Ph. D. Thesis, University of Houston (Texas) 1965; Diss. Abstr. 26, 732 (1965).
421) —, and R. S. Becker: J. Phys. Chem. 69, 3207 (1965).
422) — — J. Chem. Phys. 35, 516 (1961); ibid. 35, 2246 (1961).
423) Shustorovich, E. M., and M. E. Dyatkina: Zh. Strukt. Khim. 3, 345 (1962); J. Struct. Chem. 3, 328 (1962).
424) — — Zh. Strukt. Khim. 1, 109 (1960); J. Struct. Chem. 1, 98 (1960).
425) — — Dokl. Akad. Nauk SSSR 128, 1234 (1959); ibid. 131, 113 (1960); ibid. 133, 141 (1960); Engl. Transl.: Proc. Acad. Sci. USSR, Sect. Chem. 131, 215 (1960); ibid. 133, 767 (1960).
426) Siefert, E. E., and R. J. Angelici: J. Organometal. Chem. (Amsterdam) 8, 374 (1967).
427) Smith, J. J., and B. Meyer: J. Chem. Phys. 48, 5436 (1968).
428) Söfftge, M.: Doctoral Dissertation, Technical University Braunschweig 1967.
429) Speyer, E., u. H. Wolf: Chem. Ber. 60 B, 1424 (1927).
430) Spilners, I. J.: J. Organometal. Chem. (Amsterdam) 11, 381 (1968).
431) Srinivasan, R.: J. Am. Chem. Soc. 86, 3318 (1964).
432) — J. Am. Chem. Soc. 85, 4045 (1963).
433) — J. Am. Chem. Soc. 85, 3048 (1963).
434) Srinivasan, S. C., D. R. Wiles, and I. C. Yang: Inorg. Nucl. Chem. Letters 2, 399 (1966).
435) Stanclift, W. E., and D. G. Hendricker: Inorg. Chem. 7, 1242 (1968).
436) Stephenson, L. M., u. G. S. Hammond: Angew. Chem. 81, 279 (1969).
437) Sternberg, H. W., R. Markby, and I. Wender: J. Am. Chem. Soc. 80, 1009 (1958).
438) Stolz, I. W., H. Haas, and R. K. Sheline: J. Am. Chem. Soc. 87, 716 (1965).
439) —, G. R. Dobson, and R. K. Sheline: Inorg. Chem. 2, 1264 (1963).
440) — — — Inorg. Chem. 2, 323 (1963).
441) — — — J. Am. Chem. Soc. 85, 1013 (1963).
442) — — — J. Am. Chem. Soc. 84, 3589 (1962).
443) Strohmeier, W., u. F.-J. Müller: Chem. Ber. 100, 2812 (1967).
444) —, J. F. Guttenberger u. F.-J. Müller: Z. Naturforsch. B 22, 1091 (1967).
445) —, u. H. Grübel: Z. Naturforsch. B 22, 553 (1967).
446) — — Z. Naturforsch. B 22, 115 (1967).
447) — — Z. Naturforsch. B 22, 98 (1967).
448) —, u. P. Hartmann: Z. Naturforsch. B 21, 1119 (1966).

449) — — Z. Naturforsch. B *21*, 997 (1966).
450) —, *J. F. Guttenberger, H. Blumenthal* u. *G. Albert:* Chem. Ber. *99*, 3419 (1966).
451) —, *G. Popp* u. *J. F. Guttenberger:* Chem. Ber. *99*, 165 (1966).
452) —, and *C. Barbeau:* Makromol. Chem. *81*, 86 (1965).
453) —, u. *P. Hartmann:* Z. Naturforsch. B *20*, 513 (1965).
454) —, u. *H. Grübel:* Z. Naturforsch. B *20*, 11 (1965).
455) —, *J. F. Guttenberger* u. *G. Popp:* Chem. Ber. *98*, 2248 (1965).
456) —, u. *H. Hellmann:* Chem. Ber. *98*, 1598 (1965).
457) — Ger. Patent 1197457 (1965); Chem. Abstr. *63*, 18151g (1965).
458) —, u. *H. Hellmann:* Chem. Ber. *97*, 1877 (1964).
459) —, u. *J. F. Guttenberger:* Chem. Ber. *97*, 1871 (1964).
460) — — Chem. Ber. *97*, 1256 (1964).
461) —, u. *D. v. Hobe:* Z. Naturforsch. B *19*, 959 (1964).
462) —, u. *P. Hartmann:* Z. Naturforsch. B *19*, 882 (1964).
463) — — Z. Naturforsch. B *19*, 655 (1964).
464) —, *J. F. Guttenberger* u. *H. Hellmann:* Z. Naturforsch. B *19*, 353 (1964).
465) —, u. *H. Hellmann:* Z. Naturforsch. B *19*, 164 (1964).
466) — Angew. Chem. *76*, 873 (1964).
467) — Angew. Chem. *75*, 453 (1963).
468) — Ber. Bunsenges. Physik. Chem. *67*, 892 (1963).
469) —, u. *D. v. Hobe:* Z. Naturforsch. B *18*, 981 (1963).
470) — — Z. Naturforsch. B *18*, 770 (1963).
471) —, u. *H. Hellmann:* Z. Naturforsch. B *18*, 769 (1963).
472) —, u. *J. F. Guttenberger:* Z. Naturforsch. B *18*, 667 (1963).
473) —, u. *D. v. Hobe:* Z. Naturforsch. B *18*, 580 (1963).
474) —, u. *J. F. Guttenberger:* Z. Naturforsch. B *18*, 80 (1963).
475) —, *C. Barbeau* u. *D. v. Hobe:* Chem. Ber. *96*, 3254 (1963).
476) —, u. *H. Hellmann:* Chem. Ber. *96*, 2859 (1963).
477) —, u. *J. F. Guttenberger:* Chem. Ber. *96*, 2112 (1963).
478) —, u. *D. v. Hobe:* Z. Physik. Chem. (Frankfurt) *34*, 393 (1962).
479) —, u. *G. Schönauer:* Chem. Ber. *95*, 1767 (1962).
480) —, *H. Laporte* u. *D. v. Hobe:* Chem. Ber. *95*, 455 (1962).
481) —, u. *C. Barbeau:* Z. Naturforsch. B *17*, 848 (1962).
482) —, *D. v. Hobe, G. Schönauer* u. *H. Laporte:* Z. Naturforsch. B *17*, 502 (1962).
483) —, u. *R. Müller:* Z. Physik. Chem. (Frankfurt) *28*, 112 (1961).
484) —, *Kl. Gerlach:* Z. Phys. Chem. (Frankfurt) *27*, 439 (1961).
485) —, u. *D. v. Hobe:* Z. Naturforsch. B *16*, 402 (1961).
486) — Chem. Ber. *94*, 3337 (1961).
487) —, u. *D. v. Hobe:* Chem. Ber. *94*, 2031 (1961).
488) —, u. *G. Schönauer:* Chem. Ber. *94*, 1346 (1961).
489) —, u. *D. v. Hobe:* Chem. Ber. *94*, 761 (1961).
490) —, u. *Kl. Gerlach:* Chem. Ber. *94*, 398 (1961).
491) — — u. *D. v. Hobe:* Chem. Ber. *94*, 164 (1961).
492) — — Chem. Ber. *93*, 2087 (1960).
493) —, *G. Matthias* u. *D. v. Hobe:* Z. Naturforsch. B *15*, 813 (1960).
494) —, u. *Kl. Gerlach:* Z. Naturforsch. B *15*, 675 (1960).
495) — — Z. Naturforsch. B *15*, 622 (1960).
496) — — u. *G. Matthias:* Z. Naturforsch. B *15*, 621 (1960).
497) — — Z. Naturforsch. B *15*, 413 (1960).
498) *Sutin, N.,* and *R. W. Dodson:* J. Inorg. Nucl. Chem. *6*, 91 (1958).
499) *Sutton, D.:* Electronic Spectra of Transition Metal Complexes. London: McGraw-Hill 1968.

500) Southern New England Ultraviolet Co., Middletown, Connecticut, U.S.A.

501) *Tarr, A. M.*, and *D. M. Wiles:* Can. J. Chem. *46*, 2725 (1968).

502) *Thompson, D. T.:* J. Organometal. Chem. (Amsterdam) *4*, 74 (1965).

503) *Thompson, H. W.*, and *A. P. Garratt:* J. Chem. Soc. 524 (1934).

504) *Tondello, E., L. Oleari, L. Di Sipio*, and *G. De Michelis:* Proc. 11th Int. Conf. Coord. Chem., Haifa-Jerusalem (1968), p. 528, Ed. *M. Cais*, Amsterdam: Elsevier 1968.

505) *Traunmüller, R., O. E. Polansky, P. Heimbach* u. *G. Wilke:* Lecture on Koordinationschemisches Symposium, Jena 1969.

506) — — — — Chem. Phys. Letters, to be published.

507) *Trecker, D. J., R. S. Foote, J. P. Henry*, and *J. E. McKeon:* J. Am. Chem. Soc. *88*, 3021 (1966).

508) —, *J. P. Henry*, and *J. E. McKeon:* J. Am. Chem. Soc. *87*, 3261 (1965).

509) *Treichel, P. M., R. L. Shubkin, K. W. Barnett*, and *D. Reichard:* Inorg. Chem. *5*, 1177 (1966).

510) —, *E. Pitcher, R. B. King*, and *F. G. A. Stone:* J. Am. Chem. Soc. *83*, 2593 (1961).

511) *Tsumura, R.*, and *N. Hagihara:* Bull. Chem. Soc. Japan *38*, 1901 (1965).

512) *Turro, N. J.:* Molecular Photochemistry. New York: Benjamin 1965.

513) *Tyerman, W. J. R., M. Kato, P. Kebarle, S. Masamune, O. P. Strausz*, and *H. E. Gunning:* Chem. Commun. 497 (1967).

514) *Valentine, jr., D.:* In: Advances in Photochemistry, Vol. 6, p. 123, Ed. *W. A. Noyes, jr., G. S. Hammond*, and *J. N. Pitts, jr.* New York: Interscience 1968.

515) *Vandenbroucke, jr., A. C., D. G. Hendricker, R. E. McCarley*, and *J. G. Verkade:* Inorg. Chem. *7*, 1825 (1968).

516) *Verkade, J. G., R. E. McCarley, D. G. Hendricker*, and *R. W. King:* Inorg. Chem. *4*, 228 (1965).

517) *Volger, H. C., H. Hogeveen*, and *M. M. P. Gaasbeek:* J. Am. Chem. Soc. *91*, 218 (1969). — *Hogeveen, H.*, and *H. C. Volger:* ibid. *89*, 2486 (1967).

518) *Wehry, E. L.:* Quart. Rev. (London) *21*, 213 (1967); in: Fluorescence, *G. G. Guilbault*, Editor, New York: Dekker 1967, p. 37, esp. 95 ff.

519) *Wender, I.*, and *P. Pino* (Ed.): Organic Syntheses via Metal Carbonyls, Vol. I. New York: Interscience 1968.

520) *Werner, H.*, and *J. H. Richards:* J. Am. Chem. Soc. *90*, 4976 (1968).

521) *Whitesides, T. H.:* J. Am. Chem. Soc. *91*, 2395 (1969).

522) *Whitesides, G. M., G. L. Goe*, and *A. C. Cope:* J. Am. Chem. Soc. *89*, 7136 (1967).

523) *Wilford, J. B., P. M. Treichel*, and *F. G. A. Stone:* J. Organometal. Chem. (Amsterdam) *2*, 119 (1964).

524) — — — Proc. Chem. Soc. (London) 218 (1963).

525) *Wilkinson, G.*, and *F. A. Cotton:* In: Progress in Inorganic Chemistry, Vol. I, pp. 1—124, Ed. *F. A. Cotton.* New York: Interscience 1959.

526) *Yamada, S., H. Yamazaki, H. Nishikawa*, and *R. Tsuchida:* Bull. Chem. Soc. Japan *33*, 481 (1960).

527) *Yamazaki, M.:* J. Chem. Phys. *24*, 1260 (1956).

528) *Yang, I. C.*, and *D. R. Wiles:* Can. J. Chem. *45*, 1357 (1967).

529) *Yavorskii, B. M., N. S. Kochetkova, G. B. Zaslavskaya* u. *A. N. Nesmeyanov:* Dokl. Akad. Nauk SSSR *149*, 111 (1963); Proc. Acad. Sci. USSR, Sect. Chem. *149*, 202 (1963).

530) *Zahn, U.:* Radiochim. Acta *7*, 170 (1967).

531) *Zahner, J. C.*, and *H. G. Drickamer:* J. Chem. Phys. *35*, 375 (1961).

532) *Zeiss, H., P. J. Wheatley*, and *H. J. S. Winkler:* Benzenoid-Metal Complexes. Structural Determinations and Chemistry. New York: The Ronald Press 1966.

533) *Ziegler, M. L.*, and *R. K. Sheline:* Inorg. Chem. *4*, 1230 (1965).
534) —, *H. Haas* u. *R. K. Sheline:* Chem. Ber. *98*, 2454 (1965).
535) *Parker, C. A.:* Photoluminescense of Solutions. New York: Elsevier 1968.
536) *Novak, J. R.*, and *M. W. Windsor:* Proc. Roy. Soc. (London) A *308*, 95 (1968).
537) *Bamford, C. H.*, and *C. A. Finch:* Proc. Roy. Soc. (London) A *268*, 553 (1962); Z. Naturforsch. B *17*, 804 (1962); Proc. Chem. Soc. (London) 110 (1962); Z. Naturforsch. B *17*, 500 (1962); Trans. Faraday Soc. *59*, 118 (1963). — *Bamford, C. H., M. S. Blackie*, and *C. A. Finch:* Chem. Ind. (London) 1763 (1962).
538) —, *G. C. Eastmond*, and *W. R. Maltman:* Trans. Faraday Soc. *60*, 1432 (1964).
539) *Wilke, G.:* Angew. Chem. *75*, 10 (1963).
540) *Ward, C. B. R.:* Dissert. Abstr. *16*, 2327 (1956).
541) *Nesmejanow, A. N., E. G. Perewalowa* u. *L. P. Jurjewa:* Chem. Ber. *93*, 2729 (1960).
542) *Fritz, H. P.*, u. *L. Schäfer:* Z. Naturforsch. B *19*, 169 (1964).
543) *Riemschneider, R.*, and *G. Helm:* Chem. Ber. *89*, 155 (1956).
544) *Fluck, E.:* In: Chemical Applications of Mössbauer Spectroscopy, p. 296 and references therein. Ed. *V. I. Goldanksii* and *R. H. Herber.* New York: Academic Press 1968.
545) *Maitlis, P. M.*, u. *J. D. Brown:* Z. Naturforsch. B *20*, 597 (1965).
546) *Beckwith, A. L. J.*, and *R. J. Leydon:* Tetrahedron *20*, 791 (1964).
547) *Spinks, J. W. T.*, and *R. J. Woods:* An Introduction to Radiation Chemistry. New York: Wiley 1964.
548) *Swallow, A. J.:* Radiation Chemistry of Organic Compounds. New York: Pergamon 1960.
549) *Lind, S. C.:* Radiation Chemistry of Gases. American Chemical Society Monograph No. 151. New York: Reinhold Publ. Corp. 1961.
550) *Phillips, G. O.:* Energy Transfer in Radiation Processes. New York: Elsevier 1966.
551) *Chapiro, A.:* Radiation Chemistry of Polymeric Systems, Vol. XV of High Polymers. New York: Interscience 1962.
552) Radiation Chemistry, Vol. I and II. Adv. in Chemistry Series 81 and 82, Ed. *E. J. Hart.* Washington, D.C.: Amer. Chem. Soc. 1968.
553) *Krauch, C. H., S. Farid, D. Hess, J. Kuhls* u. *W. Metzner:* Angew. Chem. *76*, 593 (1964).
554) *King, R. B., L. W. Houk*, and *K. H. Pannell:* Inorg. Chem. *8*, 1042 (1969).
555) —, and *R. N. Kapoor:* in press.
556) — — J. Organometal. Chem. (Amsterdam) *18*, 357 (1967).
557) —, and *A. Efraty:* in press.
558) —, and *M. Ishaq:* in press.
559) *Strohmeier, W.*, u. *F.-J. Müller:* Chem. Ber. *102*, 3608 (1969).
560) — — Chem. Ber. *102*, 3613 (1969).
561) *Shriver, D. F.:* The Manipulation of Air-Sensitive Compounds. New York: McGraw-Hill 1969.
562) *Henglein, A., W. Schnabel* u. *J. Wendenburg:* Einführung in die Strahlenchemie. Weinheim: Verlag Chemie 1969.
563) *Tsutsui, M.:* Characterization of Organometallic Compounds, Part I and II. New York: Interscience Publ. John Wiley 1969.
564) *Brookes, A., S. A. R. Knox*, and *F. G. A. Stone:* In: Progress in Organometallic Chemistry, *M. I. Bruce* and *F. G. A. Stone*, Editors. Proc. 4th Int. Conf. Organometal. Chem., Bristol 1969, A 6.

Knox, S. A. R., C. M. Mitchell, and *F. G. A. Stone:* J. Organometal. Chem. (Amsterdam) *16*, P67 (1969).

565) *Abramovitch, R. A., C. I. Azogu*, and *R. G. Sutherland:* In: Progress in Organnt metallic Chemistry, *M. I. Bruce* and *F. G. A. Stone*, Editors. Proc. 4th Io-Conf. Organometal. Chem., Bristol 1969, G 4; Chem. Commun., in press.

566) *King, R. B.*, and *R. N. Kapoor:* In: Progress in Organometallic Chemi try, *M. I. Bruce* and *F. G. A. Stone*, Editors. Proc. 4th Int. Conf. Organomsetal. Chem., Bristol 1969, K 3.

567) *Kaska, W. C., D. K. Mitchell*, and *W. D. Korte:* In: Progress in Organometallic Chemistry, *M. I. Bruce* and *F. G. A. Stone*, Editors. Proc. 4th Int. Conf. Organometal. Chem., Bristol 1969, K 4.

568) *Noack, K.*, and *M. Ruch:* In: Progress in Organometallic Chemistry, *M. I. Bruce* and *F. G. A. Stone*, Editors. Proc. 4 th Int. Conf. Organometal. Chem., Bristol 1969, N 5; J. Organometal. Chem. (Amsterdam) *17*, 309 (1969).

569) *Bichler, R. E. J.*, and *H. C. Clark:* In: Progress in Organometallic Chemistry, *M. I. Bruce* and *F. G. A. Stone*, Editors. Proc. 4th Int. Conf. Organometal. Chem., Bristol 1969, O 4.

570) *Rest, A. J.*, and *J. J. Turner:* In: Progress in Organometallic Chemistry, *M. I. Bruce* and *F. G. A. Stone*, Editors. Proc. 4th Int. Conf. Organometal. Chem., Bristol 1969, N 6; Chem. Commun. 1026 (1969).

571) *Jennings, P. W.* (Bozeman, Montana): personal communication.

572) *Hoyano, J. K., M. Elder*, and *W. A. G. Graham:* J. Am. Chem. Soc. *91*, 4568 (1969).

573) *Jetz, W.*, and *W. A. G. Graham:* J. Am. Chem. Soc. *91*, 3375 (1969).

574) *Fischer, E. O.*, and *E. Louis:* J. Organometal. Chem. (Amsterdam) *18*, P26 (1969).

575) *Treichel, P. M.*, and *J. J. Benedict:* J. Organometal. Chem. (Amsterdam) *17*, P37 (1969).

576) *Brunner, H.:* J. Organometal. Chem. (Amsterdam) *16*, 119 (1969).

577) *Birchall, J. M., F. L. Bowden, R. N. Haszeldine*, and *A. B. P. Lever:* J. Chem. Soc. A, **747** (1967).

578) *Watts, W. E.:* Organometal. Chem. Rev. *2*, 231 (1967).

579) *Meinwald, J.*, and *B. E. Kaplan:* J. Am. Chem. Soc. *89*, 2611 (1967).

580) *Ohnesorge, W. E.:* Fluorescence of Metal Chelate Compounds. In: Fluorescence and Phosphorescence Analysis, *D. M. Hercules* (Editor), New York: Interscience 1966.

581) *Valentine, Jr., D.:* in: Annual Survey of Photochemistry, Vol. 1, Part 4. New York: Wiley Interscience 1969.

Eingegangen am 21. Juli 1969

SPRINGER-VERLAG
BERLIN HEIDELBERG GMBH

Preparative Organic Photochemistry

Second completely revised edition of „Präparative Organische Photochemie" by A. Schönberg with a contribution by G. O. Schenck

By **Alexander Schönberg,** Technische Universität Berlin
In cooperation with Günther Otto Schenck
and Otto-Albrecht Neumüller, Abteilung Strahlenchemie,
Max-Planck-Institut für Kohlenforschung, Mülheim/Ruhr

With 4 figures
and 51 tables
XXIV, 608 pages. 1968
Cloth DM 148,—

This new edition in English of Schönberg's well-known monograph reflects the remarkable progress that has been made since the appearance of the first German edition in 1958. The book has been greatly enlarged, as a vast number of new reactions had to be included. Nomenclature has been standardized throughout, and the bibliography has undergone a thorough revision. All references have been double checked and are therefore absolutely reliable. The typographical make-up and the presentation of chemical formulae are much improved.

The aim of the book is unchanged. Theories and reaction mechanisms are treated extensively in books already available. This monograph is an exhaustive compilation of photochemical reactions that are of preparative interest to the organic chemist. In every case sufficient experimental details are given to make this book usable as a manual of photochemical laboratory techniques. It will be indispensable for any worker in the field, and is highly recommended for use in university courses.

■ **Prospectus on request**

ISBN 978-3-540-04489-5   ISBN 978-3-540-35963-0 (eBook)
DOI 10.1007/978-3-540-35963-0

Library of Congress Catalog Card Number 51-5497.
Titel-Nr. 7716